Commission of the European Communities

Non-destructive assay of radioactive waste
Topical meeting

Proceedings of a topical meeting organized
by the Commission of the European Communities,
Directorate-General 'Science, Research and
Development' and CEA/CEN Cadarache in the
framework of the R&D programme on radioactive
waste management and disposal, held in Cadarache,
France, 20-22 November 1989

Edited by

C. Eid
Commission of the European Communities
200, rue de la Loi
B-1049 Brussels

P. Bernard
CEA/CEN
F-Cadarache

Directorate-General
Science, Research and Development

1990   EUR 12890 EN

Published by the
COMMISSION OF THE EUROPEAN COMMUNITIES

Directorate-General
Telecommunications, Information Industries and Innovation

L-2920 Luxembourg

LEGAL NOTICE

Cataloguing data can be found at the end of this publication

Luxembourg, Office for Official Publications of the European Communities, 1990

ISBN 92-826-1636-3   Catalogue number: CD-NA-12890-EN-C

Printed in Belgium

FOREWORD

R. SIMON
CEC-Brussels

The subjects of our meeting are "The measurement methods that accompany radioactive waste management from the source of the waste stream to its disposal". Non destrutive assay finds applications in :

- the initial sorting of the waste
- the declassification after decontamination
- the dosing of waste during processing and conditioning
- the certification of the compliance with the disposal criteria.

In fuel reprocessing and the subsequent fuel manufacturing additional considerations like criticality safety and fissile material accounting rely on the correct measurement of transuranic nuclides in solid and liquid waste streams.

During recent discussions at the IAEA some countries were advocating annual limits for measured waste discards of 0.1 Kg Pu for each fuel cycle facility.

I am quite curious to know how they intend to measure 100 g of Pu in about 2000 m^3 of raw waste or 4000 m^3 of conditioned waste as such quantities roughly are produced by a 500 MTHM commercial reprocessing plant.

This control of low level radioactive nuclide concentration is clearly one of the challenges that the scientific community has to meet. Improving the control of radioactivity in processes and in waste is one of the issues directly affecting the health and safety of lab workers, production staff and the general public.
I hope this meeting will contribute to tackle this important common commitment.

WELCOME ADDRESS

M. MEGY
CEA-CEN Cadarache

I would first like to welcome you to Cadarache, on behalf of CEA and the Cadarache centre and I would like to congratulate the organisers for the punctual start to this meeting.

It gives me great pleasure in welcoming you, not only from a work point of view, but also personally, as I am meeting again today many of you with whom I have had the pleasure of working with in the past in the area of radioactive waste, the subject which brings us together again today. I also believe that the subject we shall be discussing is not only important and news worthy but is also important in the long term. In spite of the fact that the measurement of alpha waste appears to be technically difficult, it could have an important impact on the way problems connected with the final phase of the nuclear combustible cycle, in particular deep geological storage, are treated. But this is very technical and I won't go into that here. Nevertheless, I would like to emphasise how important we feel that such subjects be treated in a context that brings together a maximum of parteners on an international level.
In fact, it has been proven that scientifically, "synergie" is the source of much production, but also because it is fundamental that we be more coherent in our results, in our way of solving problems and explaining them, not only in a technical way to scientists, but also in a much wider way to all those interested in our activities.

I would like to finish by saying that the Cadarache centre and its surroundings seems to me to be an excellent area in which to hold such a meeting. I hope that in the next three days you will have many fruitful discussions.

SUMMARY

The nuclear fuel cycle generates at different steps of its development a large variety of waste containing Pu. After treatment and conditioning the final destination of this waste is either to be disposed by shallow land burial or in underground geological repositories.
In all cases the decision concerning the method of disposal is determined by the quantity of Pu contained in the waste to be disposed of. For this reason and taking into account the rigorous requirements of the safety authorities concerning the protection of people and the environment and the large investments which are necessary for the development and operation of such sites, it is most important to determine accurately the Pu contents in the waste.

This seminar has notably demonstrated that some monitoring techniques have reached industrial development and that sustained efforts are being made in the different laboratories to improve the accuracy of the results and gain confidence in the general performance of the different systems implemented.

It has also provided the opportunity for scientists working on the various aspects of the waste monitoring to present their most salient results and to discuss the possibilities of making headway in all areas of the non destructive assay of radioactive waste.

C. EID
(Scientific Secretary)

The topical meeting on NON DESTRUCTIVE ASSAY OF RADIOACTIVE WASTE was held in Cadarache, France, from 20 to 22 November 1989. The meeting was organised by the CEC DGXII/D-2, Directorate-General for Science, Research and Development and the French Commissariat à l'Energie Atomique CEA and hosted by the CEN Cadarache.

The purpose of the meeting was to review the experience gained to date in this field in some laboratories of different countries. Twenty- eight papers were presented to ninety-five delegates who represented ten countries and international organisations.

The presentation were divided into six sessions covering :

- active methods
- passive methods
- integrated and/or combined systems
- performance
- correlation
- operating experience

A short discussion period followed each paper presented. A final discussion was held at the end of the meeting on the state of the art in the field.

The CEC wishes to thank the authors and session chairman for their contribution to the success of the meeting. It also wishes to express its appreciation to SGN, the Commissariat à l'Energie Atomique and in particular to Dr Patrice Bernard for arranging and hosting this meeting.

C O N T E N T S

PART A

DEVELOPMENT OF MEASUREMENT

TECHNIQUES

SESSION A1 - ACTIVE METHODS

Chairman : R. Dierckx - J.R.C. Ispra

Opening statement

In this session three papers were presented.

T.V. Molesworth discusses the development of an integrated system, combining active and passive methods to monitor on a long-term basis the waste stream produced. The measuring should not need about 25 minutes. Simultaneous analysis of the different measuring systems should give a quite complete description of the waste distribution and quantity in the waste item. D.L.S. Findlay explained in the second part of this paper the active gamma and neutron interrogation techniques, based on a LINAC. The gamma and neutrons produced by appropriate targets, create reactions by the fission or (, n) reaction, the counting of which gives information on the fissile and fertile material content.

B.H. Armitage presented in detail two active methods : the dieaway technique and the Cf-shuffler. Used in combination with the segmented gamma scanner, measurements on standard samples show a linear increase of count rate with declared mass for both methods, except for extreme spatial position of the waste. A good correlation between the results of the two active techniques were found. Calibration with drums with representative matrices will ameliorate the absolute determination of the fissile content.

P.M. Rinard discussed a fully automized, self-diagnostic Cf-shuffler to control the 235U content in a liquid waste stream, to be operated in a hot cell for three months without manual intervention. A correction for different flow rates is worked out. The system was calibrated showing a linear response up to 0.5 g/l. Measurement at 0.034 g/l 235U, the alarm quantity, resulted in a reproducibility of about 10 %.

DEVELOPMENT OF AN INTEGRATED ASSAY FACILITY

T.V. MOLESWORTH
Taylor Woodrow Management and Engineering Limited., Hayes, U.K
M. BAILEY, D.J.S. FINDLAY, T.V. PARSONS, M.R. SENE and M.T. SWINHOE
Nuclear Physics and Instrumentation Division, Harwell Laboratory U.K.

Summary

The I.R.I.S. concept proposed the use of passive examination and active interrogation techniques in an integrated assay facility. A linac would generate the interrogating gamma and neutron beams. Insufficiently detailed knowledge about active neutron and gamma interrogation of 500 litre drums of cement immobilised intermediate level waste led to a research programme which is now in its main experimental stage. Measurements of interrogation responses are being made using simulated waste drums containing actinide samples and calibration sources, in an experimental assay assembly. Results show that responses are generally consistent with theory, but that improvements are needed in some areas. A preliminary appraisal of the engineering and economic aspects of integrated assay shows that correct operational sequencing is required to achieve the short cycle time needed for high throughput. The main engineering features of a facility have been identified.

1. INTRODUCTION

In 1985, an industrial consortium (Taylor Woodrow, Rolls-Royce, Fisher Controls*, and Radiation Dynamics**) prepared a conceptual design of an integrated non-destructive assay system for 500 litre drums of cement immobilised intermediate level waste (ILW) [1, 2]. The system, known as "I.R.I.S." (Immobilised Radwaste Inspection System), proposed to use passive gamma scanning, passive neutron scanning, gamma radiography, active neutron interrogation, and active gamma interrogation in combination to derive the maximum useful information about the waste [3].

A particular feature of I.R.I.S. was the use of a single electron linear accelerator (linac) with suitable targets, to provide neutron and gamma beams for the active interrogations and the radiography. Use of a linac neutron source was also expected to benefit fissile actinide detection sensitivity, because of the much greater neutron output, compared with a compact D-T 14 MeV neutron generator tube.

In 2010, the estimated cumulative volume of conditioned ILW in the UK is 138,000 m^3 [4]. A 500 litre drum has been chosen as the standard minimum size ILW container, and therefore there is a potential total of 276,000 drums to be dealt with. Public opinion might result in a demand for independent, 100 per cent assay of drums, before final disposal.

* now Siemens
** now Ray Technologies

Hence the I.R.I.S. concept envisaged assay of all waste drums, requiring a short cycle time of 15-20 minutes, with almost continuous operation.

Lack of detailed information about active interrogation, particularly as applied to 500 litre drums of cemented ILW, made the effective design of an I.R.I.S. unit quite impracticable. Previous work on neutron and gamma-ray interrogation systems using linacs has been carried out at Los Alamos, but on 200 litre non-immobilised low level waste [5]. The systems now being considered draw on experience obtained from earlier U.K. work [6, 7].

A three year, three stage research programme was therefore started in 1987, to provide more information, in particular about the two active interrogation techniques, and their application in an integrated assay facility of the I.R.I.S. type. The work is supported by the U.K. Department of the Environment, UK Nirex Limited., the Inspectorate of Nuclear Installations, the industrial consortium, and the Commission of the European Communities. It forms part of the European Atomic Energy Community's cost sharing research programme on "Development of Test Methods for Quality Assurance".

The first stage of the project consisted largely of desk studies and calculations, supported by limited experimental work. The findings were encouraging and a decision was made to go ahead with the second stage of the project.

Following on from the calculated predictions of performance described in [8], a full-scale experimental assembly has been constructed at Harwell to allow measurements to be made on simulated waste drums. Some preliminary results are described in this paper.

2. EXPERIMENTAL ARRANGEMENTS

A schematic drawing of the experimental assembly which has been set up in the Low Energy (LE) Cell of the Harwell electron linac HELIOS is shown in Figure 1. The assembly has been designed to accommodate both active neutron and active gamma-ray interrogation. In the neutron mode, fast neutrons from a composite tantalum/beryllium target penetrate the cement drum and after thermalisation produce a thermal flux throughout the drum. This thermal flux induces fission of fissile isotopes within the drum, and a small fraction of the resultant fast fission neutrons is counted in fast neutron detectors outside the drum. In principle, the net counts from the detectors are proportional to the fissile actinide content of the drum.

In the gamma mode, bremsstrahlung with a ~7 MeV endpoint from a ~1g.cm^{-2} tantalum radiator penetrates the drum and induces photofission and photoneutron reactions in all actinide isotopes within the drum. After thermalisation, the resultant fast neutrons are counted in neutron detectors outside the drum. The neutrons photoproduced by the bremsstrahlung from the deuterium naturally present in the water in the cement can be measured separately and then subtracted by reducing the bremsstrahlung endpoint energy to ~5 MeV. The latter is well below the characteristic fission barrier and neutron separation energies of ~6 MeV in actinide nuclei. The net counts are in principle proportional to the total actinide content of the drum. For both neutron and gamma interrogation, the source of interrogating radiation and the cluster of neutron detectors are on opposite sides of the drum.

When the experimental assembly is operated in its neutron mode, the neutron target is positioned in the mid-plane of the drum close to its curved surface, the centre of the target being 500 mm from the vertical axis of the 375 mm (nominal) radius drum. To minimise the gamma-flash from the target (the intense burst of direct bremsstrahlung photons and scattered gamma-rays produced by each pulse of the electron beam), the axis of the electron beam, and therefore of the bremsstrahlung beam driving the beryllium neutron-producing component of the target, is set at 90° to the direction of a line from the target to the neutron detectors.

When in gamma mode, the bremsstrahlung radiator is 1000 mm from the drum axis, and the axis of the electron beam and therefore of the bremsstrahlung beam intercepts the drum axis. Since for the present experiments the same electron beam line is used in both neutron and gamma modes, the whole assembly has to be rotated and translated when changing modes. To facilitate this, the assembly is mounted on large air-pads. The tantalum/beryllium target consists of a 3 mm total thickness tantalum bremsstrahlung radiator (configured as five 0.6 mm water cooled sheets) followed by water cooled beryllium blocks (totalling 200 x 200 x 50 mm) driven by a 15 MeV electron beam from HELIOS. The output has been calculated as described in [3] to be 1 x 10^9 neutrons/ μ C.

The bremsstrahlung radiator can also be used separately for gamma interrogation, and has been designed to be easily re-configurable for different total tantalum thicknesses. The bremsstrahlung radiator is preceded by an electrically insulated water cooled aluminium collimator so that the position of the electron beam spot may be continuously monitored during operation. Positioning the source and detectors on opposite sides of the drum produces a rough balance between the effects of variation of attenuation of the interrogating radiation and the variation in neutron detection efficiency. It is therefore very important that neutrons produced in or near the source do not leak around the drum to the detectors, and both drum and detectors are enclosed in borated concrete shielding. In addition, because of (i) the intense gamma-flash from the linac, and (ii) possible radiation dose rates of ˜1000 rads.hour^{-1} which are to be expected near the surfaces of 500 litre ILW drums [1], the counters have to be protected by lead shielding. The thickness of lead surrounding the counters in the present arrangements is ˜100 mm. The assembly has been designed to accommodate a wide range of counters. The detector channels at 90, 180 and 270° are intended for fast neutron detector packages for neutron interrogation, and the channels at 135 and 225° for thermal neutron detectors for gamma interrogation. Several different detector packages have been prepared, incorporating 50 mm diameter ^{3}He counters, "four-packs" of 25 mm ^{3}He counters (each of which should recover from overload more quickly than a single 50 mm counter), and 50 and 25 mm ^{10}BF$_3$ counters.

The data recorded from the experimental assembly are the time spectra (relative to the electron beam pulses) of neutrons counted in the detector packages, and the electron beam charge delivered to the targets. A block diagram of the data acquisition system is shown in Figure 2. The heart of the system is a multi-input, multi-shot time digitiser as used routinely for neutron time-of-flight experiments for nuclear physics and nuclear data purposes on HELIOS. The digitiser has been configured with eight separate inputs each with 1 microsecond time channels extending out to 2 ms, and is connected to a DEC LSI-11/73 computer. For each linac pulse, the digitiser clock is started with the electron gun trigger pulse, and the digitiser records the times of occurrence of neutron detector signals. The counter signal processing electronics preceding the digitiser has been arranged so that signals from up to sixteen counters can be multiplexed as required to provide a maximum of eight separate inputs to the time digitiser. (For example, the four separate outputs from a single neutron detector package incorporating four separate 25 mm counters - a "four-pack" - could be fanned into one digitiser input). The signals from the counters are also recorded in blind CAMAC scalers and (for visual checking of counter operation) in Harwell 2130 scalers. The electron beam charge deposited in the targets is digitised in a Brookhaven model 1000 current integrator. A mean electron beam current window is set using the analogue output from the integrator to ensure that the data acquisition system operates only within specified beam current limits. Linac gun trigger pulses and 100 pps clock pulses are also recorded. All the data recorded may be displayed during data collection on the 11/73 terminal.

3. EXPERIMENTAL RESULTS

Measurements have so far been made on a 500 litre drum containing cemented simulated CAGR waste with the experimental assembly in its neutron interrogation mode. A system of holes and plugs in the drum allow actinide samples and calibration sources to be placed at known positions within the drum. The neutron detection efficiencies for the various counters used were measured using a calibrated ^{252}Cf source.

Some results are shown in Figure 3. The neutron output from the target is monitored using a small $^{10}BF_3$ counter some 3m from the target. The counter is surrounded first by ~50 mm of lead and then by 1 mm of cadmium. The time spectrum from this counter, shown in Figure 3, reflects the long decay time of the epi-cadmium neutron flux in the LE Cell. The electronics are gated off for 100 μs following the gamma-flash, and the scatter of the time spectrum data points between 100 and ~200 μs is due to overloading of the electronics caused by the gamma flash. The thermal neutron flux in the drum was measured using a 13 mm diameter $^{10}BF_3$ counter inserted into a 20 mm hole running the length of the drum. A typical time spectrum is shown in Figure 3, and the characteristic 1/e decay time is 135μs, close to the 110μs expected [3].

Also shown in Figure 3 is the fast neutron time spectrum from one of the four 4-atmosphere 25 mm diameter 300 mm active length counters in a "four-pack" in the 180° detector channel for a 6g sample of ^{235}U at the centre of the drum. The initial fast decay up to ~200 μs is due to neutrons directly from the target, and from ~400 μs onwards the decay time is the same as that of the thermal flux within the drum (as would be expected for fast neutrons produced from thermal fission within the drum). The component between ~200 and ~400μs is probably due to neutron leakage from insufficiently tight shielding. However, if the component beyond ~400 μs is evaluated from 200 μs onwards, giving 5 x 10^{-4} for the total fast neutron counts beyond 200 μs, then upon division by the total neutron target output of 1 x 10^{14} for 0.10C delivered to the target, an overall interrogation response of ~1 x 10^{-10} counts per gram of ^{235}U per source neutron is obtained. This is lower than the ~3 x 10^{-10} predicted in [8], but is consistent with the unexpected presence of some neutron absorbing elements in the cement drum not included in the calculations as deduced in [6].

4. SOME ENGINEERING AND ECONOMIC ASPECTS OF INTEGRATED ASSAY

To develop a satisfactory operational facility, its functional requirements must be known. These include the throughput rate (which determines the cycle time), and the assay information to be obtained. The necessary examination processes can then be specified, and the engineering design carried out. The main components of a facility are shown by Table I.

TABLE I: - Main components of an integrated assay facility

Individual stations each accommodating one or more assay processes
Linear accelerator(s) with beam switching arrangements
Gamma and neutron targets
Collimating devices
Shielded enclosure including foundations and roof
Mechanical handling equipment for movement and positioning of drums
Data collection and processing equipment
Control systems
Monitoring equipment
Ventilation
Safety interlocks and barriers
Services, including lighting

Decisions have to be made regarding the number and sequence of assay stations, the number of linacs, use of single or multi-purpose assay stations, use of shield doors or labryinth shielding, and permitted levels of surface contamination on drums. Ancillary features might include package recognition, weighing, contamination checking, and post-examination drum handling.

The basic mode of operation is that each waste package to be inspected passes through a series of assay stations in sequence. As a package moves from any one station, it is replaced by a succeeding package. The cycle time of the complete system is equal to the longest residence time at any one station.

In a high throughput assay unit, the sequencing of operations should be such as to achieve the lowest cycle time. A convenient method of describing and analysing a system is with an operational sequence diagram, an example being shown at Figure 4. The following factors have to be taken into consideration: movement time between stations, station loading and unloading times, examination times, and special time allowances, e.g: for decay of any induced radioactivity.

More than one examination technique can be used at a single station e.g: active neutron and gamma interrogation, and passive neutron scanning, as envisaged in the original I.R.I.S. concept. However, it may be difficult to obtain a satisfactory composite design.

The requisite interrogating beam energies can be obtained from a single linac, although outputs may be more problematical. However, in order to use a single linac for different purposes at more than one station, the electron beams need to be re-directed by physically moving the linac and/or by magnetic switching. Although cheaper, there are various disadvantages.

Investigation of all the various facets of assay facility design and operation is far from complete, but some preliminary findings are of interest.

For a high throughput unit, it is considered that drums can best be moved by powered rollers with through-wall drives. Examination stations will require drum rotation, some with drum elevating capability. Accurate positioning of drums to achieve precise and predictable geometry is likely to be essential. A labyrinth is preferred to shield doors. If possible, drums should be free from external contamination.

With appropriate assumptions about the various individual operation times, the lowest achievable cycle time was found to be 25 minutes, for a 4 station, 2 linac system. This had the highest capital (£11m) and highest annual operating costs (£1.3m). On a continuous basis, the unit was capable of assaying over 19,000 drums per year at a cost each of £182.

If a high rate of throughput is not required, because, say, only a small percentage of drums is assayed, then everything becomes far easier. However, neither the capital nor annual operating cost will be reduced in proportion, and the cost per drum assayed will be increased substantially.

To date, no specific guidelines regarding assay have been laid down by regulatory or other concerned authorities in the U.K. However, there are indications that adequate quality assurance of the waste package production processes will be demanded. Systems will be installed that will, in theory, prevent the occurrence of packages with unidentified contents.

5. CONCLUSIONS AND FUTURE WORK

The measurements presented here constitute the first test of an experimental assembly constructed primarily to check the predictions of [8]. The results indicate that on the whole the system is operating satisfactorily, but they also highlight areas where improvements are necessary if the full potential is to be realised. In particular, efforts will be made to reduce neutron leakage around the drum to a minimum.

When such improvements have been made, the variation of the interrogation response throughout the drum will be measured. The assembly will then be fully tested in its gamma interrogation mode. In addition, measurements on a Magnox waste drum and a PCM drum will be made.

The experimental work is expected to provide a sound technical basis for assay system design, but to complete such design work, definition of assay requirements is needed.

REFERENCES

1 FAIRHALL, G.A. Proc. BNES Int. Conf. Radioactive Waste Management, Brighton, May 1989, 79.

2. LOCKWOOD, N. Ibid., 85.

3. DAVIES, I.LL. et al. IAEA SM 303/7. Int. Sympos. Management of Low and Intermediate Level Radioactive Wastes, Stockholm, 1988.

4. SMITH, M.J.S. et al. Nirex Report No. 48, Jan. 1988. "Radioactive Waste Arisings in the U.K."

5. FRANKS, L.A. et al. Nucl. Instr. Meth. <u>193</u>, (1982), 571.

6. PACKER, T.W. and SWINHOE, M.T. Harwell Report AERE R 13137 (1988).

7. FINDLAY, D.J.S. Harwell Report AERE R 12863 (1989).

8. MOLESWORTH, T.V. et al. Proc. BNES Int. Conf. Radioactive Waste Management, Brighton, May 1989, 178.

<u>Disclaimer</u>

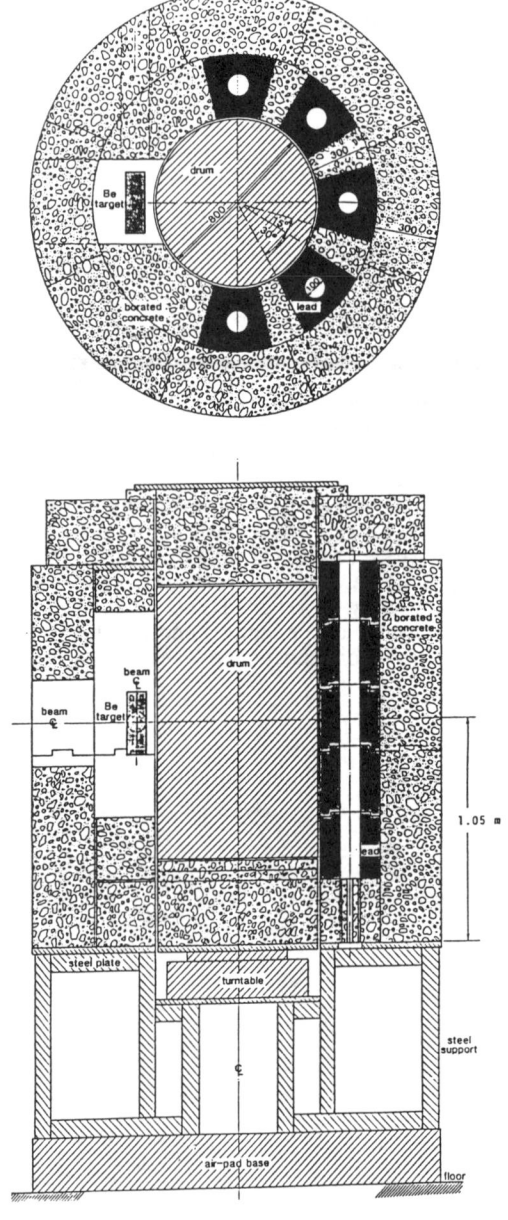

Fig. 1　Schematic diagram of horizontal and vertical sections of the experimental assembly, shown set up for neutron interrogation. The concrete "shims" on which the drum and the top shielding plug rest are arranged to accommodate a 1300 mm high drum.

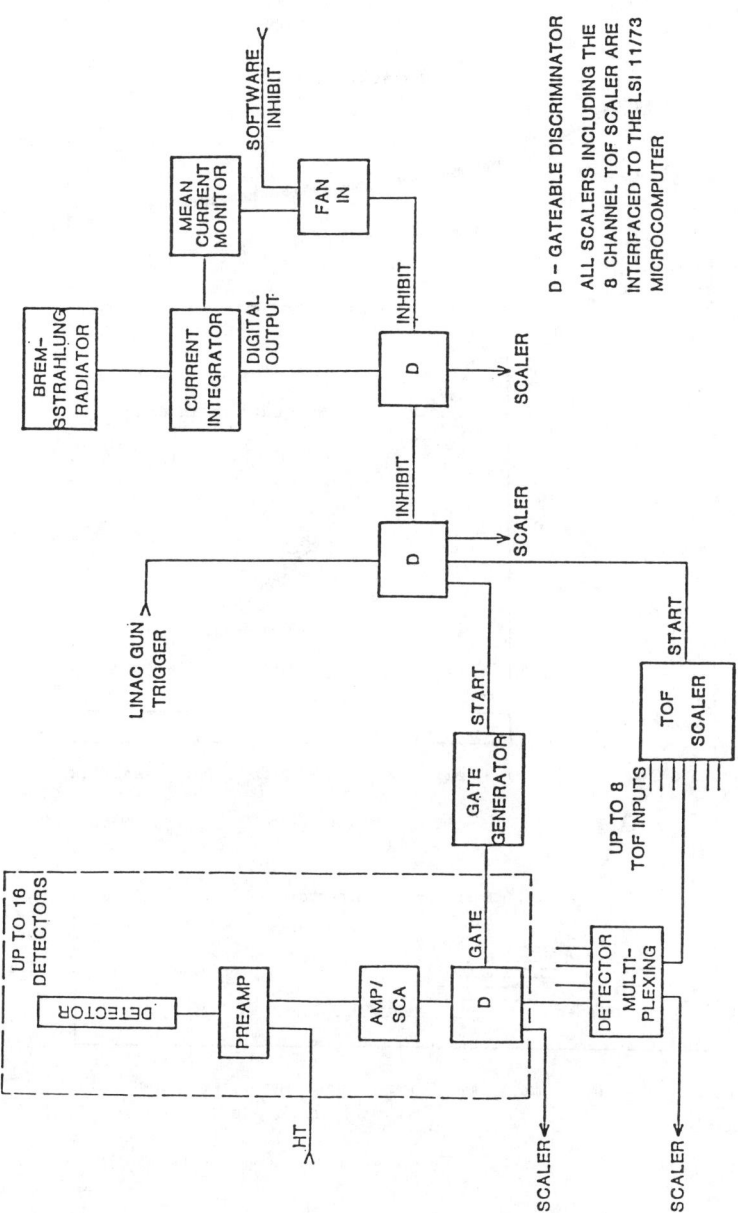

Fig. 2 Block diagram of the data acquisition system.   The "tof scaler" (time-of-flight) is the time digitiser referred to in the text.

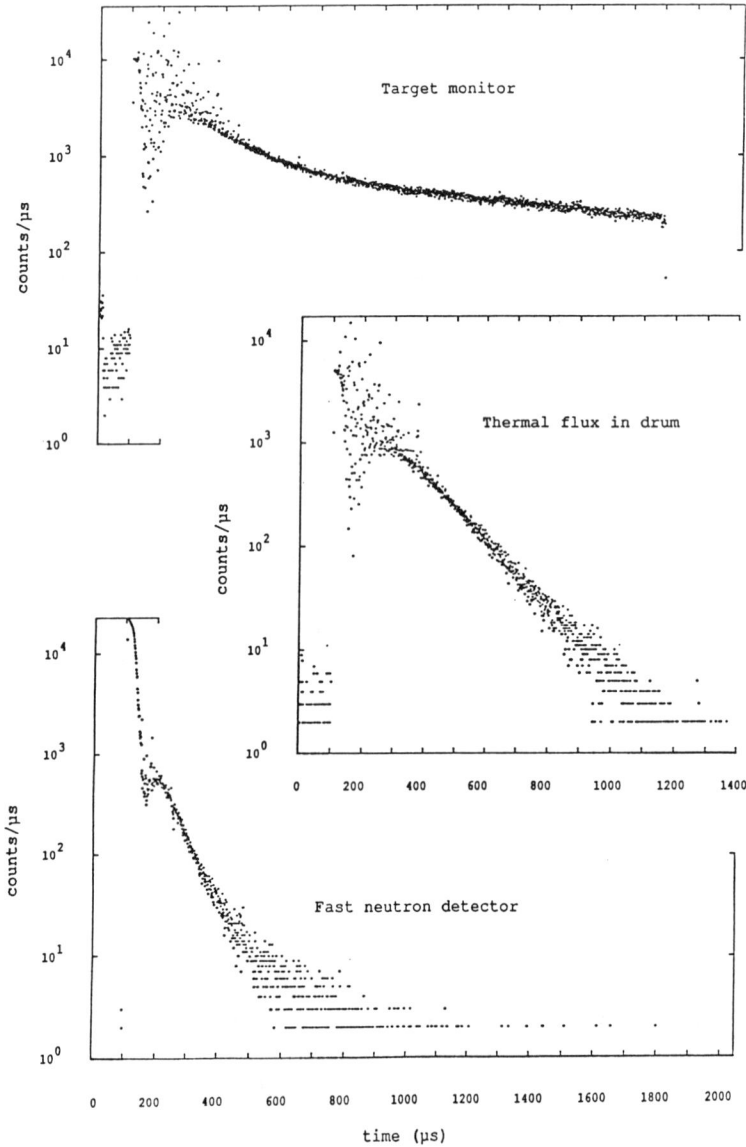

Fig. 3   Time spectra of neutron counts from three different neutron detector packages. Not all data points are shown.

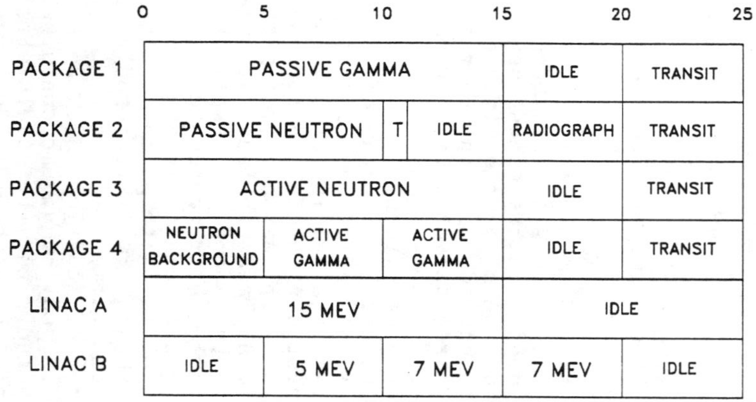

Fig.4   Typical operational sequence diagram

	0	5	10	15	20	25 MIN
PACKAGE 1	PASSIVE GAMMA			IDLE	TRANSIT	
PACKAGE 2	PASSIVE NEUTRON	T	IDLE	RADIOGRAPH	TRANSIT	
PACKAGE 3	ACTIVE NEUTRON		ACTIVE GAMMA	IDLE	TRANSIT	
PACKAGE 4	NEUTRON BACKGROUND	ACTIVE GAMMA	ACTIVE GAMMA	IDLE	TRANSIT	
LINAC A	15 MEV		IDLE			
LINAC B	IDLE	5 MEV	7 MEV	7 MEV	IDLE	

At any point in time, a vertical section indicates what is happening to each of the packages that is present in the system.

The time needed for the whole examination process is determined by the availability of the linac or by the activities taking place at one of the stations (e.g. package 2 in System 4). This gives the longest time that package must stay at any one position, including any intermediate holding positions if required.

The sequence followed by any one package can be traced by moving from left to right through each package line in turn.

A transit time of 1 minute is assumed for lifting the drum between passive neutron examination and radiography — in addition to the "normal" transit time of 5 minutes between stations. However, no time is allowed for switching targets between active gamma and active neutron interrogation.

DEVELOPMENT OF ACTIVE NEUTRON INTERROGATION TECHNIQUES

AT HARWELL

B.H.ARMITAGE, P.M.J. CHARD, T.W PACKER, M.T.SWINHOE and D.B.SYME
AEA Technology, Harwell, United Kingdom

Summary

Active neutron interrogation techniques capable of measuring the fissile content of a range of waste drum sizes and contents have been developed at Harwell. This paper describes measurements which have been made to investigate the behaviour of these assay systems for the difficult case of concreted waste in a heterogeneous matrix. The drums have been measured using a Cf shuffler and a differential die-away system, with supporting information obtained from a segmented gamma- scanner. Good correspondence has been observed between the two different neutron interrogation techniques. It was concluded that the measurement of highly heterogenious wastes is likely to be more effective if calibration can be undertaken with representative artificial matrices. Further measurement and analysis remains to be undertaken.

1. INTRODUCTION

The development at Harwell of active neutron interrogation techniques for assay of fissile material has been influenced by the diversity of waste forms present in the U.K., and also by the fact that 235U or fissile Pu may be dispersed throughout the volume of a disposal container, or concentrated into lumps or aggregates. Therefore the emphasis has been on a generic programme to allow measurements in the widest set of conditions. The focus of effort has been on the differential die- away technique using compact pulsed neutron sources, and more recently attention has been directed to neutron interrogation with a Cf shuffler source.

At Harwell we have made an extensive study of the effects of the matrix on differential die-away measurements. Such effects can lead to very large variations in response from a given amount of fissile material in large volume drums. Our work has involved the investigation of 21 different matrices in the 200 l drum size [1,2]. It has lead to a procedure whereby during the assay additional measurements are made both to characterize the matrix type and to apply the appropriate compensation. The result of this procedure is that a drum containing an unknown matrix can be assayed to an accuracy of 25% provided the matrix lies within the range of the 21 considered, and provided the matrix is reasonably homogeneous.

In addition a short programme of matrix characterisation has been undertaken with a Cf shuffler [3]. Here various waste forms have been examined in smaller volume containers (60 l).

This report is concerned with the field testing of our active neutron interrogation systems with 200 l waste drums. It is also concerned with the use of the segmented gamma- scanner in order to provide information on heterogeneity and the location of U and Pu within drums.

2.   MEASUREMENTS ON WASTE DRUMS

For the field tests of differential die-away (DDA) and shuffler measuring systems, a number of 200 l waste drums containing 235U were identified.   The documentation on these drums (Table I)indicated that in the majority of cases the 235U

Drum No.	Declared contents		Comments
	235-U g	239-Pu g	
1	97		
2	57		
3	99		
4	99		
5	23	14	
6	97		
7	57	10	
8	2	13	
9	42	1	
10	27		
11	96		
12	98		
13	50		non-shreddable items
14	66		93% 235-UO2 pellets
15	77		
16	94		
17	27		

TABLE I: Properties of measured radioactive waste drums.

was immobilised in a single concrete-filled tin (volume 2.5 l).   In turn, most of these tins were placed in a 60 l can, and the can was subsequently placed in a 200 l drum.   The remainder of the drum was generally filled with shredded waste. Additionally, several of the drums contained smaller quantities of Pu.   The overall average density of the drums was about 0.5 g/cm3.   These drums present a difficult problem for neutron assay in terms of neutron penetration and heterogeneity.

The drums were measured using a modified segmented gamma-scanner [4].   This gave two pieces of information for each drum: (1) the location of the 235U in the drum based on emission of 186 KeV gammma- radiation and (2) the mean attenuation of the drum for several different gamma- energies (from a 75Se source) as a function of height.   The results confirmed the presence of higher attenuation close to the 235U (agreeing with the presence of concrete) and also showed that 10 of the 17 drums had the 235U within 5-30 cm of the bottom of the drum.   The results also showed that the mean attenuation through the drum varied with height in a similar manner for all the drums examined.   It was observed that the attenuation was approximately constant in the lower part, but decreased to about one third of this value in the upper part of the drum.

Additional information on the position of the U (and hence the concrete) within the drum was obtained from the 186 KeV gamma-ray intensity at a number of different drum orientations as the drum was rotated. Hence, if the concrete occupies a position along the axis of the drum, which is otherwise filled with uniform waste, a constant U count should be observed.

Table II lists the results of the measurements. In the third column of the table, the 186 KeV gamma- radiation was measured with the drum in continuous rotation and at a height corresponding to the observed position of the 235U. The ratio of the highest to the lowest 186 KeV intensity (fourth column), observed with eight equally spaced azimuthal measurements, gives an indication of the distance of the U from the centre of the drum. It is observed that except for Drums 1 and 14 the U gamma-ray count scales with the declared contents, and Drums 4 and 10 have U far from the axis of the drum.

Drum No.	Height of U in drum (cm)	U gamma count relative to declared contents	Azimuthal variation of U count (high/low)	Position of U in drum (B= bottom,T=top C= central axis)
1	65-75	7	3	T,C
2	60-75	26	4	T
3	60-75	25	6	T
4	10-25	21	57	B
5	15-25	32		B
6	40-55	26	3	T,C
7	5-15	36		B
8	*			
9	20-30	25	6	B
10	8-15	22	35	B
11	20-30	23	5	B
12	10-30	18	5	B
13	60-80	43	2	T,C
14	40	2		
15	8-20	26	6	B
16	10-20	38	3	B,C
17	20-30	17	4	B

* no measurement

TABLE II: Information on the position of U within the drums
--------- derived from segmented gamma- scanner measurements,

2.1 Shuffler Assay

The measurements were made using an all-polyethylene passive neutron detection chamber [5], adapted for use as a Cf shuffler. The measured limit of detection for this system (using in this case a low intensity (5×10^7) 252-Cf source) in a plant environment is approximately 6 g 235U for drums of low passive output during a measurement period of 20 minutes.

Because the construction of this chamber was different from that used in the earlier experiments with the californium shuffler (mentioned above), it was not possible to use an absolute calibration for these drums.

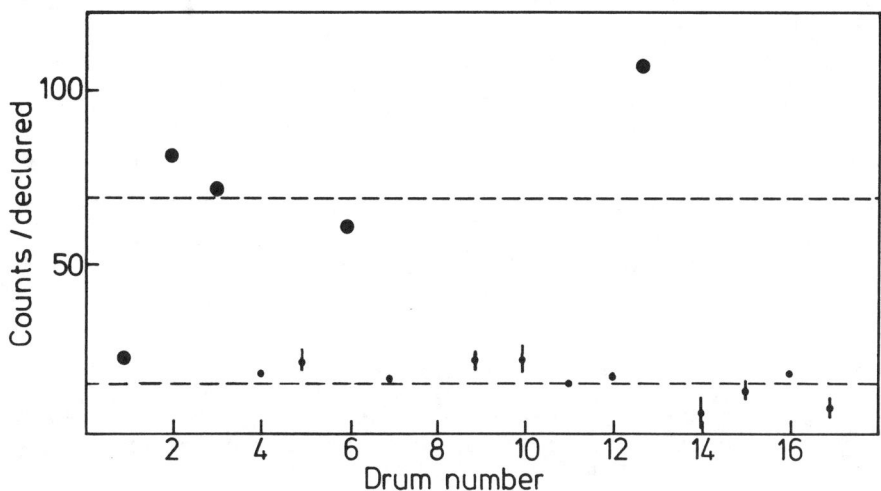

FIGURE I: Display of the measured response relative to the declared drum contents. The irradiation position of the Cf source is at height of the centre of the drum. The large circles are for drums where the concrete is not near the bottom.

In the first set of measurements the neutron source was moved to the mid-height of the drum for four seconds, and withdrawn, and after a delay of about 1 second, counting took place for 4 seconds. Each drum was measured at four equally spaced azimuthal orientations for 20 irradiation/counting cycles. The difference between these individual results was often large, (greater than a factor of two).

The mean response for each drum, expressed as the number of measured counts divided by the declared 235-U equivalent, is given in Table III The results are also presented in Figure I, where the large circles represent drums where the concrete (and the 235-U) is not near the bottom of the drum. Here it can be seen that the drums with the uranium at the bottom lie approximately on a straight line of constant ratio while the remainder of the drums have a significantly greater relative response. However there are two exceptions. The records for Drum 14 show that it contains 93% 235-U in pellet form, so that the low result is probably due to neutron self-shielding. No explanation for the relatively low result for Drum 1 (where the uranium is not at the bottom of the drum) has, as yet been found. A similarly low result was also obtained for this drum with the U gamma-ray measurements.

The highest relative result of all is obtained from Drum 13. The explanation for this is also unknown, but it is noted that it contains non-shreddable items.

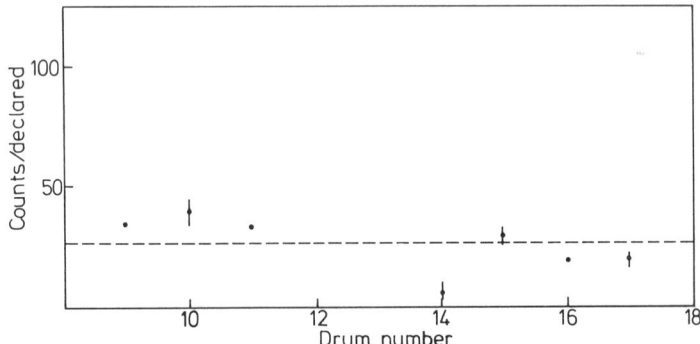

FIGURE II: Display of the measured shuffler response
----------   relative to the declared contents. The Cf source
              irradiation position is at one-third of the drum
              height. The large circles are for drums where the
              concrete is not near the bottom.

     In the Cf shuffler matrix characterisation work referred to
in the introduction, two BF3 monitors (one bare for thermal
neutron counting, and one enclosed in polyethylene and cadmium
for fast neutron counting) were placed under the drum. Those
measurements indicated that the monitors were capable   of
tracking the bulk properties (neutron absorption and neutron
moderation) of artificial matrices. The present measurements,
however suggested that the bulk properties of the waste drum
matrices differed little one from another.
     Additional measurements were made with the irradiation
position lowered from the centre of the drum to a position about
one third up from the bottom. The results which are shown in
Figure II indicated that the absolute response increased by
about 50% for those drums with uranium at the bottom of the
drum, and decreased by about 10% for the remainder of the drums
chosen for remeasurement. Again, self-shielding effects are
observed with Drum 14 containing 235-U in pellet form.
     The results indicate that the shuffler technique is capable
of reliable assay of drums with 'difficult' contents provided
different categories of drum can be distinguished by additional
information (in this case SGS measurements and/or plant records
). In plant conditions a calibration drum would be used to give
absolute values.
     Further work on the analysis of the drums is required.
This will involve calibration of the shuffler with known fissile
samples and known matrices. It will also include Monte-Carlo
calculations of the response using a model based on the recorded
contents of the drums.

2.2 Neutron die-away assay.

     The differential die-away measurements were made in a

graphite and polyethylene chamber designed to take a single 200
l drum [5]. The measurement consisted of observing the response
of eight fast neutron detector packages within the chamber
walls. Calibration was based on the response of a 0.54g 93%
235U sample in the form of a 0.025mm foil at the centre of an
empty 200 l drum. The sensitivity of the system is such that it
is capable of observing fissile material at the mg level. The
thermalised interrogating flux was monitored by a bare BF3
detector attached to the chamber wall. An additional BF3
monitor was located in direct contact with the underside of the
drum. This detector (the external matrix monitor) acts as a
monitor of the drum contents or matrix and a Cd shield beneath
it ensures that it is sensitive only to thermal neutrons
emerging from the matrix.

Measurements were made with each drum at four equally
spaced azimuthal orientations. This was necessary in order to
take into account the asymmetry of the assay system due to the
presence of the pulsed neutron source at one corner of the
chamber. It was noted that in general rotation of the waste
drums resulted in little change in response.

Matrix correction for the waste drums was made by following
the prescription described earlier [1]. In summary, the
procedure consisted of making a correction for the fact that
fast fission neutrons cannot be counted for the first 0.4ms
following the intense burst of 14 MeV neutrons. To correct for
this the external matrix monitor response (which is unaffected
by the 14 MeV burst) integrated for the first 5ms after the
pulse, is divided by the corresponding quantity integrated
between 0.4ms and 5ms following the burst.

A further matrix correction is based on the fact that the
counting efficiency for fast fission neutrons is dependent on
the moderating properties of the matrix. As explained in [1]
this is taken into account by the second matrix correction $E^{-0.2}$
where E is obtained by measuring the count rate obtained from a
Cf source under the waste drum relative to the same quantity
obtained with an empty drum.

Here the combined matrix correction factor C is given by

$$C = E^{-0.2} \frac{EMM(0-5)}{EMM(0.4-5)}$$

Further guidance on the properties of the matrices can be
obtained from the measured lifetimes of thermal neutrons as
obtained in the M and EMM detectors. Thus, in the earlier
report [1] it was concluded that matrices for which T(M)/T(EMM)
is between 2.5 and 4.5 are characterised as being highly
moderating and highly absorbing. Matrices for which the ratio
is greater than 4.5 were found in ref.[1] to be subject to large
assay errors. The reason for this is that the response varies
greatly depending on whether the fissiles are in the inner or
outer regions of the drum. Thus, in one example it was found
that the response in the outer regions is about 11 times that at
the centre. Hence,there is an expectation that better agreement
with the declared contents will occur for waste drums that fall
within the range of matrices examined in ref.[1].

The results of the assay are given in Table IV. It can be
seen that of the thirteen drums examined by DDA eight have
matrices outside the limits previously set for adequate matrix

compensation. It must be emphasised, however, that the matrix compensation procedures have been developed for homogeneous matrices. We have as yet made no systematic attempt to examine the effects of gross inhomogeneity in matrices. However, it is clear that the EMM will be most sensitive to the composition of the matrix material in its immediate vicinity. It is noted that the EMM values do not appear to be sensitive to the position of the concrete in the drums. This must be due to the fact that, for the drums where the concrete is near the bottom, the gamma-measurements indicate that typically there is 15cm of matrix material between the concrete and the bottom of the drum

Drum No.	T(EMM) (ms)	T(M) (ms)	T(M)/ T(EMM)	Matrix Cat. 4	C	DDA meas-ured 235U equiv-alent (g)	DDA meas-ured/ declared
1	0.08	0.52	6.5	no	3.72	24	0.25*
2	0.05	0.51	10.2	no	2.73	35	0.61*
3	0.07	0.52	7.4	no	3.26	69	0.70*
4	0.08	0.51	6.4	no	2.85	19	0.19
5	0.11	0.53	4.8	no	3.07	4.8	0.11
6	0.08	0.55	6.9	no	3.05	52	0.54*
7	0.14	0.49	3.5	yes	2.46	26	0.36
8	0.05	0.49	9.8	no	2.08	7.2	0.32
9	0.14	0.52	3.7	yes	1.94	8.7	0.20
10	0.15	0.50	3.3	yes	2.69	9.8	0.36
11	0.06	0.50	8.3	no	2.48	9.4	0.10
12	0.12	0.51	4.2	yes	2.37	18	0.18
13	0.12	0.52	4.3	yes	2.62	43	0.86*

* concrete not near the bottom of drum.

TABLE IV: Results of DDA measurements

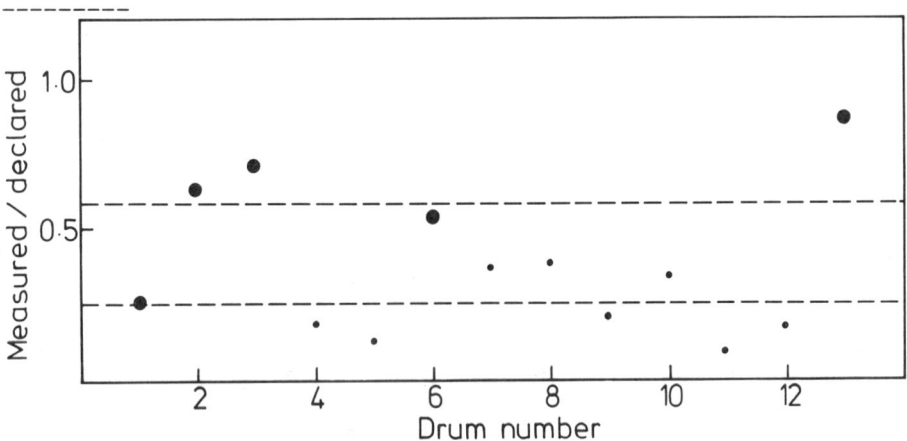

FIGURE III: Display of measured DDA response relative to the declared contents. The large circles are for drums where the concrete is near the bottom.

It is observed that the best absolute agreement with the declared contents occurs when the concrete is not at the bottom of the drum. The explanation for this may be that in such cases the concrete is surrounded by matrix material more susceptible to neutron penetration. Evidence that this may have occurred has already been found in the gamma- attenuation measurements, where the attenuation in the upper part of the drums was found to be less than in the lower part. However Drum 1 is an exception, and the depressed response may be due to the presence of the concrete at an inner position in the drum (as suggested by the gamma- measurements) and also to the matrix being a more severe case than those examined in ref [1]. The best agreement with the declared contents occurs for Drum 13 whose matrix appears to be within the limits previously examined.

3. COMPARISON OF RESULTS FROM NEUTRON INTERROGATION TECHNIQUES

The relationship between the neutron interrogation results obtained with the two instruments is shown in Figure IV, where the shuffler counts are plotted against the DDA response. As can be seen there is a good measure of agreement, although not all the drums have as yet been measured by both techniques.
The results of the die-away measurements indicate that the matrix compensation procedures of ref.[1] are not adequate to provide a reliable assay for the heterogeneous concrete-

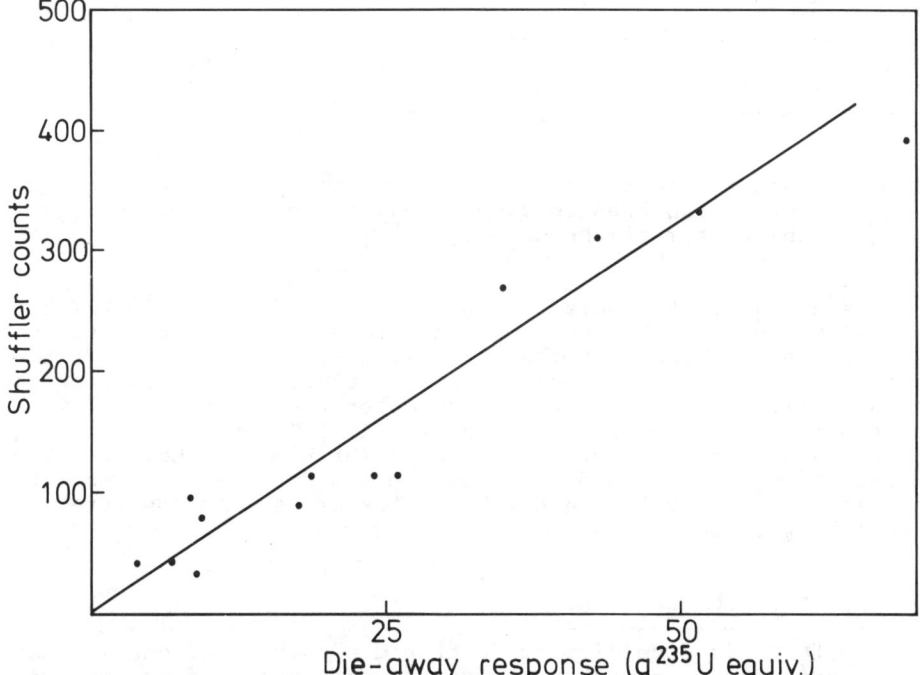

FIGURE IV: Plot of shuffler counts against DDA response.

containing waste drums considered. A more effective approach is to use the existing documentation on the waste drums to construct artificial matrices, one with concrete near the bottom of the drum and one with the concrete at a central position. Confirmation of the appropriate calibration drum will then be obtainable from the segmented gamma- scanner data.

Drum No.	Shuffler counts /declared g		DDA measured /declared g	
1.		22*		0.25*
2.		82*		0.61*
3.		72*		0.70*
4.	19		0.19	
5.	22		0.11	
6.		61*		0.54*
7.	17		0.36	
8.			0.32	
9.	20		0.20	
10.	21		0.36	
11.	14		0.10	
12.	15		0.18	
13		109*		0.86*
14	6			
15	14			
16	10			
17	8			
Mean	15	69*	0.23	0.59*
Sigma (%)	34	41	43	34

TABLE V: Measured shuffler and DDA responses per declared g. The asterix applies to drums where the concrete is not near the bottom of the drum.

Assuming that these calibrations lead to a reliable normalisation of the neutron interrogation data, the neutron die-away results indicate that the standard deviation of the assay errors for the eight drums with concrete near the bottom is 43% while the corresponding figure for the five drums with the concrete in a central position is 34% (TABLE V).
If a similar procedure is adopted for the Cf shuffler the corresponding standard deviations are 34% for the 11 drums with concrete near the bottom and 41% for the drums with concrete at a central position.

4.   MONTE CARLO MODELLING

Monte Carlo modelling of neutron die-away interrogation has been undertaken in order to seek confirmation of an enhanced response for drums where the concrete is not close to at the bottom. The details of the model were based on the known mass of the waste drums (circa 100kg) and the SGS data which indicated that the gamma-attenuation was typically about three

times less in the upper region of the drum than in the remainder.
The calculations consisted of comparing the response from two drums each containing 2.5 l of concrete, one at a height of 15cm from the bottom and the other at a height of 60cm. In each case the remainder of the drum was filled with a mixture of shreddable and non-shreddable waste, at a density of 0.4 g/cc up to a height of 60cm, and at a density of 0.13 g/cc up to the top. The results indicated that the response increased by a factor of about 2.5 when the concrete was at the higher position. This is in good agreement with the ratio (2.57) of the mean values of 0.59 and 0.23 (see Table V) obtained for the measured neutron die-away responses with the concrete in the two locations.

4. CONCLUSIONS

Consideration of the results obtained from the Cf shuffler and from the neutron die-away interrogation leads to a number of general conclusions:

1. There is good correspondence between the measured response obtained with neutron interrogation between the Cf shuffler and neutron die-away.

2. The interpretation of waste assay measurements by neutron interrogation (both neutron die-away and shuffler) is greatly aided by the use of a SGS. For this particular type of shreddable matrix filling the SGS measurements agree well with the declared contents.

3. The measurement of highly heterogeneous wastes is likely to be more effective if calibration can be undertaken with representative artificial matrices.

Measurement and analysis are continuing to complete the study of this set of drums. At a later stage it is intended to unpack the present drums in order to be able to more clearly relate the measured results to the actual contents.

References
----
1. B.H.Armitage and A.C.Sherwood (1987), 9th ESARDA Symposium on Safeguards and Nuclear Material Management, London

2. B.H.Armitage and M.J.Cogbill (1988), AERE Report R-13065.

3. T.W.Packer, M.T.Swinhoe and D.B.Syme (1989), 11th ESARDA Symposium on Safeguards and Nuclear Material Management, Luxembourg.

4. K.P.Lambert, J.W.Leake and G.Wakefield (1986), AERE Report R-11615.

5. T.E.Sampson, B.H.Armitage and T.L.Morgan (1985), 7th ESARDA Symposium on Safeguards and Nuclear Material Management, Liege.

MONITORING A LIQUID WASTE STREAM
WITH A DELAYED-NEUTRON INSTRUMENT

P. M. Rinard, T. Van Lyssel, K. E. Kroncke,
C. M. Schneider, and S. C. Bourret
Los Alamos National Laboratory, USA

Summary

A flowing raffinate stream is to be continuously assayed by a delayed-neutron instrument to detect concentrations of ^{235}U that could cause a criticality problem in a holding tank. The instrument is to assay a concentration of 0.034 (g ^{235}U)/L in 100 s with a precision of 10% (1 σ) and to operate unattended for a few months at a time, so it can detect and adjust for changes in the neutron background, the flow rate, and for electronic drifts and malfunctions. In laboratory tests with conditions slightly different from what may be found in the plant, repeated assays on a solution with 0.034 (g ^{235}U)/L flowing at 80 L/h through the 2-L assay tank had relative precisions of 9-11%.

1. THE MEASUREMENT PROBLEM AND TECHNIQUE

An instrument has been built to monitor the concentration of ^{235}U in a flowing raffinate stream so that a criticality accident in the holding tank into which the stream flows can be avoided. The major measurement criterion is to assay a solution holding 0.034 (g ^{235}U)/L with a 1-σ precision of 10% within 100 s. The flow rate may range from 0 to 102 L/h, although the rates are usually between 60 and 100 L/h. The instrument is to run unattended for as long as 3 months. Warnings and alarms are to be generated when concentrations exceed certain limits.

It is not possible to measure the gamma rays emitted by the ^{235}U because of the intense gamma-ray background from fission products in the solution. Exposure rates from these fission products may be as large as 10 R/h on the surface of the assay chamber.

The neutron emission rate from spontaneous fissions is too small to meet the assay criteron because of the low concentration of ^{235}U and the low rate of spontaneous fissioning.

The assay technique chosen for this instrument is thus an active neutron interrogation that counts delayed neutrons; such an instrument is often called a shuffler [1,2]. Previous work at Los Alamos with a shuffler and static solutions formed the basis for this project [3,4]. The detector can be made insensitive to the gamma rays and can detect enough delayed neutrons to be sufficiently sensitive in the time allowed.

2. OVERVIEW OF THE INSTRUMENT

To generate fissions that produce delayed neutrons, the raffinate in an assay chamber is irradiated with neutrons from a ^{252}Cf source. The delayed neutrons are counted after the ^{252}Cf is removed to a shielded position. The source is shuffled back and forth a number of times to form a single assay. The irradiation and count times can be selected to yield the minimum uncertainty in the assay result, within the constraint of the time specified to complete the assay.

The ^{252}Cf source is doubly encapsulated, attached to a flexible cable, and positioned by a stepping motor. Proximity switches sense the presence of the cable at three locations, two for overtravel protection, and a third to verify that the source is properly within its shield while delayed neutrons are counted.

The assay chamber and neutron detector tubes must be inside a large hot cell with the raffinate line. The detector head will be mounted on a wall about 10 m above the floor (Fig. 1). The assay chamber is a cylindrical annulus with a volume of about 2 L. The liquid flows into the chamber from the bottom and out through the top. A ring of ^{3}He detector tubes (mounted in polyethylene) encircles the chamber behind 2.5 cm of lead to keep the gamma-ray dose rate in the tubes below 1 R/h. Figure 2 is a drawing of these components.

The detection geometry and ^{252}Cf source size were determined through Monte Carlo calculations using the Los Alamos code MCNP [5].

A flow meter in the raffinate line will inform the instrument of the current flow rate and a correction will be made for flow rates that differ from a standard rate of 80 L/h.

The irradiation position for the ^{252}Cf source is in the center of the assay chamber. This uses the source most efficiently and requires the minimum source. A guide tube for the ^{252}Cf source rises from the center of the chamber and enters a wall of the hot cell. The tube continues through the wall inside a polyethylene plug and then into the corridor where the stepping motor and most of the electronics are located.

The source will be kept within the wall except when it is to irradiate the liquid. Additional shielding will be placed on the wall to further reduce the dose rates to personnel

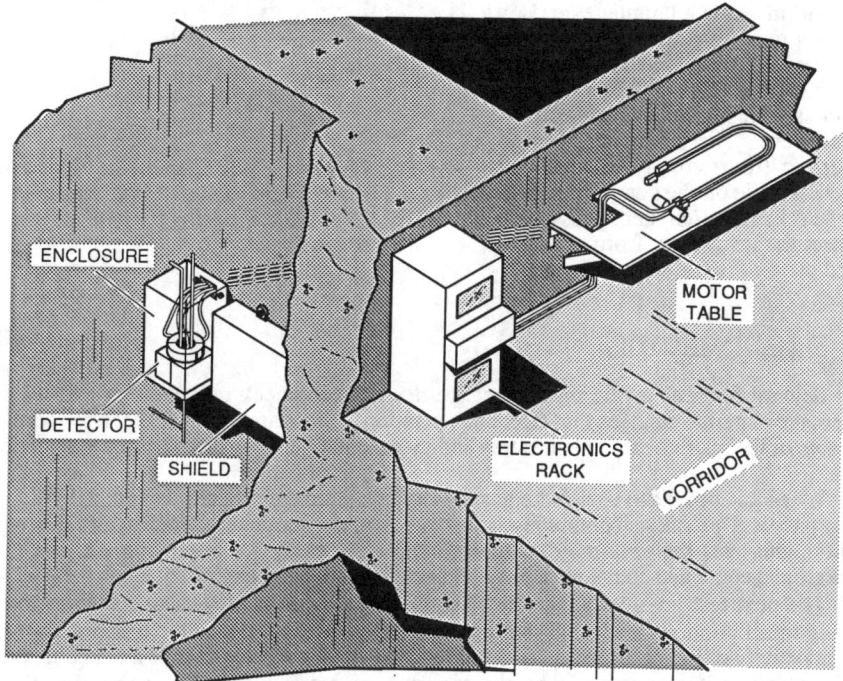

Fig. 1. The assay chamber of the instrument is mounted on a hot cell wall with a raffinate line entering from the bottom and leaving from the top. The guide tubes for capsules containing ^{252}Cf and ^{235}U pass through a plug in the wall into a corridor; electrical cables travel the same route. An electronics rack and stepping motors for the capsules are in the hall.

working in the vicinity of the instrument during shutdown periods of the plant. The shielding is designed to keep dose rates below 0.1 mrem/h in the corridor and below 50 mrem/h on the surface of the wall shields.

The major pieces of electronics are high voltage supplies for the detector tubes, a multichannel scaler board (designed and built at Los Alamos), a computer, and a stepping-motor indexer. The computer communicates with the indexer over a serial line and then with other components through ports in the indexer (which are available to the user). A transfer system interface unit, built at Los Alamos, electrically interfaces the stepping-motor indexer to the proximity switches, status lamps, and alarm relays.

Software for the computer was written in the C language at Los Alamos to control the operation and perform analyses. The user can configure the operation of the instrument by the settings in an extensive list of parameters. The parameters can be modified from within the code and saved as a file on the computer's disk.

A key switch and passwords protect the instrument from unauthorized use.

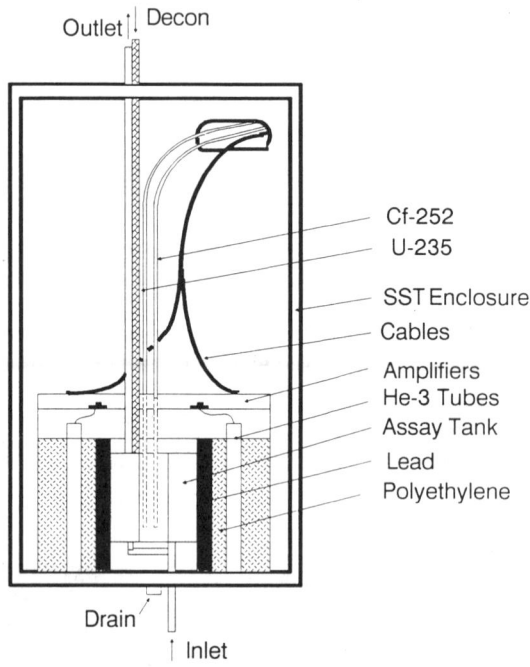

Fig. 2. This is a view of the enclosure with the assay chamber and surrounding components, as seen when facing the wall on which they are mounted.

3. SELF-DIAGNOSTICS

Several features of the instrument are designed to check its operation and watch for external or internal changes that affect its operation. These are important features because of the role of the instrument in plant safety and its need to operate unattended for long periods of time.

When an authorized user (who has the key switch in the proper position and has given a valid password) requests continuous assays to begin, the computer first exercises the stepping motors and tests the proximity switches, lamps, and relays. The instrument then takes a background count and compares it to the expected range in the parameter file.

The next test involves a capsule of solid ^{235}U, which is also used at regular intervals during the continuous assays. The normal concentration of ^{235}U in the liquid is well below the sensitivity of this instrument, so almost all assays are expected to give a result of zero. To make sure that a series of zero assays does not mean a failure of the instrument, the ^{235}U capsule is placed inside the assay chamber using a flexible cable and stepping motor, duplicating the mechanism used for the ^{252}Cf capsule. If an assay with this ^{235}U capsule included is outside limits set by the operator in the parameter file, the operator is informed of a fault.

This normalization process with the ^{235}U sample is performed before continuous assays are started. If the result is out of the expected limits, continuous assays will not be started.

After all these test results are favorable, assays are performed continuously. They will continue until the key switch is turned to the stop position.

The continuous series of assays is interrupted briefly (at intervals set by the operator in the parameter file) to check the background count rate and repeat the normalization assay with the ^{235}U sample.

If a background count rate is outside the preset limits, the operator is informed but assays continue with the new background rate. If a subsequent check of the background shows that it has returned to within the preset limits, the instrument uses the background rate set in the parameter file.

Similarly, if the normalization assay is outside its preset limits, the operator is notified and an appropriate normalization factor is applied to the assays until a subsequent normalization assay is within the preset limits.

The ^{3}He detector tubes are divided into two banks; each has its own power supplies. If an electrical failure should occur in one bank, the computer will detect it because the counts in the two banks are compared at the completion of every count time. A failure sends an alert to the operator, but assays continue with the good bank.

Each detector bank also includes a small ^{10}B-lined detector tube. These tubes monitor the irradiation flux from the ^{252}Cf source, so they are close to the assay chamber and not shielded by the lead. Boron-10-lined detectors are much less sensitive to gamma rays than are the ^{3}He detectors and are unaffected by the radiation from the fission products. The usual role of flux monitors in shufflers is to respond to neutron spectrum changes caused by varying amounts of moderators (such as water) in the materials being assayed. This is not expected to be a problem with this instrument, but the adjustment capability is provided. While the count rates from the ^{10}B tubes are within the limits set in the parameter file, no correction to the ^{3}He count rates is made for changes in the irradiation flux. The operator may even choose not to use these flux monitors through a simple entry in the parameter file.

Assay results are supplied to a plant computer through a serial communication line and an analog voltage line. If no transmission is received over the serial communication line for a few minutes, the plant computer will assume there is a problem and alert the operators.

3. DATA ANALYSIS

Several corrections are made to a raw count rate from the ^{3}He detectors before a calibration curve is used to convert the count rate to a ^{235}U concentration.

The first correction is for the decay of the ^{252}Cf source relative to a reference date. The current background, caused almost entirely by the ^{252}Cf source itself, is also corrected for the decay of the source and subtracted from the delayed-neutron count rate.

If the operator has elected to use the flux monitors, a correction for irradiation variations is made if the measured count rate in the flux monitor is outside the limits given in the parameter file.

The flow rate of the liquid through the assay chamber has a small but readily measured effect on the assay result. A high flow rate washes away delayed-neutron precursors quickly and would produce a lower concentration if no correction were made. For example, a solution moving at 80 L/h would give an uncorrected assay result that is 14% low. Changing the flow from 80 L/h to 100 L/h without the computer's knowledge leads to an additional 3% error in the assay result. The flow-rate correction is shown in Fig. 3.

During an assay the computer records the times actually used while moving the ^{252}Cf source, while irradiating the liquid, and while counting delayed neutrons. A correction is calculated for the minor deviations of these times from the standard times given in the parameter file. If a motion problem causes a timing correction larger than the limit set in the parameter file, the operator is informed of a problem. This correction is usually less than 1%.

4. LABORATORY EXPERIENCE

4.1 Laboratory Conditions

The instrument has been tested at Los Alamos using a small flow loop containing 4-M nitric acid solutions with various ^{235}U concentrations. The mass of the ^{252}Cf source during

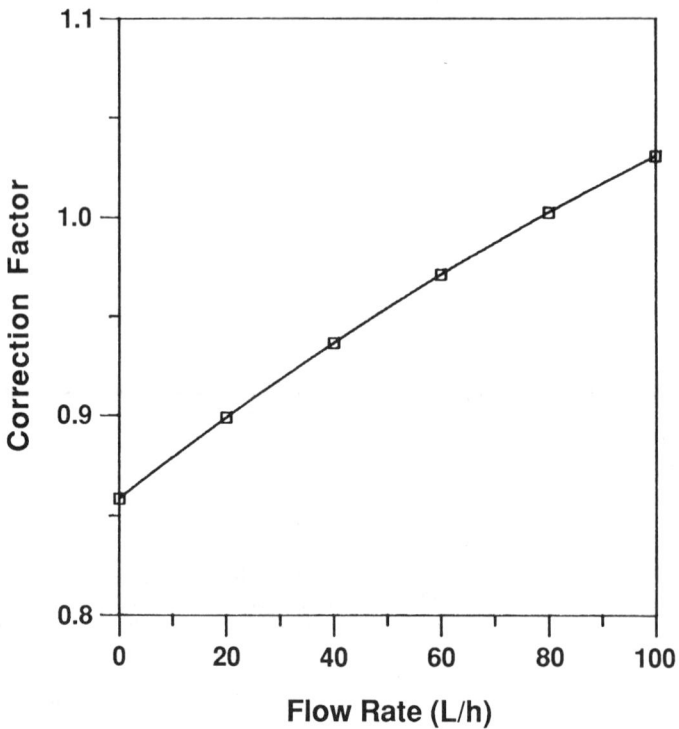

Fig. 3. This correction factor is applied to the delayed-neutron count rates, according to the flow rate of the raffinate stream. The correction is relative to a rate of 80 L/h, the rate most commonly expected.

these tests was about 61 μg. The source was stored within a shield of polyethylene, but still produced a background count of about 190 counts/s.

The precision of an assay was optimized by using five shuffles of the source, each of which irradiated the solution for 11 s; delayed neutrons were counted for 7 s. About 1 s was needed for each 209-cm movement of the ^{252}Cf capsule. Each assay is thus completed in the required time of 100 s.

4.2 Calibration

A provisional calibration curve was developed for these solutions and is shown in Fig. 4. The uncertainties in the concentrations were specified by chemists who analyzed the solutions. The curve is linear and essentially passes through the origin within the precision of the data; the slope is 727.29 (counts/s)/(g/L).

The concentrations below 0.05 g/L are of primary interest. A concentration of 0.48 g/L was also measured because it is an alarm point for approximate assays calculated after each individual shuffle (that is, every 20 s); this concentration is not shown in Fig. 4 for clarity in the region of lower concentrations. The multiple data points at each concentration are measurements of the same solution at flow rates of 40, 60, 80, and 100 L/h. The flow rates at the plant are expected to be between 60 and 100 L/h, so the count rates are adjusted to the flow rate of 80 L/h (using Fig. 3) to minimize the magnitude of the adjustments.

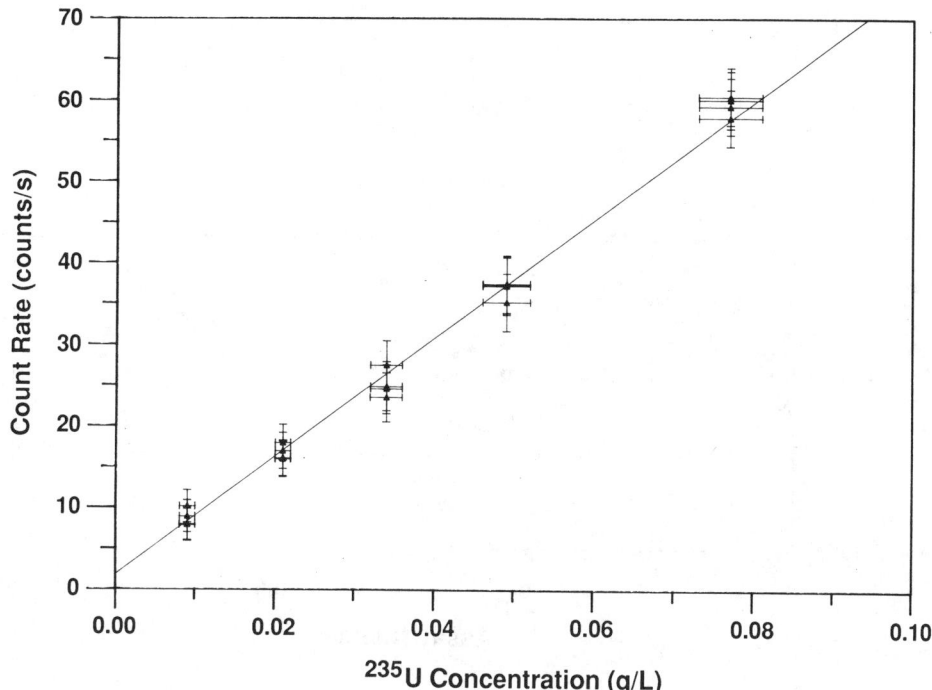

Fig. 4. A provisional calibration curve was developed for the laboratory conditions. Uncertainties in the concentrations are taken from chemical analyses. At each concentration there are corrected count rates from flow rates of 40, 60, 80, and 100 L/h. The concentrations of major concern are below 0.05 g/L; data points at 0.48 g/L were also taken but could not be included in this figure without obscuring the more important data points at lower concentrations.

4.3 Precision at 0.034 g/L

The precision at the important concentration of 0.034 g/L was given special attention, although solutions and conditions in the plant may produce a different precision.

A series of 137 consecutive assays at this concentration gave the results shown in Fig. 5, using the calibration of Fig. 4. The average assay value is 0.0348 g/L with a standard deviation of 0.0031 g/L (or 9.0%). A chemical analysis showed the concentration of this solution to be 0.034 ± 0.002 g/L. Other sets of such measurements gave concentrations of 0.0341 and 0.0323 g/L with standard deviations of 10% and 11%.

The precision expected from counting statistics alone is calculated to be 9%. The standard deviation of the measurements also includes contributions from fluctuations in the flow rate, ^{252}Cf positioning variations, and electronics drifts, all of which seem to be small and relatively unimportant.

4.4 Normalization Checks

The normalization sample has 1 g of ^{235}U. After 13 repeated normalizations, the average equivalent concentration was 0.357 with a standard deviation of 0.006 g/L (or 1.8%). Such a large value will assure the plant that the shuffler is working properly.

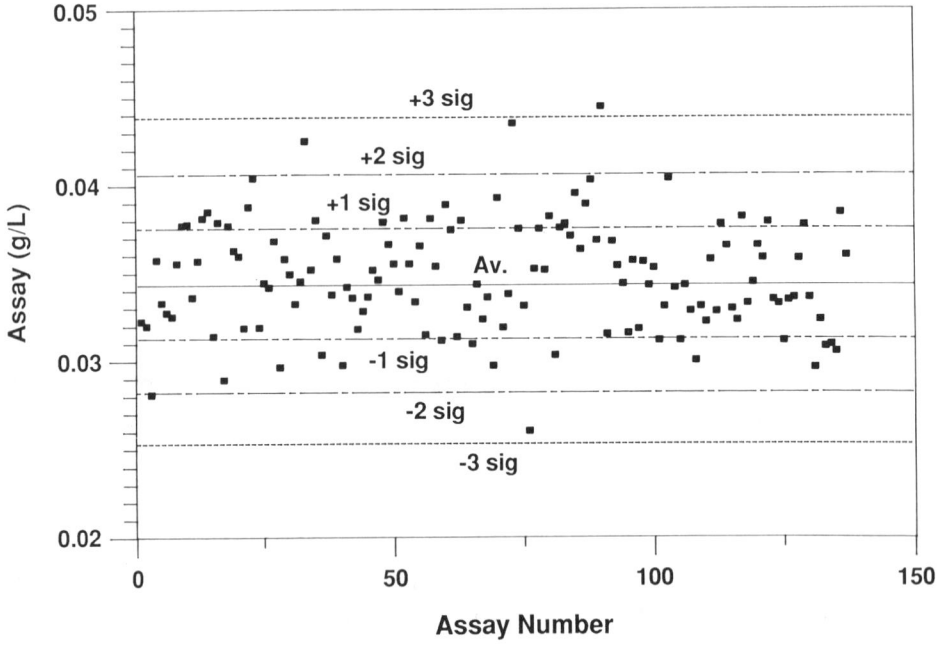

Fig. 5. *With a solution of 0.034 g/L flowing through the instrument at 80 L/h, 137 consecutive assays were taken. The sequence of results is shown along with its average value and uncertainty bands.*

5. PLANT INSTALLATION

Installation is scheduled to begin in the fall of 1989, and a few months will be needed to prepare the hot cell and mount the equipment.

A test loop will be permanently connected to the instrument; it can be calibrated periodically with known solutions that more closely match the raffinate liquid.

For several months, the instrument will assay the raffinate output of the plant on a trial basis. If its performance is satisfactory, it will become a routine monitor for criticality protection.

ACKNOWLEDGMENT

The authors are indebted, for the use of the ^{252}Cf source, to the U.S. Department of Energy's Californium Industrial Loan Program as administered by the Office of Nuclear Materials Production through the facilities of the Oak Ridge National Laboratory.

REFERENCES

1. Rogers, D. R., *Handbook of Nuclear Safeguards Measurement Method* (NUREG/CR-2078, MLM-2855, Division of Facility Operations, Office of Nuclear Regulatory Research, U.S. Nuclear Regulatory Commission, Washington, DC, 1983).

2.   Gozani, T., *Active Nondestructive Assay of Nuclear Materials* (NUREG/CR-0602, SAI-MLM-2858, Division of Siting, Health and Safeguards, Office of Safeguards Development, U.S. Nuclear Regulatory Commission, Washington, DC, 1981).

3.   Crane, T. W., "Liquid Sample Shuffler," Los Alamos National Laboratory report LA-10291-MS (January 1985).

4.   Crane, T. W., "Liquid Sample Shuffler for Active Gamma Solutions," Los Alamos National Laboratory report LA-10925-MS (February 1987).

5.   Briesmeister, J. F., editor, "MCNP - A General Monte Carlo Code for Neutron and Photon Transport, Version 3A," Los Alamos National Laboratory report LA-7396-M, Rev. 2, Manual (September 1986).

SESSION A2 - PASSIVE METHODS

Chairman : R. Odoj - KFA-Jülich

In seven presentations of the session passive non destructive assays
were explained and the results were demonstrated. The presentations
dealed with neutron measurements as well as gamma-spectrometric
investigations. The different investigations showed clearly, that both
methods are used for different nuclide contents in the waste and are
complementary.

The presentation of R. Carchon and P. de Baere described a feasibility
study for enlarging an existing neutron system for 281 drums to a
measuring system for 2201 barrels. The resulting neutron well counter
with VCD electronics is very flexible and can easily be built up in
half a day. The measured samples had a Pu weight ranging from 0.25 -
5g. It is concluded from the investigation that the minimum detection
limit is approximately 30 mg Pu-total. To make a correct interpretation of the PU-total amount in the waste barrel, the isotopic
composition has to be known.

The paper of B.H. Pedersen and W. Hage on the effects of the spatial
distribution of plutonium in waste matrices dealt with the examination
of the pair and triple correlations for the assay of Pu contaminated
low beta- and gamma-radioactive waste. Following the mathematical
model it could be proved with Pu-samples ranging from 2-20 g that the
mass dependence is well treated by the triple correlation method in the
case of a voided matrix. The pair correlation method leads to better
results in a concrete matrix. Best results were obtained inserting the
α-ratio i.e. the ratio of the (α,n) reaction rate to the spontaneous
fission neutron emission rate. The time correlation analyser proves to
be a reliable instrument capable of applying both pair and triple
correlation analysis methods.

The determination of the radionuclide inventory in waste packages by
gamma-spectroscopie methods was covered in 2 presentations of J.A.
Suarez Gonzalez del Rey and G. Pina from Spain on one side and B.R.
Martens and P. Filss from Germany on the other hand.

The first presentation by Suarez et.al. described the basic mechanical
components of the experimental device and the results obtained in the
quantitative calibration of the equipment. Differences in the calibration of the equipment with the QCY-44 standard from the British
Calibration Service and with a Eu-152 standard led to some modified
abundance values. The measured efficiency curve showed a small variation of the efficiency with matrix density depending on the gamma-
radiation energy. The conversion of the efficiencies obtained with
real packages of the some density show a good similarity. Loss factors
of efficiency for real scale packages due to the attenuation in the
matrix can be calculated from the results obtained with smaller
packages.

In the second paper by Martens and Filss on the radionuclide inventory for waste packages using gamma-ray-detectors of low beam width a mathematical derivation was given for the countrate of a point source in different positions of a waste barrel. The results obtained for narrow detector cones show a minor dependence of the countrate with the radial position of the source in the barrel. It is very low for the typical μ-values of waste matrices. A further improvement is achieved by combining the countrates measured at two different directions of the collimator axis toward the barrel. By a combination of a 0° and 18° measurement the standard deviation decreases from 80% to 25%.

The two presentations of G. Simonet, France and C.H. Orr and K. Allred, UK, describe measuring methods for large sized objects to be investigated.

The remote Gamma-ray mapping by G. Simonet compares the pictures of photosensitive films and radiosensitive films. The used photographic device and the real time video camera are portable equipments which were used in different French nuclear facilities to detect radio-activity in cells, rooms and halls previous to decontamination or even dismantling. The Film efficiency differs for different gamma energies so that listing, localization and estimation of the activity is possible. A very simple equipment for the detection of risk areas and to examine effluent circuits and to monitor reprocessing operations is given.

The second paper of Orr and Alldred presents a more sophisticated arrangement of neutron coincident counting combined with High Resolution Gamma Spectroscopy. Passive neutron coincidence counting in used for the detection of Pu-240 eq, while HRGS is used for U-235 detection by the 186keV gamma radiation. Crated items arising from the decanning of Magnox fuel will be monitored with this equipment in order to provide a quantitative assessment of some 40 specified radio-isotopes. The measured items range from 1 m^3 to 25 m^3. The exact physical arrangement of the equipment was outlined and the arrangement will go in to operation at Sellafield in 1990.

STUDY OF A PASSIVE NEUTRON SYSTEM FOR
THE DETERMINATION OF Pu IN 220 L WASTE BARRELS

R. CARCHON, P. DE BAERE
Nuclear Research Centre, SCK/CEN, B-2400 Mol, Belgium

Summary

A study has been made of a passive neutron system for the determination of the plutonium content in 220 l waste barrels. A hexagonal
cavity for 28 l drums was adapted for 220 l drums and a variable
dead-time electronics system was used for counting. Optimum shielding
was studied such as to obtain the lowest background possible.
Two different matrices were studied: a hydrogenous organic matrix and
an ash matrix with a density of 0.40 g/cm³. Calibration curves have
been established, using PuO_2 samples up to 5 g.

1. INTRODUCTION

    The precise determination of the Pu-content in a great variety of
containers at different stages of the fuel cycle is of great importance.
Non destructive assay (NDA) techniques are often very helpful tools to
reach this aim, because of reasons of non accessibility or because sampling, being often difficult, may not be representative for the whole.
There exist a series of possibilities but the matrix is often setting
limitations to the method that can be applied.
    Some of these measurements are already in exploitation, such as gamma
ray scanning of 220 l barrels, and passive neutron counting on 28 l drums.
As waste barrels of 220 l, including light to moderate densities currently
have to be monitored, the investigation and eventually construction of
such a system was imposed.
    The study described here reveals only a feasibility study of such a
system, starting from the existing device for 28 l drums and reassembling
the different pieces of the well counter to form a cavity fitting around a
220 l barrel.
    The method described here is based on the detection of passive
neutrons from fissile materials in the waste barrel under investigation.
It remains clear that in order to be able to make a correct interpretation
of the total Pu amount in the waste barrel, the isotopic composition has
to be known.

2. EXPERIMENTAL SET-UP AND EQUIPMENT

    The neutron detector used in this experiment was assembled from 6
neutron detector banks, formerly used on a measurement configuration for
28 l drums [1,2]. A solid hexagonal steel frame, suitable for 220 l drums,
was designed, on the circumference of which the detector banks together
with their associated charge sensitive preamplifiers were fixed.

Each detector bank, 1 m high by 18 cm wide and 6 cm thick, contains 3 LCC ^{3}He proportional counters having 1 m active length, 2.5 cm diameter and 4 bar gas pressure. The counters are embedded in a polyethylene moderator block, which is cladded with a 1 mm thick cadmium sheet. The drum side of the blocks is covered with a steel plate of 2 cm thickness, serving as gamma shield and spectrum shifter.

The distance between two opposite detector banks is 67.5 cm. On the outer side (this means outside seen from the well cavity), a 3 cm thick polyethylene plate is mounted that serves for backscattering of fast and epithermal neutrons into the counter.

The analogue output pulses from the three counters are combined in a preamplifier which is fixed on the modular block.

The electronics used was the variable dead-time counter (VDC) system, consisting of pulse amplification and discrimination circuits for the six detector units, a pulse mixing circuit and one fast and four slow counters with nominal dead-times of 16, 32, 64 and 128 µs. The dead-time losses of the different pulse counters are analysed in terms of count rates of neutrons from spontaneous fissioning nuclides among the plutonium isotopes, being coincident neutrons.

The detection efficiency of the system was measured by means of the samples described in paragraph 3, and amounted to 5.3 %, representing a reduction of 2 % as compared to the original system developed for 28 l drums.

Shielding was studied and optimized in order to obtain the lowest background possible. The detector system was set up in a paraffin castle of 20 cm thickness in all directions. Tests have shown that in this way the background was lowered by a factor of 10. Because of the thickness of the paraffin shielding, no additional cadmium layer was necessary.

3. SAMPLE DESCRIPTION

The samples used for carrying out the proper measurements were contained in small aluminium boxes, and encapsulated in a double plastic bag.

The characteristics of the samples are given in table I, displaying

TABLE I: Sample data

Sample No.	Updated Isotopic Composition (18.05,88)						Pu weight (g)	^{240}Pueff mass (g)
	^{238}Pu (wt.%)	^{239}Pu (wt.%)	^{240}Pu (wt.%)	^{241}Pu (wt.%)	^{242}Pu (wt.%)	^{241}Am (ppm)		
M15/1	0.759	79.187	14.847	3.791	1.417	39437	0.252	0.049
M15/2	0.759	79.187	14.847	3.791	1.417	39437	0.504	0.098
M15/3	0.759	79.187	14.847	3.791	1.417	39437	0.757	0.146
M15/4	0.759	79.187	14.847	3.791	1.417	39437	1.009	0.195
M15/5	0.759	79.187	14.847	3.791	1.417	39437	2.523	0.487

for each sample the plutonium weight, the Pu isotopic composition and Am content updated to the measurement day, and the ^{240}Pu effective mass, ^{240}Pueff, calculated according to the following formula [3]:

$$^{240}Pueff = 2.66 \ f^{238} + f^{240} + 1.64 \ f^{242}$$

where the f^m's are the weight fractions of the plutonium isotopes with mass m.

The samples also contained 93 % enriched uranium in a ratio $\frac{Pu}{Pu + U} \approx 0.30$. Combinations of the samples were used in order to build up a mass range between 0.25 and 5 g total Pu.

4. MEASUREMENT PROCEDURES

Before starting the measurement series, an exact determination of the dead-times has been made with the help of an Am-Be pure (α,n) source of 1 Ci $(3.7 \; 10^{10}Bq)$ strength. The dead-times were calculated as being 15.47 μs, 31.55 μs, 63.54 μs and 127.55 μs. They were used throughout the calculations.

Measurements were steered from a Commodore 3032 CBM 32K microcomputer interfaced to the VDC electronics, starting every campaign with a background registration for a certain preset time.

The neutron background was measured several times each day and the last measurement was stored in memory in order to be subtracted from every measurement value.

The different coincidence count rate values Xi (i= 1 to 4) are calculated based on the registered count rates Ci (i = 0 to 4) in the different counters, according to the following formula:

$$Xi = Co - \frac{Ci}{1 - Ci\tau i} \qquad (i = 1 \text{ to } 4)$$

wherein τ^i is the appropriate dead time for counter i.
Co is the count rate of the total counts scaler
Ci is the count rate of the scaler with dead-time τi

These values are printed out and an appropriate ^{240}Pu effective mass is calculated. A statistical treatment is made of the 4 estimates, to yield the most probable value. However, it is envisaged to use the count rate value of the 32 μs scaler only, as this is best matching the die-away time of the detector assembly.

At low count rates, the Xi values are not very dependent on the precision with which the dead-times are known. On the contrary, when count rates become higher, it has been shown that in order to determine the systematic uncertainty in the VDC coincidence count rate Xi to less than 1 %, the dead-time τi of the slow scaler must be known to 0.03 % [4].

5. MEASUREMENT RESULTS

5.1. Background measurements
As explained in the previous paragraph, the background has been measured on several occasions to account for changes in measurement conditions.

This value amounted to 3 counts per second for all counters, and has been used in the determination of the detection limit of the equipment.

5.2. PuO$_2$ calibrations
Calibrations have been set up, using the samples described in paragraph 3. Combinations of these samples were also made in order to cover a mass range up to 5 g total Pu.

Two different matrices were investigated that were considered appropriate in relation to neutron counting in a passive way: the first was composed of hydrogenous materials such as paper, gloves, plastic materials, with a density of 0.1 or less, the second was made up of calcined clay balls of diameter 1 cm and density 0.4, simulating ashes of the incineration in which burnable waste was treated at 600°C.

The coincidence count rates obtained in this exercise are summarized in tables II and III concerning the light matrix and the ash matrix respectively. A least squares straight line was fitted to the different X_i values and the results are reproduced in table IV. Figures 1 and 2 give the graphical representation of these results.

The 1σ error indicated is calculated according to the following formula:

$$\Delta X_i = \frac{1}{\sqrt{t}} \left[\sqrt{C_o} - \frac{\sqrt{C_o} - \sqrt{C_o - C_i}}{1 - C_i \tau_i} \right]$$

5.3. Detection limit calculation

The detection limit is defined as that quantity of material that yields a signal three standard deviations above background [5].

For each matrix, four values are obtained for each X_i, of which an arithmetic mean has been taken. For the light matrix, a mean value is obtained amounting to 0.073 g Pu or 0.014 g ^{240}Pu$_{eff}$ and in case of the ash matrix, these values amount to 0.058 g total Pu or 0.011 g ^{240}Pu$_{eff}$ mass.

If these values are compared with those obtained with the original set-up, which was intended for 28 l waste boxes, that minimum detection limit was stated at approximately 30 mg total Pu.

6. CONCLUSIONS

A simple passive neutron detection system is obtained that is applicable to the measurement of Pu-contaminated waste, packed in 220 l barrels.

Flexibility is reached in its use and it can easily be built up in half a day, from the modules installed on a similar device for 28 l waste drums.

The system has already proven its usefulness. An effort could be made to improve detection efficiency and to decrease the detection limit. But even within these restrictions, the instrument is able to provide useful information. We could recommend to use the passive neutron detection system as part of an overall quality assurance policy.

ACKNOWLEDGEMENTS

Thanks are due to P. Van Iseghem and P. Vandewauwer for useful discussions and to M. Alen for preparing the manuscript for publication.

REFERENCES

1. BERG R., BIRKHOFF G., BONDAR L., BUSCA G., LEY J. and SWENNEN R. (1975). EUR 5158 e. On the determination of the Pu^{240} in solid waste containers by spontaneous fission neutron measurements application to reprocessing plant waste.

2. BIRKHOFF G., BONDAR L. and COPPO N. (1972). EUR 4801 e. Variable dead time neutron counter for tamper resistant measurements of spontaneous fission neutrons.

3. U.S. NUCLEAR REGULATORY COMMISSION (May 1984). REGULATORY GUIDE 5.34. Nondestructive assay for plutonium in scrap material by spontaneous fission detection.

4. LEES E.W. and HOOTON B.W. (1978). AERE-R 9168, AERE-R 9367, AERE-R 9701. Variable dead time counters.

5. CRANE T.W. (1980). LA-8294-MS. Measurement of Uranium and Plutonium in Solid Waste by Passive Photon or Neutron Counting and Isotopic Neutron Source Interrogation

TABLE II: Data for the light matrix

^{240}Pueff mass(mg)	X1(s^{-1})	X2(s^{-1})	X3(s^{-1})	X4(s^{-1})
49	0.036±0.003	0.061±0.004	0.081±0.005	0.105±0.006
97	0.067±0.006	0.110±0.008	0.152±0.010	0.186±0.011
146	0.099±0.009	0.154±0.012	0.234±0.016	0.279±0.018
194	0.115±0.012	0.192±0.015	0.267±0.018	0.327±0.021
291	0.143±0.016	0.244±0.021	0.389±0.027	0.442±0.030
437	0.238±0.024	0.367±0.031	0.534±0.038	0.697±0.046
485	0.265±0.023	0.412±0.029	0.650±0.038	0.754±0.043
679	0.391±0.041	0.638±0.054	0.882±0.068	1.109±0.084
825	0.453±0.051	0.719±0.067	1.096±0.088	1.302±0.108
968	0.470±0.055	0.837±0.075	1.195±0.096	1.448±0.121

TABLE III: Data for the ash matrix

^{240}Pueff mass (mg)	X1(s^{-1})	X2(s^{-1})	X3(s^{-1})	X4(s^{-1})
49	0.039±0.003	0.060±0.003	0.087±0.004	0.105±0.004
97	0.061±0.006	0.103±0.008	0.155±0.009	0.192±0.011
146	0.083±0.008	0.146±0.010	0.216±0.013	0.264±0.015
194	0.113+0.011	0.207±0.014	0.286±0.017	0.325±0.020
291	0.143±0.015	0.251±0.020	0.382±0.025	0.468±0.030
437	0.231±0.023	0.386±0.030	0.583±0.039	0.714±0.047
485	0.301+0.027	0.505±0.036	0.690±0.044	0.895±0.053
679	0.372±0.034	0.597±0.045	0.834±0.056	1.067±0.070
825	0.462±0.046	0.735±0.061	0.996±0.076	1.164±0.095
968	0.510±0.018	0.870±0.024	1.304±0.031	1.561±0.039

TABLE IV: Results of the fitting functions

light matrix	ash matrix
^{240}Pueff = 1.959X1 - 0.029	^{240}Pueff = 1.855X1 - 0.012
^{240}Pueff = 1.169X2 - 0.019	^{240}Pueff = 1.135X2 - 0.021
^{240}Pueff = 0.804X3 - 0.023	^{240}Pueff = 0.793X3 - 0.022
^{240}Pueff = 0.664X4 - 0.024	^{240}Pueff = 0.654X4 - 0.024

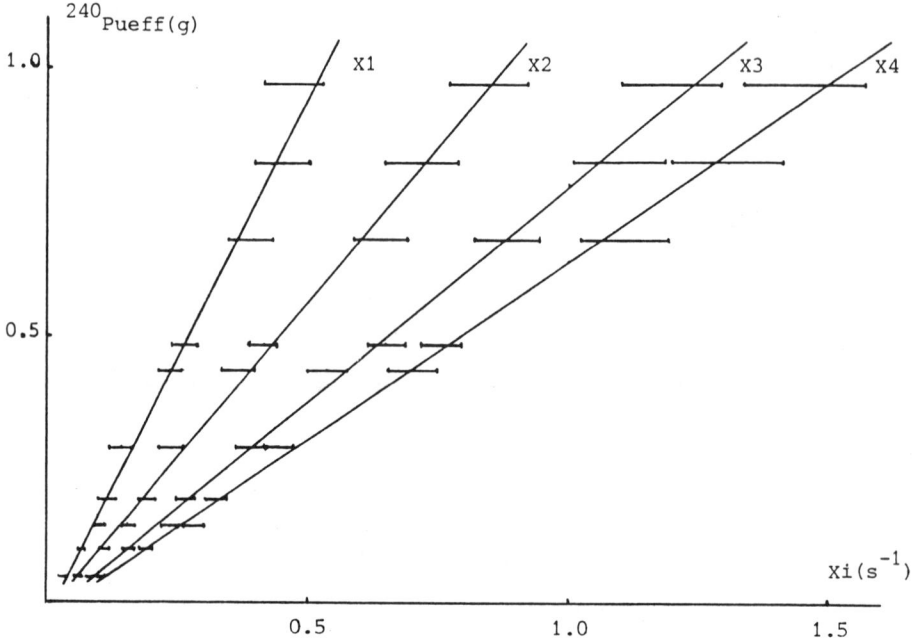

FIGURE 1. Calibration curves for the light matrix

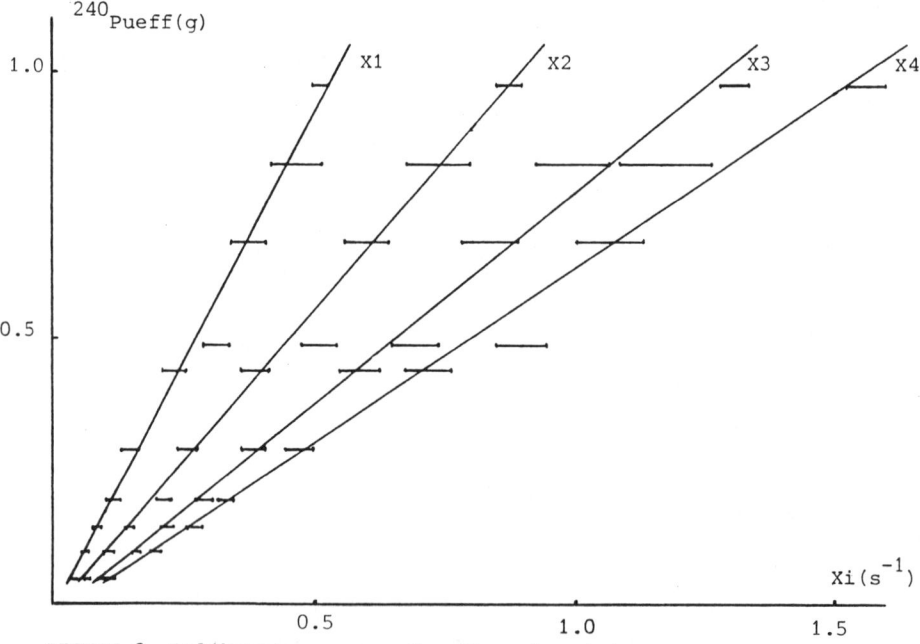

FIGURE 2. Calibration curves for the ash matrix

REMOTE GAMMA-RAY MAPPING : ALADIN
NUCLEAR INSTALLATION REMOTE ACTIVITY LOCATION DEVICE

G. Simonet

CEA - IRDI - DERPE, Saclay Reactors Department
91191 - GIF sur YVETTE (France)

ABSTRACT

The mapping of gamma activity is an important safety element and an essential prior step in many cases.

Classical problems :

o Organization of intervention in a "hot area".
o Drawing up of suitable estimates and decontamination procedures.
o Strategy of dismantling, decommissioning basic nuclear installation
o Containment monitoring, checking of basic materials and storage areas, systematic monitoring of risk areas and protections, detection of leaks.
o Examination of effluent circuits, monitoring of reprocessing operations.
o Waste management, quantification for conditioning.

Possible incidents to be processed :

o Localization of a radioactive source (site, hospital).
o Alarm to be identified.
o Technical fault, handling error, etc.

A localization technique of radioactive sources has been developed at DERPE CEN/Saclay. It is based on photographic camera principle. It essentially comprises a radiation proof "black-box", with an aperture and the view capture placed opposite.

(1) In the first static version, a conventional photographic-type emulsion photographs the area while a second radiation sensitive emulsion gives a more or less contrasted and spread spot corresponding to each source.

(2) In the second version, the landscape is identified in real time by a video camera. A light-transparent scintillator screen converts gamma-rays to light, visible by the camera.
In both cases, the gamma transparent obturator is open only during the picture shot.
The superimposition of the acquisitions, made in strictly identical geometric conditions, let immediately know the location of each source of radiation in the observed area.
Picture processing is digitally performed in order to facilitate the presentation and to improve the accuracy of the dosimetry, taking into account physical data and calibration parameters.

1. PRESENTATION

Nuclear Installation Remote Activity Localization Device (Appareil de Localisation d'Activité à Distance en Installations Nucléaires or "Aladin, a marvelous lamp") allowing to display radioactivity.

The industrial development of nuclear activities requires the drawing up of precise directives for interventions; that involves use of thorough techniques providing complete information : list, localization, and estimation of activities.

1.1 Principle

If a picture of radioactive sources can be recorded, such sources will be pointed out : as the radioactive radiation prints particular screens, it is therefore possible to extend the analogy, of light to gamma radiation, of the darkroom to a radiation-shielded container, and of the pinhole to a collimator.

It is sufficient to take the photography of a certain landscape and the radiography of the associated radiation, using two media (photographic and radiographic). After acquisition, the two records are superimposed in order to provide an immediate and strictly objective view of the locations of each emissive areas. The equipment is particularly efficient for high dose rates and multiple sources (as they are dissociated).

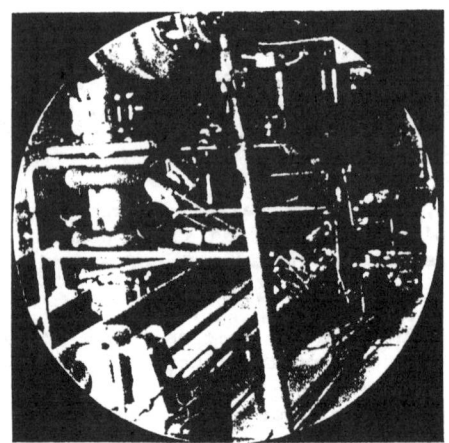 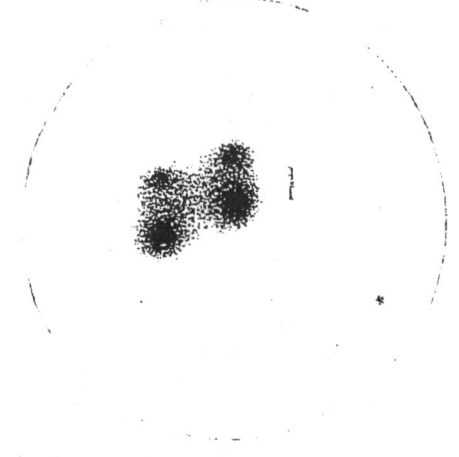

Fig. 1 - Valve gallery : photosensitive film, radiosensitive film.

1.2 Feasibility

Figure 1 shows an example of equipment use in EL.3 reactor's valve gallery. Before shutdown of the reactor, maintenance interventions on the valve controls and other mechanisms increase. After three clad failures, activity is such that access is only possible for a few minutes.

Gamma ray mapping enables to localizate the main activity at the base of the four expansion bellows on the main pipework. Interventions are then possible after placing local protective shields. Active valves being located, subsequent dismantling is facilitated. Relevant valves are cut out with an explosive cord.

1.3 Present resources

a) scintillators or intrinsic detectors (medical scanner type)
b) ionization chamber (babyline type).

The present facilities, suited to laboratory measurements, are quite satisfactory with regard to information on activities they receive. But they become inefficient when there are several sources, due to lack of discrimination about origins of emitted activities. Therefore, it is essential to finalize the proposed equipment, which complements the range of radioprotection equipment.

2. EQUIPMENT

2.1 "DARKROOM", PINHOLE-COLLIMATOR, OBTURATOR

The Darkroom-Pinhole assembly allows to get a clear picture regardless the distance. That avoids both space and adjustments related to optical systems. Furthermore this principle avoids blackening problems of the lenses submitted to nuclear environments.

The propagation of gamma radiations also follows a straight line. So their impact spot localizes the origin direction, if they are canalized through an efficient pinhole, that means if the detector only receives information through this aperture.

The higher the density, the better the attenuation factor of gamma rays: a tungsten alloy is choosen, its density of 19 gives it a good tightness to gamma rays.

Such an alloy choice allows to minimize the shield thickness, thus space and consequently weight. The final assembly is portable and miniaturized, compatible with remote handling ports (170 mm dia.) for introduction into hot cells.

Shielded pinhole forms a double-cone collimator. A geometry giving the best compromise between conflicting and non independent parameters is looked for (large sight angle - small focal lenght and good clearness, small aperture and small exposure time ...) : the optimization requires mathematical formulation of a good contrast and a good definition.

Used sight angle is near 60°.
Commutable collimators are available.

A shutter, transparent to gamma rays, is open only during exposure time for visible light.

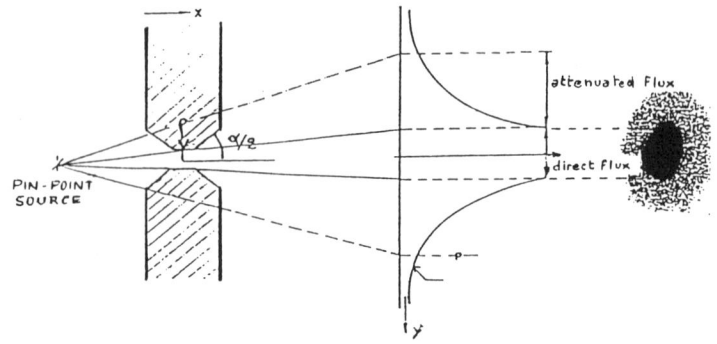

Fig. 2 - Radiation transmission.

2.2 STATIC VERSION "PHOTOGRAPHIC CAMERA"

For this option, the pictures are taken on photosensitive and radiosensitive emulsions.

2.2.1 Arrangement of Films

In order to improve the accuracy inherent to superimposition of the negatives, and to reduce interventions and exposure times, we combine the photographic film and the emulsion in the same cassette, into which simultaneous exposure takes place.

After the exposure time required for the visible photo (first film attained by the radiations, facing the pinhole), the illumination is interrupted by a shutter which is transparent to gamma radiations.

In order to cover a broad range of activities, simultaneously detected in a single exposure, films with different sensitivities are superimposed (Structurix S Gevaert, Kodak no screen, type R mono ...). See Figure 3.

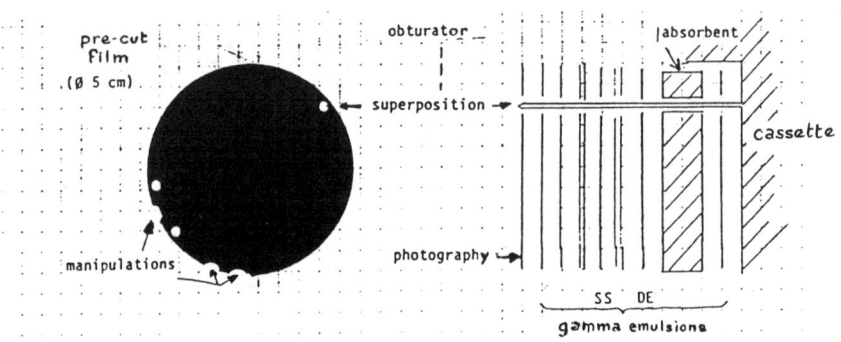

Fig. 3 - Arrangement of films.

Cutting of the film with a puncher and loading into the cassette are carried out in a darkroom. After exposure, films are brought back to the darkroom for simultaneous processing in regular baths (developer-fixation).

After drying, the prints are available for a first visual examination or for sophisticated processing.

2.2.2 Picture processing

Interaction of radiation hitting the film results, after development, to darken the silver bromide grains of the emulsion. Interpretation consists in assessing the darkening degree. The positioning of the negative on a negative reader is standardized by claws similar to the cassette ones. The transmitted light is picked up by a video camera and forwarded to the picture processing, programmed according to the needs.

The dynamics is achieved by encoding on 8 bits (256 intensity levels). The picture is defined by 512 x 512 pixels.

a) Localization of active areas

The two pictures, visible and active, are stored in two different memory planes.

Generally, the area photography is left black and white, whereas data from radiosensitive film are displayed in color for differentiating the two negatives, cumulated in one by superimposition, in order to facilitate localization of active objects.

The arbitrarily defined color code (typically rainbow) is displayed on the final unique document, and distributes colors on different active areas, giving them relative values of apparent doses.

b) Dosimetry

It should be noted that, generally, we are simply looking for an order of magnitude of dose rates. The approach of this evaluation, however, requires the knowledge of several parameters and physical data concerning the media and the process.

Films and development : reproducibility is difficult to reach with photographic techniques. The correlation is only valid if films from the same batch are used, and if their development is performed simultaneously (identical timing, temperature, bath regeneration).

In order to overcome unhomogeneities coming from digitization of the negative (illumination range of negative reader, camera resolution, film "background"), acquisition of exposed film is followed by examination, in same conditions, of a blank film with same origin. The content of each pixel of the blank will be substracted from the content of the exposed film.

Calibration : the darkening analysed on the gamma film is compared with optical density of radiography coming from a source of known activity. Available calibrated sources let experimentally obtain correlation curves.

Films : response curves of different films are studied and for each of them, the optical density evolution versus exposure is stated (intensity x exposure time). Figure 4 shows results obtained with Structurix S (SS) and Direct Exposure.

Isotopic nature of sources : film efficiency differs for different gamma energies (In Fig. 4, the SS film response is shown as dotted line for ^{241}Am, 60 keV and as plain curves for ^{60}Co, 1.3 MeV). If the isotope is not initially defined, its energy can be determined by transmission measurement : absorbant screen (w) is placed and the comparison of optical densities of obtained spots with and without attenuation lets calculate the energy.

Solid angle : the density detected by the film corresponds to transmitted activity; emitted activity from the source is extrapolated as inverse of the square of the distance.

Attenuation : eventually, an attenuation correction for inserted materials in the radiation path is performed.

Sighting angle versus collimation axis : an experimental correction for the equipment takes into account the radial position of the spot.

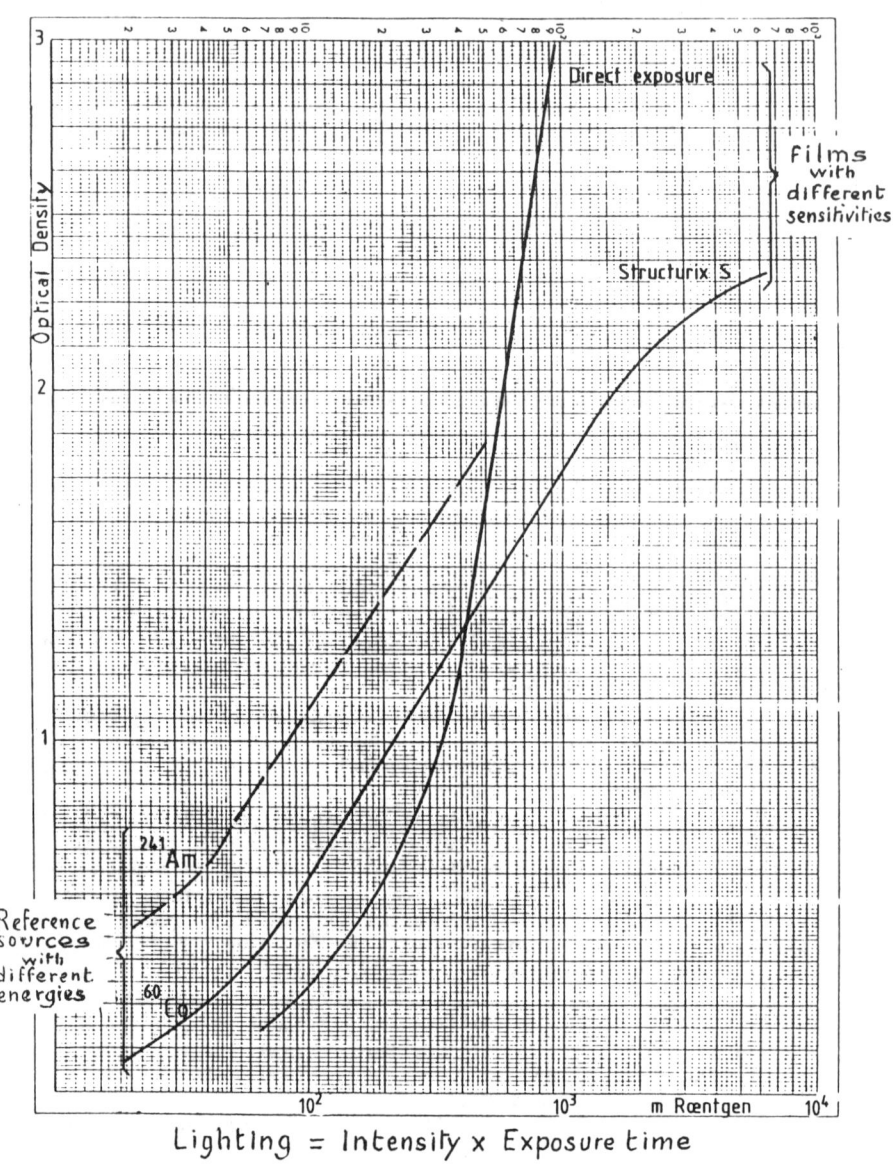

Fig. 4 - Response of various films to received doses.

Fig. 5 - Aladin and
        its accessories

2.2.3 Characteristics

Overall dimensions :
They let introduce the equipment into a 254 mm diameter port.
The weight is limited to : 27.5 kg for simple use,
                          36.6 kg with rotating support and charger.
Resolution : 1.5°. It is the resolving power allowing to distinguish
two spot sources whose received doses are similar.
Measurable limit doses : they are not strictly defined; generally, it
is possible to modify practical conditions. Detection threshold is lowered
as exposure time is extended, and distance of the equipment decreased.
The following example gives some data for a Cs pneumatic source :

Spot source activity	Distance	Exposure time	TT level environment
2 Ci	1 m	30 min	2.4 Rad/h
1 mCi	1 m	1 day	1.2 mRad/h
10 µCi	30 cm	13 day	12 µRad/h

2.3 REAL TIME VIDEO CAMERA VERSION

2.3.1 Detection

Negatives are recorded through a video camera. This camera is placed
in a pinhole-collimator assembly (§ 2.1). A luminescent screen converts
gamma radiation into light, visible by the camera.
If the screen is transparent, the photography of the scene is not
disturbed.
If detection efficiency is sufficient, the screen is thin and spatial
resolution of gamma sources is good.
So, the video camera observes, either the landscape through the light-
transparent scintillator, if obturator is open, or radioactive sources
through the obturator. The screen (preferentially BGO) is placed against
the camera whose input window is made of optical fibers. Then, it makes it
possible to get the screen f r apart from camera.
The used camera is of single photon detection type. Its sensitivity
is very high (10^{-7} lux). It is formed with an image intensifier coupled
through optical fibers to an intensified CCD camera. For instance, the
picture of a 1.7 Ci ^{60}Co source, located at 3 m, is obtained in 10 s.
The dimensions of the shielded camera are compatible with introduction
requirements for remote handling port. The electronic and power supply unit
remotely controls the camera through a connection cable, which transmits
data to acquisition system.
Note. The camera is customarized. Sometimes it is necessary to decouple
detection and electronic parts, by a remote cable. Thus shielding is lighted.
Sensitivity increases with cooled camera. Ray-hardened camera increases life.

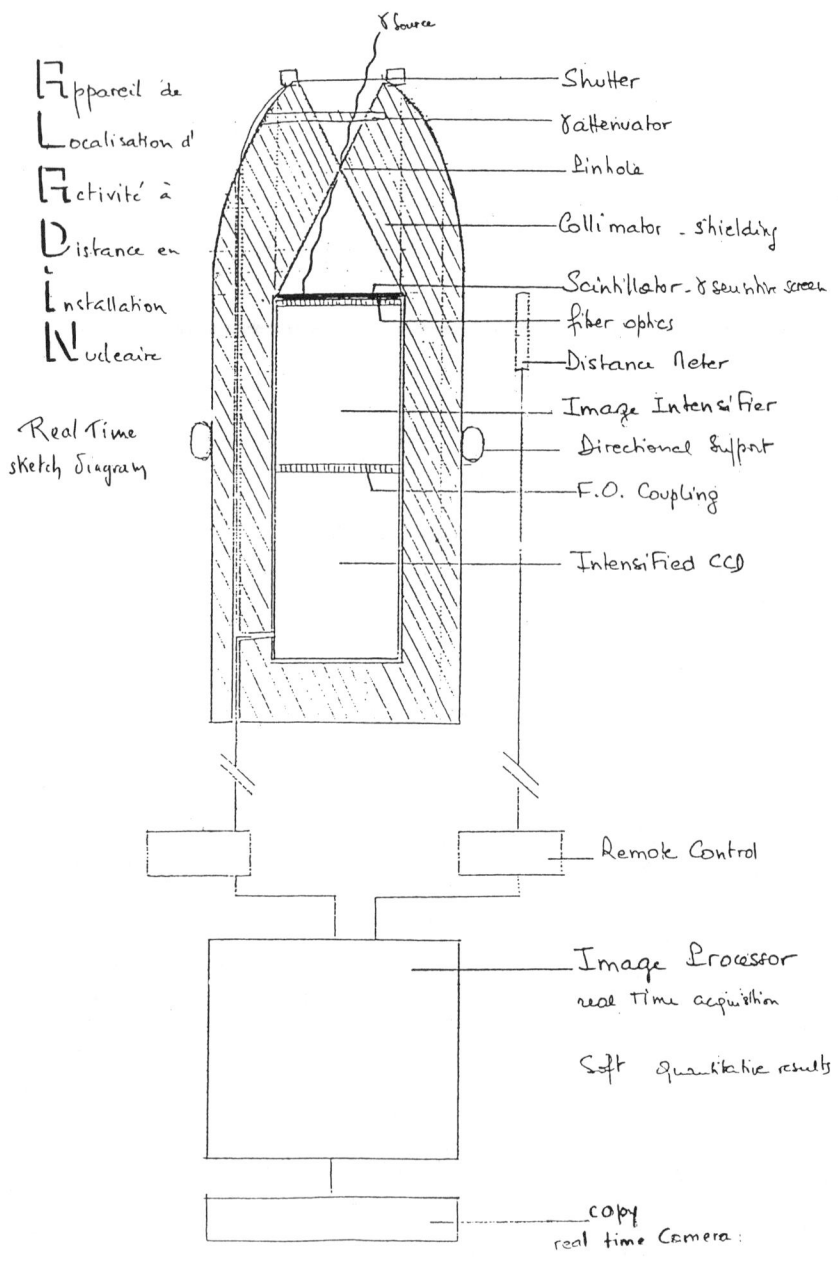

γ Source

A ppareil de
L ocalisation d'
A ctivité à
D istance en
I nstallation
N ucleaire

Real Time
sketch diagram

Shutter
γ attenuator
Pinhole
Collimator - shielding
Scintillator - γ sensitive screen
fiber optics
Distance Meter
Image Intensifier
Directional Support
F.O. Coupling

Intensified CCD

Remote Control

Image Processor
real Time acquisition

Soft Quantitative results

copy
real time Camera

Fig. 6 - Real time sketch diagram.

2.3.2 Image acquisition and processing

The real time picture formation requires, in one hand, the memorization for transient data storage (particularly for visible light), in the other hand, integration on more or less extended duration of acquired data (particularly for gamma detection).

A real time data acquisition system brings a dynamic memory for restitution of pictures, which are stored on hard or floppy disk. The processor is equipped with image processing classical softwares performing continuous integration, background correction, filtering, coloration and superimposition functions.

2.3.3 Dosimetry

The evaluation of dose emitted by sources is performed in the same way as described in § 2.2.b. Calibration is made for several camera adjustments.

The knowledge of source-camera distance, necessary for received dose to emitted dose extrapolation, can be obtained by direct measurement. Near the camera, a telemeter, for instance of laser type, has simply to be placed. It is read by the camera and directly interpreted by image processing system.

CONCLUSION

The first version equipment, photographic device type, operational since 1984, is completely independent. But, it works in blind mode for source detection and for necessary shooting time. The results are only available after development and processing.

The second version equipment, real time video camera, is intended for override these disadvantages. It is under realization. Its performances, due to high sensitivity of the camera, rapid delivery of the results, make it very interesting.

In both cases, the equipment is portable and can access every room. A peelable painting avoid surface contamination event. Personnel intervention is limited to place the equipment near hostile area, or on a robot or other teleintervention device.

The process interest comes from objectivity and important amount of delivered data in complex environments. It is the only apparatus able to remotely detect and localize the whole radioactive sources and their specific activities as well. The processing of acquired images, compared to records given from calibration sources, and corrected for physical data (particularly for source-device distance) let to evaluate the dose rate emitted by each source. A color code of activity source, on the final document, facilitates the interpretation.

Multiple interventions with the first equipment were fully satisfactory : decontamination (Tricastin), dismantling (Cadarache - Rapsodie, Brennilis - EL4, La Hague - Elan II B - AT1, Marcoule - Piver, Fontenay-aux-Roses - RM2).

Aladin brings an unrivaled way for data knowledge accessing in order to elaborate strategies related to nuclear activities.

Both versions are patented.

DETERMINATION OF THE RADIONUCLIDE INVENTORY FOR WASTE PACKAGES USING GAMMA-RAY-DETECTORS OF LOW BEAM WIDTHS

B.-R. MARTENS
Bundesamt für Strahlenschutz, Germany
and
P. FILSS
Kernforschungsanlage Jülich, Germany

Summary

An explanation is given of how the radionuclide inventory of cylindrical waste packages can be determined by spectrometric measurements of the emitted gamma-rays. A suitable detection scheme is described and theoretical results are compared with measurements performed with the aid of calibration sources. The radionuclide inventory can be determined almost independently of the spatial distribution of the radionuclides within the waste package by screening the detected gamma-rays and by turning the detector or package during the measurement cycle. Thus a volume near the surface is seen by the detector for a shorter time than a volume deep inside, which compensates in part for the different attenuation of the emitted photons. Systematic deviations can be avoided as far as possible by calibrating the scanning device. Limits are defined by a single point source and by a uniform distribution of the activity.

1.   INTRODUCTION

Radioactive wastes to be disposed of must fulfill certain requirements. The property which is of chief importance for the protection of man against the hazards of ionizing radiation is - apart from the quality of packaging and/or the quality of the waste product - the radionuclide inventory of a waste package to be disposed of. In as far as the limiting values are not exhausted, an upper estimate of this inventory may be sufficient, and this might be performed, for example, on the basis of simple dose rate measurements. The more these limits are exhausted the more it is necessary to use improved methods which should be able to determine the radionuclide inventory within certain well-defined limits.

It is the aim of this work to give some simplified algorithms which allow the computation of the unscattered photons counted by an external gamma-ray-detector. Results are given for cylindrical waste packages which, in the case of a nonhomogeneous activity distribution, have to be scanned over their lateral surface with the aid of an appropriate scanning device. Basic considerations together with detailed comparisons with experiments have been published elsewhere [1,2]. It is recommended that the theoretical considerations and computations presented here be used for
   - proper calibration of the measuring device,
   - optimization of the measuring method and for
   - assessment of unavoidable uncertainties.

2. DERIVATION OF THE PHOTONS COUNTED FOR ROTATING POINT OR VOLUME SOURCES

We begin with a point source of activity A which is a source confined to a volume whose dimensions are small compared with the distance m between the source and the detector and small compared with the attenuation length of the material for its own radiation [3]. For the photons counted by the detector during the time interval T we obtain the well known equation

$$(1) \quad N_0 = \frac{K \, A \, T}{4 \, \pi \, m^2}$$

K stands for the calibration constant of the detector describing its efficiency (effective area) and 4π is the full solid angle valid for any isotropically emitting radiation source. If the photon's path between the source and detector is filled with a pure absorbing medium, an additional factor $\exp(-\mu m)$ at the right-hand side of equation (1) must be considered, μ being the linear attenuation coefficient for gamma-rays.

Fig. 1 shows a sectional diagram of a point source P at distance r from the origin of ordinates O inside a cylindrical waste package with radius R. Any photon reaching the detector at the position m must travel the distances x inside and y outside the package. It has been proved to be a suitable detection scheme if the detected gamma-rays are screened by a collimating device and if the detector or the package turns around O as the center or axis of rotation. Assuming a cone shaped receiving pattern for the detector, only photons falling into the detector at angles between $\tau+\tau_0$ and $\tau-\tau_0$ are counted. According to Fig. 1 this cone intersects the circular path of our point source at the angles α_1 to α_4. Thus equation (1) is easily extended for this detection scheme resulting in

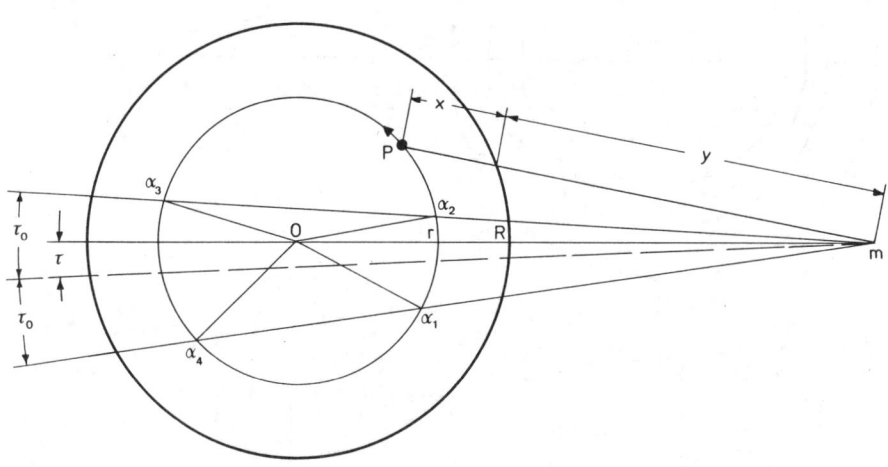

Fig. 1   Detection scheme for rotating point sources

$$(2) \quad N_p = \frac{K\,A\,T}{4\,\pi} \left\{ \frac{1}{2\,\pi} \int_{\alpha_1}^{\alpha_2} \frac{\exp(-\mu x)}{(x + y)^2}\, d\alpha + \frac{1}{2\,\pi} \int_{\alpha_3}^{\alpha_4} \frac{\exp(-\mu x)}{(x + y)^2}\, d\alpha \right\}$$

for the number of photons counted during the time T for a single rotation.

It seems worthwhile to expand equation (2) in the case of a source material with the source strength C distributed homogeneously throughout a cylinder with radius Z inside a waste package. We will consider only the case where the attenuation coefficients in the radioactive source material and in the outer shield material are identical. Here, the number of photons counted might be computed according to the quite general equation

$$(3) \quad N_V = \frac{K\,T\,C}{4\,\pi} \int_V \frac{\exp(-\mu x)}{(x + y)^2}\, dV$$

dV is the differential volume element, x and y are the photon path lengths inside and outside the waste package between the volume element dV and the detector, and the integration must be performed over the whole volume of

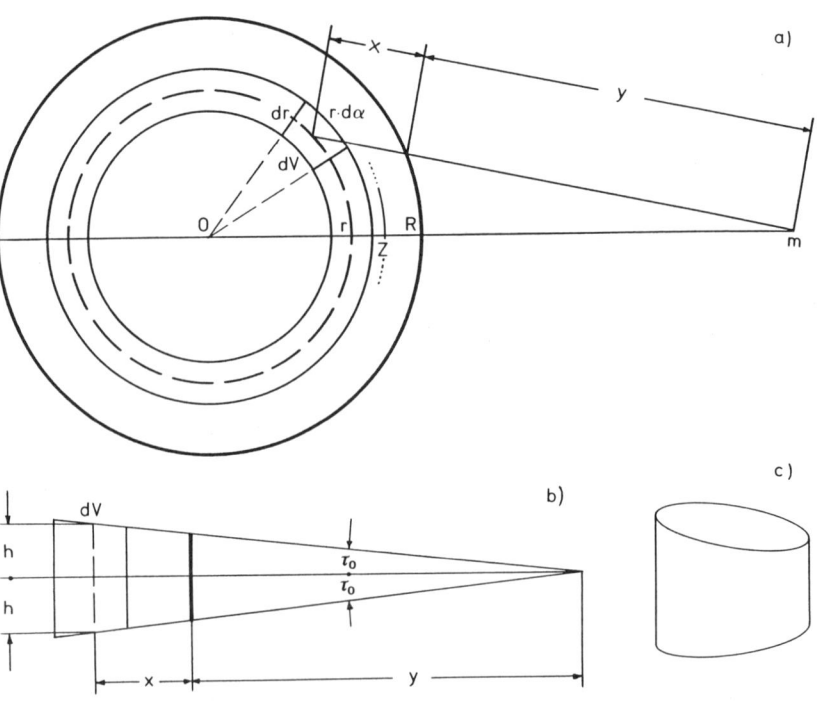

Fig. 2   Sketch for the integration of volume sources

the source seen by the detector. Considering the same detection scheme as before, Fig. 2a) shows a sectional view from the top and Fig. 2b) a lateral sketch of this arrangement. The photons counted by a low beam width detector are estimated, neglecting small additional photon path lengths if a volume element dV is defined as not differential in the direction of the axis of rotation (Fig. 2b)). If in the first step a pyramid-shaped pattern is assumed for a collimated detector with a quadratic cross section, then all these differential elements dV are sections of a volume $V=2\pi m Z^2 \tan(\tau_0)$ shaped like a truncated cylinder (Fig. 2c)). The differential volume elements are of the form $dV=2(x+y)\tan(\tau_0)r\,dr\,d\alpha$. With $C=A/V$ equation (3) may be approximated by

$$(4)\quad N_v \approx \frac{K\,A\,T}{4\,m\,\pi^2}\,\frac{1}{Z^2}\int_0^Z r\,dr\left\{\int_{\alpha_1}^{\alpha_2}\frac{\exp(-\mu x)}{x+y}\,d\alpha + \int_{\alpha_3}^{\alpha_4}\frac{\exp(-\mu x)}{x+y}\,d\alpha\right\}$$

An additional factor $\pi/4$ should be used at the right-hand side of (4) if the detector has a circular receiving pattern instead of the square pattern used for the derivation of (4). This simple correction holds true as long as the photon flux density is approximately constant over the cross section of the receiving pattern.

Equations (2) and (4) are easily evaluated. If the trigonometry is taken into account, this leads to:

(5) $x + y = m\cos(\tau) - r\cos(\alpha+\tau)$

(6) $x = \sqrt{R^2 - (m\sin(\tau))^2} - r\cos(\alpha+\tau)$

(7) $\tau = \arctan(r\sin(\alpha)/(m - r\cos(\alpha)))$

For the limits α_1 up to α_4 we get

(8) $\alpha_1 = \arcsin(m\sin(\tau-\tau_0)/r) - (\tau - \tau_0)$

(9) $\alpha_2 = \arcsin(m\sin(\tau+\tau_0)/r) - (\tau + \tau_0)$

(10) $\alpha_3 = \pi - \alpha_2 - 2(\tau + \tau_0)$

(11) $\alpha_4 = \pi - \alpha_1 - 2(\tau - \tau_0)$

For the computation of formulas (2) and (4) it seemed expedient to split off a common factor

(12) $H = N_0/2\,\pi = K\,A\,T/(8\,m^2\,\pi^2)$

When $n_p = N_p/H$ and $n_v = N_v/H$ are introduced it may easily be verified that these quantities depend only on the relative lengths x/m, y/m, Z/m, R/m and r/m and on the attenuation constant μm. The evaluations in the next chapter might therefore be used for any arrangement to which the product of μ and m and the above mentioned reduced dimensions correspond.

3. EVALUATION OF THE THEORETICAL RESULTS

Fig. 3 shows reduced photon counts n_p as a function of the radial position r of a rotating point source with the attenuation factor μ as a parameter. These graphs can be interpreted as possible deviations if the activity of a source of unknown position in the detection plane is to be determined. If the mean value of each graph is taken as a basis for the calibration of our scanning device, an overestimationof the activity occurs for a point source near the center of rotation for low values of μ while an underestimation occurs for large values of μ. Small deviations occur for $\mu \approx 1/R$. This result has been obtained in a previous publication [1] on the basis of a simplified algorithm.

Fig. 4 shows reduced counts n_p as before for an attenuation factor $\mu=0.114$ and for two different angles τ between the direction 0 m and the axis of the receiving cone. Uncertainties in the determination of the activity can be reduced by a suitable linear combination of the counts obtained at different angles τ. For example, if a corrected photon count

(13) $n_c = n_p(\tau=0°) - 0.5n_p(\tau=18°)$

is used for the determination of the activity, any possible deviations are substantially reduced (see Fig. 4; the standard deviation decreases from $\approx 80\%$ to $\approx 25\%$ and a greater reduction is gained for the maximum deviations). Thus it is recommended that such simple corrections be used in the case of nonhomogeneously distributed sources (e.g. hot spots) inside a waste package.

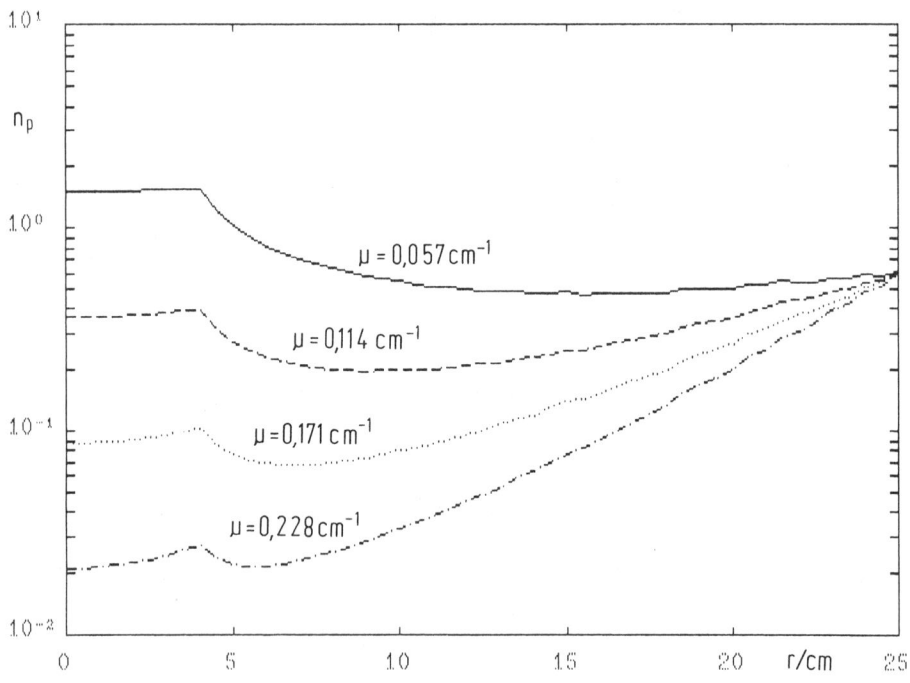

Fig. 3   Reduced photon counts for a rotating point source as a function of its circular path radius (R=25 cm, m=60 cm, $\tau=0°$, $\tau_0=4°$)

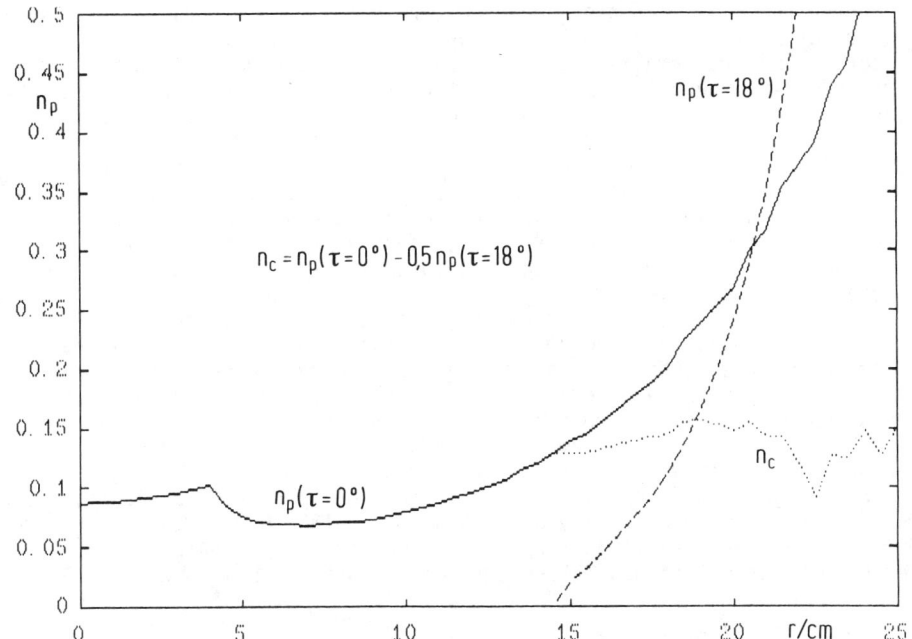

Fig. 4   Reduced photon counts for a rotating point source for two
different angles τ and corrected counts n_c (R=25 cm, m=60   cm,
τ_0=4°, μ=0.171 cm^{-1})

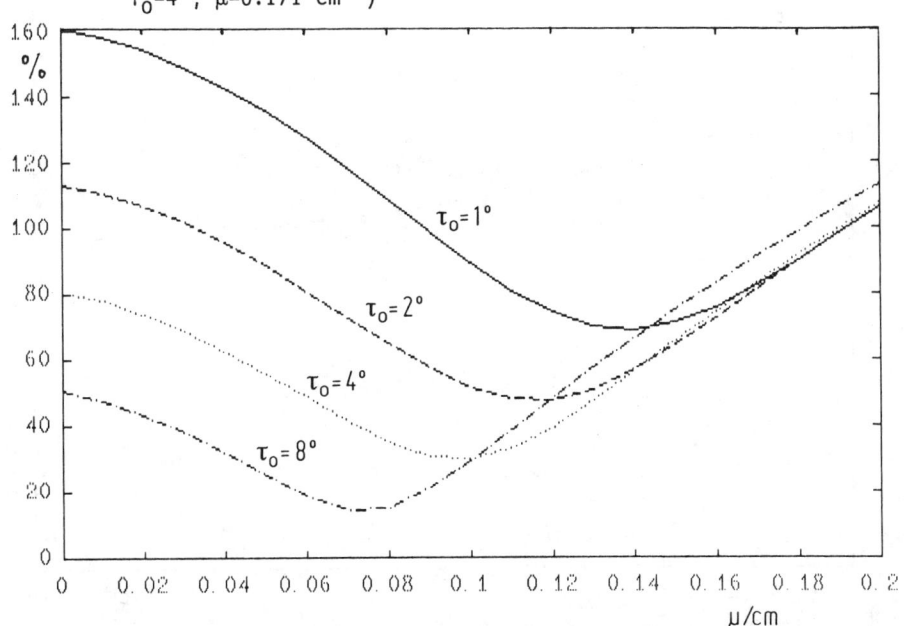

Fig. 5   Standard deviation for rotating point sources in dependence on
the attenuation coefficient μ for different beam widths (R=25 cm,
m=60 cm, τ=0°)

Another question of interest which can be answered by numeric evaluation of equation (2) is how to find an optimal beam width for the collimated detector, aiming at small deviations. Calculating the quotient between the standard deviation and the mean value of n_p for an unknown point source, we obtain the results shown in Fig. 5. In dependence on the attenuation coefficient μ we find broad minima for the relative uncertainties defining the best possible beam width for the corresponding value of μ if the previously mentioned corrections have not been performed. We find very small optimal beam widths for high densities of the matrix material (e.g. $2\tau_0 \approx 2°$ for a Co 60 source embedded in a cement matrix possessing an approximate density of 2 g per cubic centimeter). Thus a compromise between the uncertainty of the results and the time necessary for a measurement ought to be made.

Fig. 6 shows reduced photon counts n_v as a function of the radius Z of a cylindrical homogeneous volume source inside a cylindrical waste package and again of the attenuation coefficient μ as a parameter. It should be noted that the underlying equation (4) of chapter 2 is valid only if the attenuation coefficient is constant over the whole waste package and that a factor $\pi/4$ should be used assuming a cone-shaped pattern for the collimated detector. These computations should be compared with measurements already published [2]. In this previous work cylindrical specimens fabricated from cement with a constant activity concentration (C=1330 Bq per cubic centimeter for Co 60) were examined for calibration and testing purposes using a detector with an efficiency K=1.86 cm² for Co 60 and a beam width $2\tau_0=8°$.

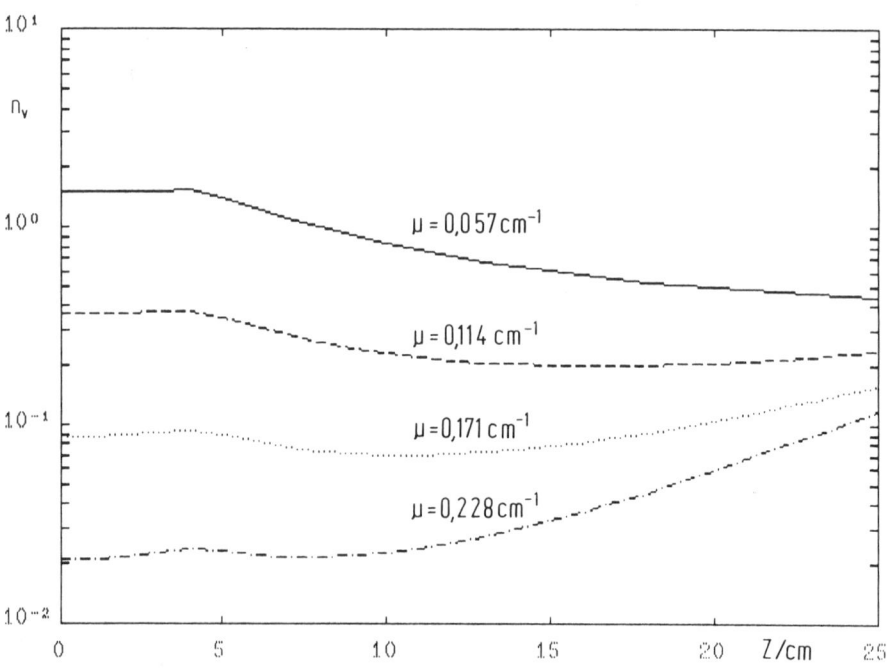

Fig. 6   Reduced photon counts n_v in dependence on the radius Z for different attenuation constants μ (R=25 cm, m=60 cm, τ=0°, τ=4°)

Table I lists the various matrix densities, attenuation coefficients, reduced photon counts n_V, the measured count rates and the calculated count

density/g cm^{-3}	1	2	3	4
attenuation coefficient/cm^{-1}	0.057	0.114	0.171	1.228
reduced photon counts	1.166	0.772	0.555	0.426
measured count rates/s^{-1}	42	26	18	16
calculated count rates/s^{-1}	36.0	23.9	17.1	13.2

TABLE I: Comparison of measured and calculated count rates for homogeneous volume sources embedded in different cement matrices (Co 60, m = 35 cm, Z = 10 cm, $\tau = 0°$, $\tau_0 = 4°$, K = 1.86 cm²)

rates $N_V/T = (\pi H/4)n_V/T$. According to the previous chapter, the scaling factor amounts to

$$(14) \quad \frac{\pi H}{4 T} = \frac{\pi K V C}{32 \ m^2 \ \pi^2} = \frac{K Z^2 \tan(\tau_0) C}{16 \ m} = 30.89 \ s^{-1}$$

It is concluded that the comparison between equation (4) and the measurements proves satisfactory. The uncertainties for a cylindrical source of unknown diameter inside a cylindrical waste package are smaller in comparison with the uncertainties for a single point source of unknown position. A correction procedure might easily be evaluated even for cylindrical volume sources, combining the counts gained at different angles τ.

Finally it should be mentioned that similar methods can be applied for rectangular waste packages. A suitable detection scheme for this is a scanner driven with constant velocity along the lateral face of a container. Using a detector with a cone-shaped pattern as above, we find that sources deep inside the container are seen by the detector for longer than sources near to the surface and the different photon attenuations are therefore partly compensated.

References

1.   Filß, P.; Odoj, R.; Enge, R.: Möglichkeiten und Grenzen der gammaspektrometrischen Aktivitätsbestimmung in Abfallfässern, 20. Jahrestagung des Fachverbandes für Strahlenschutz e. V., 67, Basel (1987)

2.   Filß, P.: Specific activity of large volume sources determined by a collimated external gamma detector, Kerntechnik 54 No.3, 198, München (1989)

3.   Jaeger, R. G. (ed.) Engineering compendium on radiation shielding (vol.I), 363, Berlin (1968)

DEVELOPMENT OF A NON DESTRUCTIVE EXPERIMENTAL METHOD FOR THE QUANTIFICATION OF GAMMA EMITTER RADIONUCLIDES IN PACKAGES

J.A. SUAREZ G. DEL REY
G. PIÑA LUCAS
C.I.E.M.A.T. Spain
E.N.R.E.S.A. Spain

———————

This work is carried out and partially financed within the frame of the European Community Research Programs.

———————

Summary

The main scientific and mechanical components of the experimental device are described.

Previous assays showed very good results for qualitative identification of radionuclides in drums containing liquid radioactive wastes incorporated in cement (homogeneous), solid radioactive wastes (heterogeneous) and radioactive sources.

Energies from 50 to 1400 KeV have been studied and only the identification of Am-241 gave bad results.

It is very important the selection and preparation of a secondary standard to be used for efficiency determination of the system as previous stage to the quantification of radionuclides.

The standard radionuclide selected has been Eu-152.

The efficiency calibration determined in function of the abundance of different energies of Eu-152 is different to the obtained with standard cocktail (QCY-44) used for gamma efficiency calibration.

To adjust Eu-152 efficiency calibration, a standard QCY-44 is necessary to modifie abundance values given on bibliographie and it is posible to use Eu-152 as secondary standard. The procedure performed to prepare this secondary standard of Eu-152 is described.

Finally the results obtained in the quantitative calibration on the equipment for the determination of radionuclides in homogeneous packages and the first experimental data corresponding to the tests achieved for the quantitative determination in heterogeneous packages are presented.

1. INTRODUCTION

The purpose of the investigation is to describe the ac - tual situation of the experimental study and the adjusting of the equipment used during 1988 and part of 1989 which consti- tutes the start-up of the tests programmed in order to carry out research and development tasks foreseen in the present work.

The experiments are carried out facing the problem from its essential part studying everything that occurs during the spectrometric exploration of packages of different inner com- pounds prepared in our installations, in order to reduce practical way the errors produced in the quantitative deter- mination preventing partially the use of empiric calculation factors.

With this work it is expected to obtain an experimental technique for the radiological characterization of liquid and solid radioactive waste conditionned in 200 l. drums. This tech nique is necessary in the field of the radioactive waste mana gement, within the frame of the Spanish Energetic Plan.

2. EXPERIMENTAL PLANNING

The tasks which are considered necessary in order to achie ve the proposed objetives are as follow:
- Planning and documentation.
- Preliminary test on viability.
- Development of measurement and mechanic systems.
- Complementary tests.
- Mounting and adjusting of the whole system.
- Optimization of the gamma-spetrum equipment.
- Optimization of the rotation-elevation mechanic system.
- Quantitative detection.
- Preparing of radioactive secondary standard for its incorporation in the packages.
- Colimation.
- Preparing of homogeneous packages.
- Quantitative detection in homogeneous packages.
- Preparing of light heterogeneous packages.
- Quantitative detection in light heterogeneous packages without compaction and compacted.
- Preparing of heavy heterogeneous packages.
- Quantitative detection in heterogeneous packages with inner shielding.
- Geometric localization of heterogeneous sources incor porated into a cement matrix.
- Quantitative detection in heavy heterogeneous packages.
- Development of calculation codes.

3. TASKS CARRIED OUT

3.1. Planning and documentation

There is a descriptive document of the Project where the Objetives, Process, Tasks Definitions, Action Definitions and Verifications, Chronogram, Decisions Diagram and Budget are detailled.

3.2. Preliminary tests on viability

Preliminary studies on the question started in 1982 with the consecutive presentation of several reports and identifi cation tests in order to obtain the efficiency loss counting factors resulted the various geometries presented by the packa ges to characterize.

3.3. Development of measurement ond mechanic systems

Specifications have been written down and it has been bought a NIMs, MCA ond Computer System with Reverse-Electrode Coaxial Ge Detector in Horizontal Dipstick.

The mechanic System is mainly composed of elevating plat form traction mechanism and rotating motor.

3.4. Complementary tests

Necessary experimental studies have been carried out to check that the instruments bought will fulfil the requirements expected.

3.5. Mounting and adjusting of the whole system

It has been carried according to schedule.

3.6. Optimization of the gamma-spectrum equipment

It has been carried according to schedule.

3.7. Optimization of the mechanic rotation-elevation system

It has been carried according to schedule.

3.8. Analytic detection

The experimental results obtained show that there is not much difficulty in the qualitative determination of the radio nuclides present in the different types of packages tested. The most un favorable case will be the one of a radioactive source placed in the center of drum which will be the most logical po sition which can present this kind of packages. The correspon ding gamma-spectrum has photopeaks higher than 295 KeV are de tected with a good resolution as shown in Fig. 1.

3.9. Preparing of radioactive secondary standard for its in - corporation in packages.

The quantitative detection obliges us to use a radioacti ve standard or standards with different energies in order to prepare packages with a known gamma activity valid for calibra tions in the necessary efficiencies for each type of packages.

From the possible suitable radionuclides, Eu-152 presents an energy range urge enogh to make the calibrations oscilating between 122 and 1.408 KeV.

In the case of Eu-152 it is observed that numeric values of its gamma-emission intensity, corresponding to the differ ent energies of this one, are not too much concident in the different publications the last years [1][2](3)(4)(5)(6)(7).

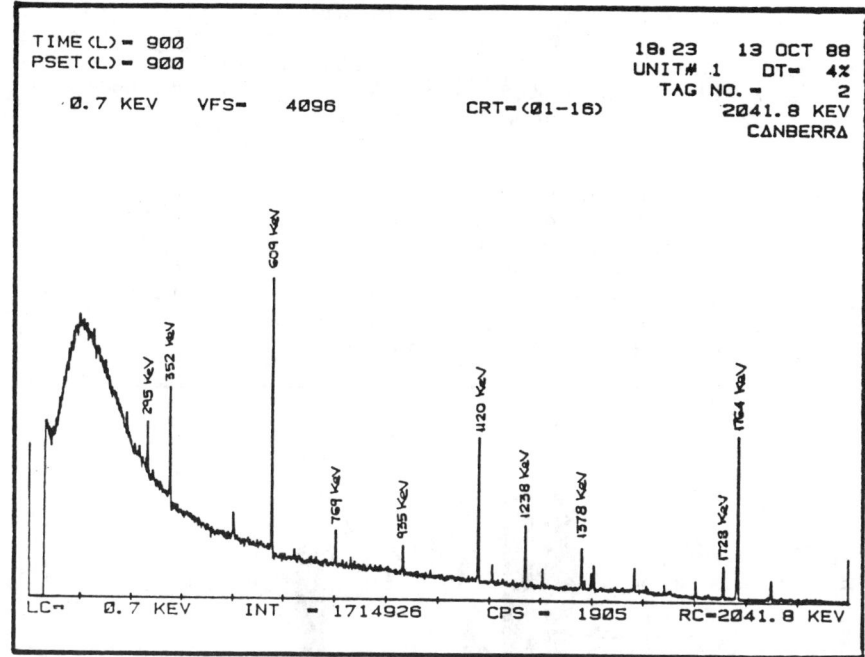

FIGURE 1. Gamma spectrum of Ra-226 placed in the center of
drum.

In our laboratory we calibrate the gamma-spectrum equip
ment with the QCY-44 standard from the British Calibration Ser
vice and the results are not coincident when using Eu-152 stan
dard.

For this reason we have adjusted some values of the gamma
emission intensities of Eu-152 so that it produces the same re
sults than the QCY-44 standard, obtaining in this way the posi
bility to made the calibration in efficiencies with Eu-152.

3.10. Colimation

It has been designed and mechanized a colimator with a
changeable piece on its front side so that its diameters can
be changed.

3.11. Preparing of homogeneous packages

With the achievement of this task we obtained standard
homogeneous packages and packages of specific activity exac-
tely known for quantitative analyses of real packages with ra
dioactive homogeneously distributed throusch the cement matrix.

The work has been faced in two aspects:
- The first one is the arrangement of packages at partial
  scala whith gives us the necessary data to know the be
  haviour of the packages at real scale due to the fact
  that the preparasion of test pieces enables a larger
  variety of test besides of an easier manageability.

- The second aspect concerns the arrangement of standard packages and packages of known specific activity at real scale in order to adjust the calibrations to the necessary efficiencies and to carry out the checking of these calibrations in the same counting geometry that will be present when making the characterization on real pro - blem-packages.

It should be stressed the difficulty to prepare the packa at partial and real scale with densities higher thats 1.96 g/cm^3, which enables to conclude that the margin of densities that can be obtained are between 1.79 and 1.91 g/cm^3. also shown by the frequencies distribution of the densities obtained with homogeneous packages prepared during 1987 and 1988 in the CIEMAT'S Waste Handling Unit, as shown in Fig.2.

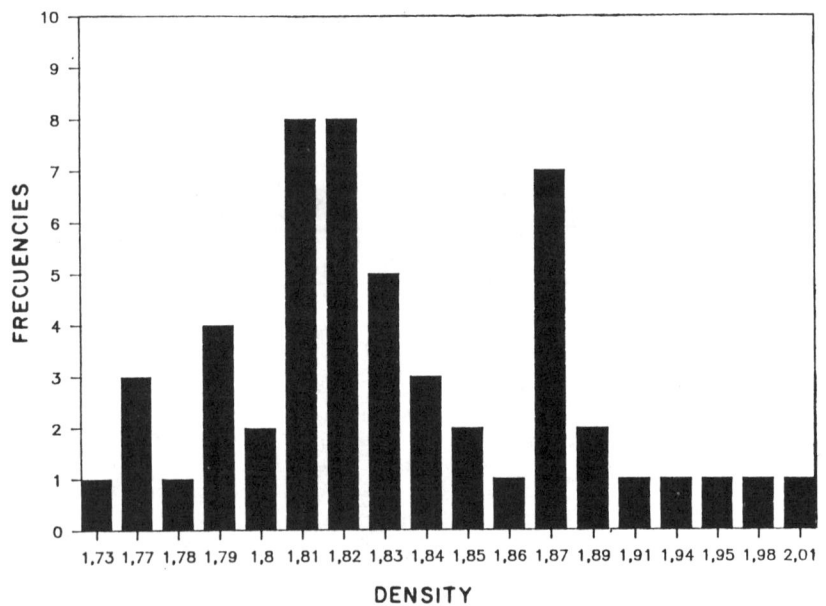

FIGURE 2. Frecuencies distribution of the densities un homoge neous packages.

3.12. Quantitative selection in homogeneous packages

3.12.1. Loss of efficiency as o function of dead time.

It is known the loss effiiency in the behaviour of gamma-spectrum systems when the counting rate is higher. The actual Ge detectors present this effect in a higher deegree which can be slightly diminished when using specially designed electronics.

The study has been done with radioactive samples of cons tant physical and geometric characteristics and a constant growing radioactive content, using an Eu-152 secondary standard. With the results obtained a function correlating the loss of efficiency of the detector with the "dead time" of the counting system has been found, as shown in Fig.3.

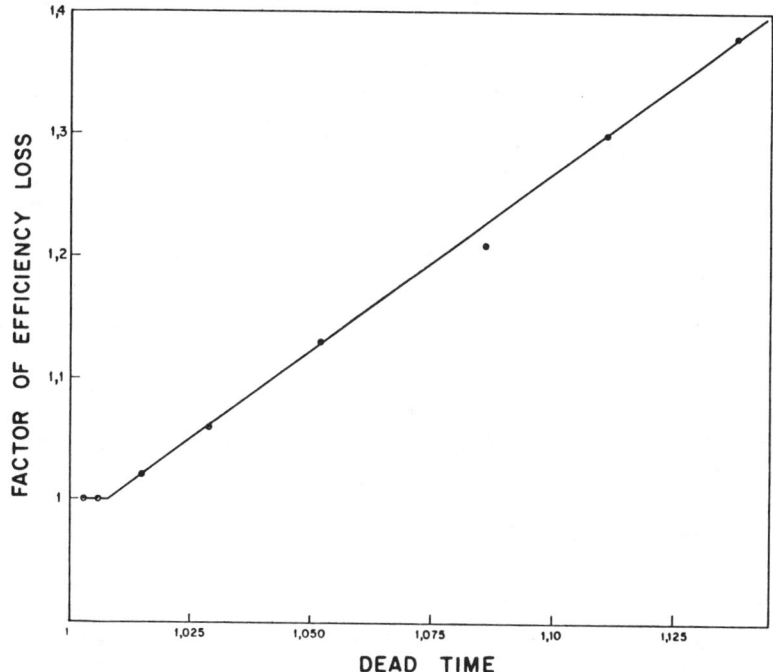

FIGURE 3. Efficiency loss Factors of the Ge Detector as a function of the dead time.

The use of loss factors correcting the mentioned efficiency loss preventing in that manner the error in the calculation of the activities that can reach very significative values.

3.12.2. Efficiencies calibration at partial scale

In order to obtain enough data and not to produce to many packages at real scale, a partial scale has been tested with 18 l. test pieces containing Eu-152 secondary standard of a density range between 1.0 and 1.96 g/cm^3. The efficiencies obtained with these packages is decreasing when the density is higher, as shown in Fig. 4, with little significant values in the range of 1.8 and 1.9 g/cm^3 with a maximum loss of 7% in the worse case corresponding to on energy of 122 KeV whith obviously is the most affected by the density. These values have undoubtedly an advantage due to the small deviation whith means that smaller errors when using the attenuation factors.

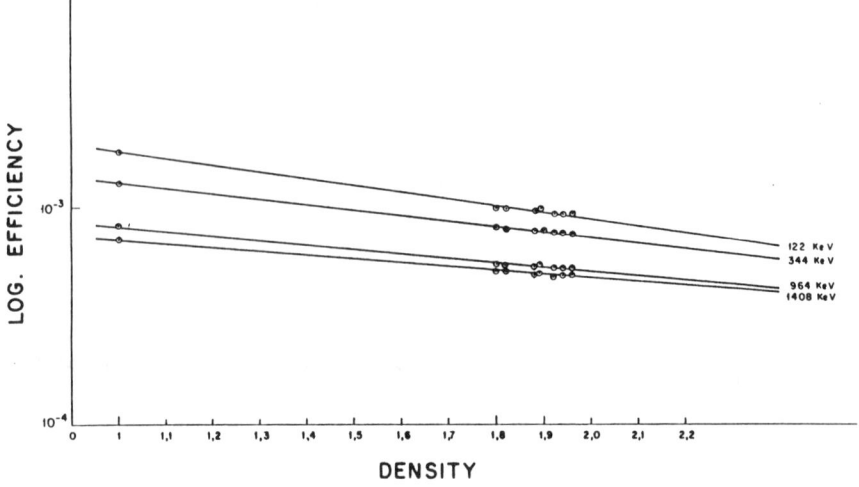

FIGURE 4. Efficiency variation of the homogeneous packages
with de matrix density.

3.12.3. Calibration in efficiencies at real scale.

Four packages at real scale have been prepared under
theoretical conditions to produce a density range between 1.7
an 2.0 g/cm^3 but the experimental values obtained are between
1.87 and 1.93 g/cm^3.

The efficiencies obtained with these packages do not pre̲
sent variations depending of the density, as shown in Table I,

TABLE I.- Efficiencies with homogeneous real scale packages
at different density.

KeV	Efficiency x 10^{-3}			
122	7'87	7'40	8'31	8'13
245	7'49	7'49	7'83	7'95
344	6'10	5'78	6'49	6'43
444	5'49	5'15	5'73	6'90
779	4'55	4'45	4'76	4'80
964	3'61	3'84	3'76	4'16
1408	3'70	3'94	3'97	4'18
Den-sidad	1'87	1'89	1'90	1'93

because the experimental error is higher that the variation expected in this density range which, as already indicated in the tests with partial scale packages, is scarcely signifi - cant.

The conversion of the efficiencies obtained with partial scale to real scale packages and the comparison with the effi ciencies obtained with real scale packages with the same den- sity value of 1'90 g/cm^3, are presenting prominent similarity, as shown in Fig. 5, from which we can deduce that the calibra tion of the detection system is done with real scale packages and the loss factors of efficiency due to attenuation of the matrix density are calculated from the results obtained with partial scale packages.

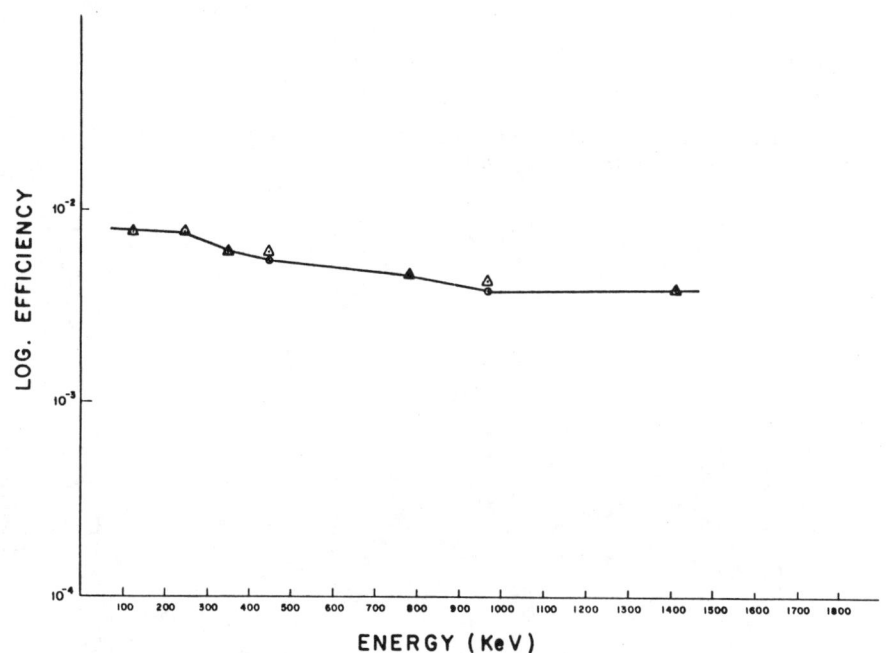

FIGURE 5. Efficiency with the homogeneous real scale packages.
Real scale ⊙ Partial scale △

REFERENCES

1. G. Erdmann, W. Soyka.- Die   - Linien der Radionuklide.
   Jül-1003-AC. September 1973.

2. Atomic Data and Nuclear Data. Tables. Vol.13.
   N°s 2-3.-1974.

3. National Bureau of Standards.- 1976. Radioactivity
   Section.

4. G. Erdmann, W. Soyka.- The gamma Roys of the radionucli
   des.- Tables for Applical Gamma Roy Spectrometry. Ver -
   lag Chemie. Weinheinn. New York 1979.

5. Debertin K. Nucl. Inst. and Meth. 158. 479-486 (1979).

6. Yoshizawa Y., Iwata Y., Imma Y., Nuch. Inst. and Meth.
   174. pg.133. (1980).

7. U. Rens, W. Westmeier. Atomic Data and Nuclear Data
   Tables. Vol.29. N°2. September 1983.

EVOLUTION OF RADIOMETRIC INSTRUMENTATION SYSTEMS ON OPERATING REPROCESSING AND WASTE HANDLING FACILITIES

C H ORR AND K ALLDRED
Physical Science and Engineering Development Group
Research and Development Department, BNFL Sellafield

SUMMARY

A wide variety of special purpose radiometric instrumentation systems are now in use in the Sellafield Magnox fuel reprocessing facility and in the associated plants and waste handling operations. These systems, which were developed by the Physical Science and Engineering Development Group have accumulated many years of successful operation. However as processes have been improved and regulatory requirements have become increasingly stringent, it has become necessary to re-assess the monitoring approach employed in a number of these devices. As examples of this process of evolution, which is being carried out in parallel with the development of other special instrumentation for the new plants now under construction at Sellafield, the paper outlines the way in which two of the existing devices are currently being replaced by more sophisticated and versatile systems. For the monitoring of crated waste, a flexible neutron coincidence monitor will incorporate high resolution gamma spectrometry to provide a more reliable overall assessment of plutonium and to assess the presence and distribution of U-235 contamination. Swarf arising from the decanning of Magnox fuel will be monitored by high resolution gamma spectrometry and the data analysis extended in order to provide a quantitative assessment of some 40 specified radioisotopes.

1 INTRODUCTION

1.1 As the nuclear industry has matured, operators have sought to achieve increasingly detailed knowledge of the quantity, characteristics and distribution of radioactive materials in their plants. This trend has been re-enforced by the tighter controls imposed by the regulators on those operations in general, on discharges of effluents to the environment, and of wastes to storage or disposal sites. At Sellafield these pressures have given rise to the installation, over the past 20 years, of a series of special purpose radiometric instrumentation systems, ranging from relatively simple stand-alone manually operated devices for initial checking of waste packages, to very complex automated systems incorporating several distinct measurements integrated totally into the plant and its operations.

1.2 Sellafield is currently the scene of a major programme of plant construction, while at the same time the existing plants are themselves the subject of wide ranging improvements and modifications. The Physical Science and Engineering Development Group - who were responsible for the development of

the earlier series of instruments - have contributed to this latter
activity by evolving greatly improved versions of several of the
more critical of the earlier generation of systems. This paper
describes two such undertakings which are currently close to
completion.

2   MEASUREMENT OF LARGE PCM ITEMS IN CRATES

2.1 The Challenge

A wide variety of large waste items ranging in size from $1m^3$ to
~ $25m^3$ are stored at Sellafield awaiting final disposal. These
include items from decommissioned plutonium plants at
Sellafield together with plutonium and uranium waste from other
UK nuclear sites. Many of these waste items were stored in the
late 1950s and early 1960s when the measurement technology for
fissile inventory was limited. More recently measurements were
carried out using a simple total neutron counter devised by
this group. Given the associated uncertainties in these early
measurements it is now intended to re-measure several hundred
such items contained in GRP overcrates, employing more
up-to-date radiometric measurement techniques and equipment.

The measurement specification for this task requests that the
monitoring system should provide:

(a) an estimate of the total plutonium content of each crated
     item.
(b) an indication of the presence of, and if possible an
     estimate of the U-235 content of each crated item.
(c) an indication of the location of plutonium (and U-235)
     within the crated item.

In order to achieve these objectives, the early total neutron
counter is to be replaced by a system based on neutron
coincidence counting. This will utilize design features which
have been successfully employed for some years on a smaller
scale for the monitoring of waste drums and small crates.

2.2 General Principles

The total plutonium content of each crated item is to be
determined via measurement of the plutonium-240 equivalent
content, Pu-240eq, by passive neutron coincidence counting
(NCC) and the percentage plutonium-240 equivalent, %Pu-240$_{eq}$,
relative to total plutonium content by high resolution gamma
spectrometry (HRGS). The total plutonium content of the crated
item, M_{Pu}, is then calculated using:

$$M_{Pu} = \frac{Pu\text{-}240eq \times 100}{\%\ Pu\text{-}240eq}$$

The presence of U-235 will be established via measurement of
its 186 keV gamma radiation using the same HRGS detectors as
for plutonium isotopic composition measurement. Interpretation
of 186 keV photopeak data in terms of a best estimate of U-235

mass will require a "special investigation" of the crated item by a specialist team. Such an investigation will involve consideration by radiometric physicists of additional parameters such as the crated item size, weight, history, presence of other isotopes etc.

Information on the distribution of plutonium is to be derived using two methods:

(a) by comparing the total neutron count rates from groups of He-3 neutron detectors at known positions in relation to the crated item.
(b) by comparing HRGS counts at known positions with respect to the crated item.

2.3 Physical Arrangement

Measurement of crated items will essentially be a two stage process in which crates are moved initially to a gamma monitoring station and then into a neutron coincidence counting (NCC) chamber on a rail guided trolley. (Figure 1)

Gamma measurement requires the continuous and controlled linear movement of each crated item, between two HRGS (germanium) detectors. The vertical position of each HRGS detector will be independently adjustable to accommodate the expected wide variety of crated item heights. In addition the solid angle view of each detector will be restricted using adjustable lead collimators and shields to reduce background contributions to the measurement spectra and aid fissile material distribution determination. Automatic begin/end count control of the HRGS electronics will be achieved by the use of infra-red sensors to detect the leading and trailing edges of the crated item as it passes the HRGS detectors.

Once past the gamma monitoring station the crated item will continue to move on the trolley until totally enclosed by the NCC chamber, at which point it will halt for the second measurement stage. One end of the chamber wall will slide open to allow the crated item to enter (and exit).

The walls and top of the NCC counting chamber are to be of modular construction with individual modules fabricated from polythene and covered with cadmium sheet on all surfaces. (Figure 2) Each module will contain up to 6 He-3 neutron detectors (100 cm active length; 5cm dia; 2 atm pressure) connected in parallel to a single (head) amplifier thus forming a discrete neutron counting unit. The base of the chamber will be formed by a concrete platform accommodating additional detector modules in a recessed channel. In total the chamber will contain 128 He-3 detectors distributed between 26 detector modules in an arrangement which will optimise and equalise detection sensitivity within the chamber.

2.4 Operations and Measurement Control

The operating sequence involves first a cycle of background and standardisation measurements on both HRGS systems and background measurements on the NCC system.

The HRGS standardisation measurement is to check that the performance of each HRGS detector and associated electronics is within acceptable limits. This involves the exposure of a Ba-133 source from a shielded position close to each detector via an electrical linear actuator to check detector gain stability, efficiency and resolution (FWHM).

The cycle of standardisation and background counts will continue until it is interrupted by the operator to carry out a waste measurement. At this point the system data processor will check that recent standardisation and background counts are valid and all standardisation sources shielded. The NCC chamber door will then be opened and the trolley movement mechanism started. Once the crated waste item is between the HRGS detectors data collection commences in a series of consecutive (~10 minute) counts during which the observed count rates in specified energy regions of interest within each spectrum are recorded. This results in a series of crate "segment" gamma counts. In addition, data will be collected continuously to give a whole crate count.

Counts in each specified region of interest will then be analysed, net photopeak count rates determined together with the associated errors and the data used to determine the percentage Pu-240eq, the presence of U-235 and distribution of fissile material in the crate.

The technique used to derive plutonium isotopic composition, and hence %Pu-240$_{eq}$, will be based on the determination of Pu-239/Pu-241 and Am-241/Pu-241 isotopic ratios from selected gamma photopeaks between 125keV and 370KeV in the HRGS spectrum. Using the Am-241/Pu-241 ratio to estimate the elapsed time since chemical separation of the fuel, the measured Pu-239/Pu-241 isotopic ratio is corrected to the separation date. For plutonium of Magnox origin this allows the isotopic abundances (Pu-238 to Pu-242) to be inferred from isotope correlation algorithms which have been derived from mass spectrometry data. Correction of the abundances to the measurement date is then carried out again using age information obtained from the Am-241/Pu-241 ratio.

Neutron counting (coincidence and total) of the crated item is to commence once the trolley is wholly within the NCC measurement chamber and the chamber doors are closed.

The neutron coincidence count is to be recorded as a series of consecutive short time period counts. A statistical analysis of these time segmented neutron coincidence counts is to be carried out to identify and reject amplifier pulses due to events other than neutron detection (eg EHT breakdown, RF interference).

The observed neutron coincidence count (once checked for statistical outliers) is to be further corrected for any moderating effects of the waste matrix using the ratio of total neutron count rates from a small number of unmoderated He-3 neutron detectors ("thermal flux probes") sited within the neutron coincidence counting cavity to the total neutron count rate from the moderated neutron detectors which provide the coincidence pulses.

Following these corrections to the neutron count, the Pu-240eq content and error will be calculated and this data combined with the % Pu-240eq and error to derive the total plutonium content and error.

2.5 Electronic Equipment

Each HRGS detector (20% coaxial) will be powered from commercially available NIM electronics mounted locally to minimise RF interference detection. Spectroscopy amplifiers having differential pulse input facilities (Ortec 672) will also be incorporated to minimise interference detection. The ADC output of each system will be linked to a dedicated MCA module mounted in a Harwell 6000 series rack. This electronic module was developed in-house as a ruggedised unit suitable for operation in plant environments. Spectral data in the MCAs can be addressed directly from a Sellafield developed Z80 based system data processor also in 6000 series format and "on-line" spectrum analysis carried out to provide Pu isotopic composition analysis, U-235 content analysis and fissile distribution assessment.

Each neutron counting module of the NCC chamber will contain a neutron pulse amplifier linked by superscreened cable to He-3 detectors. The digital pulses from each amplifier will be fed to an array of 6000 series format counter/timers to provide a facility for system checking and plutonium distribution analysis. In parallel, these pulses will also be summed into the neutron coincidence electronics for pulse correlation analysis using the fast RAM based coincidence circuitry. All operating parameters (eg coincidence gate length, predelay count times etc) and scaler outputs (counter/timers and coincidence) of the coincidence system can be addressed by the data processor mentioned above. In this way gamma and neutron data can be processed quickly to provide Pu and U-235 content and distribution information. Both amplifiers and neutron counting electronics have been developed by PS & ED Group for plant applications.

All communications between the operator and measurement system will be via a VDU and plant standard control panel. At this point information and instructions will be passed between the operator and measurement system on all aspects of routine operations, system diagnostics and measurement control.

3   THE SWARF INVENTORY MONITOR

3.1 The Challenge

The second example of the group's current programme is the
Swarf Inventory Monitor (SIM) which will use high resolution
gamma spectrometry to assess the radioisotope activity
inventory of Magnox swarf. This monitor is now in the final
stages of design and construction and is scheduled for
operation by March 1990.

Magnox swarf is the cladding and associated debris removed in
the mechanical decanning of Magnox fuel prior to its
dissolution and reprocessing. As an intermediate level waste
stream it will be encapsulated in cement in the new EP1 plant,
before being sent for final disposal. The licensing and
regulatory bodies have stipulated that each drum of
encapsulated swarf is to be monitored to determine its
radioisotope activity inventory.

Direct measurement of drums of swarf will be impractical
because of signal attenuation in the bulk material. The
preferred option, therefore, will be to measure the swarf in
small batches as it is produced. The Fuel Handling Plant
(FHP), which removes the cladding from the Magnox fuel, already
has a sophisticated gamma spectrometry system, the Uranium in
Swarf Monitor (UISM), which estimates the uranium carry-over in
such batches of swarf, before they are placed into a bin for
export to EP1. This monitor is used for process control,
alarming on high masses of uranium so that the plant operators
can search for and remove any large piece of uranium, eg a
piece from a broken fuel rod, before the swarf is discarded.
It is proposed to replace the UISM by the more powerful Swarf
Inventory Monitor, which will retain the process control
function of the UISM, but will also produce the detailed
activity inventory required by the encapsulation plant.

3.2 Radiometric Technique

A typical Magnox swarf spectrum is shown in Figure 3. The
dominant gamma activities in the swarf are due to fission
products in entrained irradiated fuel debris and to Co-60
activation in steel and nimonic items.

The algorithms from the UISM are to be preserved for the
process control measurement, but several extensions and
improvements are required to produce the activity inventory.

For the process control measurement the Cs-137 activity will be
calculated from the 662 keV peak intensity, corrected for the
detector efficiency and the self attenuation in any uranium
pieces. The correction will be derived from an empirically
calibrated linear approximation relating the absolute detection
efficiency for the 662 keV gammas to the relative efficiencies
for the 605 and 796 keV Cs-134 gammas. For low irradiation
fuels where the Cs-134 lines are weak the 696 keV and 2186 keV
gammas from Pr-144 will be used.

The fuel irradiation will be determined using the activity ratio, (Ru-106). $(Cs-137)/(Cs-134)^2$, of Fox et al [1].

Using a relationship based on data derived from the inventory code FISPIN [2], the Cs-137 specific activity per gram of uranium can be inferred from the irradiation and so the total mass of uranium in the swarf can be estimated. An alarm will be set if this mass exceeds a preset threshold.

The SIM inventory calculation will directly quantify the gamma emitters listed in Table 1. Commercial software supplied by Nuclear Data Inc. will be used to analyse the gamma spectrum and correct the measured peak intensities for the tray background and the detector efficiency.

An improved, iterative particle size determination algorithm has been developed to allow each photopeak to be corrected for self attenuation. This uses all of the major gamma lines from Cs-134, Ru-106 and Ce-144. The algorithm models two particle sizes. It assumes that one particle size group has zero thickness (no self attenuation) and determines a characteristic size for the remaining material, together with the mass distribution between the sizes. The technique remains simple and rapid, but it will cope with the common situation where larger fragments of uranium are mixed with finely divided material.

The isotopes present will be identified and quantified, and the irradiation determined from the Fox ratio. The cooling time of the swarf debris will be determined using ratios derived from Zr-95/Ce-144, Cs-137xEu-154/Cs-134, or Cs-137/Cs-134 depending on the detectability of the respective isotopes.

The activities of the isotopes listed in Table 2 will then be inferred from the Cs-137 activity and the measured irradiation and cooling, using look up curves of isotope correlations based on FISPIN data. The swarf in each batch measured will be taken from four fuel rods, with randomly varying irradiations. Hence the entrained fuel debris will have a mixture of irradiations. Each look-up curve will be biased to minimise the errors associated with the irradiation mixtures when the activities are integrated over the many tray measurements (approximately 100) associated with a single encapsulated drum of swarf.

3.3 Physical Arrangement

A schematic diagram of the monitor installation is shown in Figure 4. The swarf from four fuel rods will be deposited on a swarf sorting tray where it will be monitored by the germanium detector mounted on the cell roof. The pulse processing electronics will be sited as close as practicable to the detector, and the spectral data routed via ethernet to the microVAX data processor in an air conditioned control room. The SIM will be controlled via an RS232 link from the FHP plant control computer, and will integrate the inventory data for all the swarf destined for a single encapsulated drum. The final drum inventory will then be transmitted 500m via an RS422 link to the EP1 Inventory Computer. All data will be backed up onto tape streamer, so that a failure of the EP1 link will not disrupt the FHP operations.

The germanium detector is to be mounted on a precisely engineered movable table. This will permit rapid change-over of a failed detector, to minimise plant downtime, yet will allow the realignment of the detector to within 0.1mm. An infra-red proximity sensor will be used to sense any disturbance of the detector position during normal operations.

Ultra high count rate capability is required to obtain adequate statistics within the 60s acquisition period permitted by the plant processes, and to provide a wide dynamic range to cope with the variation in activity of monitored swarf. The germanium detector will be fitted with a Transistor Reset Preamplifier and coupled to a Nuclear Data Pulse Processing System with a gated integrator amplifier, $1.5\mu S$ successive approximation ADC, loss free counting unit and high speed multi-channel buffer. This system will cope with detected count rates of up to 400,000 events per second and also offers an accurate dead time correction system based on the Westphal [3] virtual pulse generator operating in live time correction mode. To extend the dynamic activity range even further, a lead attenuator will be automatically positioned over the collimator if the count rates from the swarf become excessive.

Previous experience with germanium detectors in process plants has shown them to be sensitive to electrical and mechanical noise. To overcome this an anti-microphonic mount for the germanium crystal is to be used and the detector preamplifier screen has been re-designed. Filters will be installed on the EHT and preamplifier supplies and superscreened cables with collet lock BNC connectors will be used for all analogue signals. The pulse processing electronics will also be modified to improve their electromagnetic screening and to ruggedise the units for plant use. The respective equipment manufacturers are to implement these modifications.

All peripheral equipment control signals will be routed via a single RS422 link to a programmable interface sited with the pulse processing electronics. This interface will provide optical isolation between the monitoring equipment and the monitor computer, further enhancing the system's immunity to the plant electrical environment.

3.4   System Control and Maintenance

The monitor will be a sophisticated instrument which will have to be maintained by the plant instrument engineers who, at this point in time, have very limited experience of gamma spectrometry. Considerable effort has been committed to provide high quality diagnostics and to make the monitor 'user friendly'.

All routine diagnostic, maintenance and calibration procedures will be performed via interactive dialogue with the monitor. This maintenance mode will feature pull down menus, 'Help' information on request, and electronic notes to accompany any warning messages.

The various data libraries required by the software will be generated from a single database, which can be amended via password access.

During its normal operations the monitor will calculate and record Page's statistics and the Shewhart control statistic for important monitor parameters such as the energy calibration and the detection efficiency. This statistical information will then be used to warn the operators of impending system failure and so permit early remedial work during windows in the plant operations. On line spares for all equipment will be maintained to allow rapid change-over in the event of a unit fault.

To check the correct operation of the system the monitor will automatically perform a self test using an standard source very eight hours, and immediately following any alarm message.

4   CONCLUSIONS

4.1   Both the enhanced crate monitor and the swarf inventory monitor will be installed during 1990. The description above is only an outline of the development work which underpins the design and development of two systems, and which is itself only a part of the wider programme of uprating and improvement of the systems in routine use on the operating plants at Sellafield.

4.2   In parallel with these projects, the group are also carrying out an extensive programme of development work in support of the design and construction of new waste processing facilities and for the THORP Oxide Fuel Reprocessing Plant [1]. Here the specified targets for the measurement systems are such that it has been necessary to utilise the full range of alternative radiometric techniques including several forms of neutron interrogation and fast neutron coincidence counting.

5   REFERENCES

1   FOX G H; MACDONALD B J AND GARDNER N: International Conference on Nuclear Fuel Reprocessing and Waste Management, RECOD 87; Paris (Aug 23-27, 1987) 1015.

2   BURSTALL R F: UK Atomic Energy Authority Paper ND-R-328(R) (1979).

3   WESTPHAL G P: J. Radioanal. Chem 70 (1982) 387.

Table I   Isotopes to be directly measured

Isotope	Half life	Isotope	Half life
Co 60	5.27 yrs	Ag 110m	249.8 days
Zn 65	244.1 days	Cs 134	2.06 yrs
Zr 95	64.0 days	Cs 137/Ba 137	30.0 yrs
Nb 95	*64.0 days	Ce 144/Pr 144	284.9 days
Ru 103/Rh 103	39.3 days	Eu 152	13.3 yrs
Ru 106/Rh 106	1.02 yrs	Eu 154	8.8 yrs
Ag 108m	127 yrs		

Table II   Isotopes activities to be inferred

Isotope	Half life years	Isotope	Half life years
C 14	5.7×10^3	U 234	2.5×10^5
Se 79	6.5×10^4	U 235	7.0×10^8
Sr 90	28.5	U 236	3.3×10^7
Zr 93	1.5×10^6	U 238	4.8×10^9
Tc 99	1.3×10^5	Np 237	2.1×10^6
I 129	1.5×10^7	Pu 238	87.7
Sb 125	2.7	Pu 239	2.4×10^4
Sn 126	1×10^5	Pu 240	6.6×10^3
Cd 113	13.7	Pu 241	14.4
Cs 135	3×10^6	Pu 242	3.7×10^5
Pm 147	2.6	Cm 244	18.11
Sm 151	90	Am 241	433
Eu 155	4.96	Am 242	152
		Am 243	7.4×10^3

* In equilibrium with Zr 95

FIGURE 1   GENERAL ARRANGEMENT OF THE CRATED WASTE MONITOR

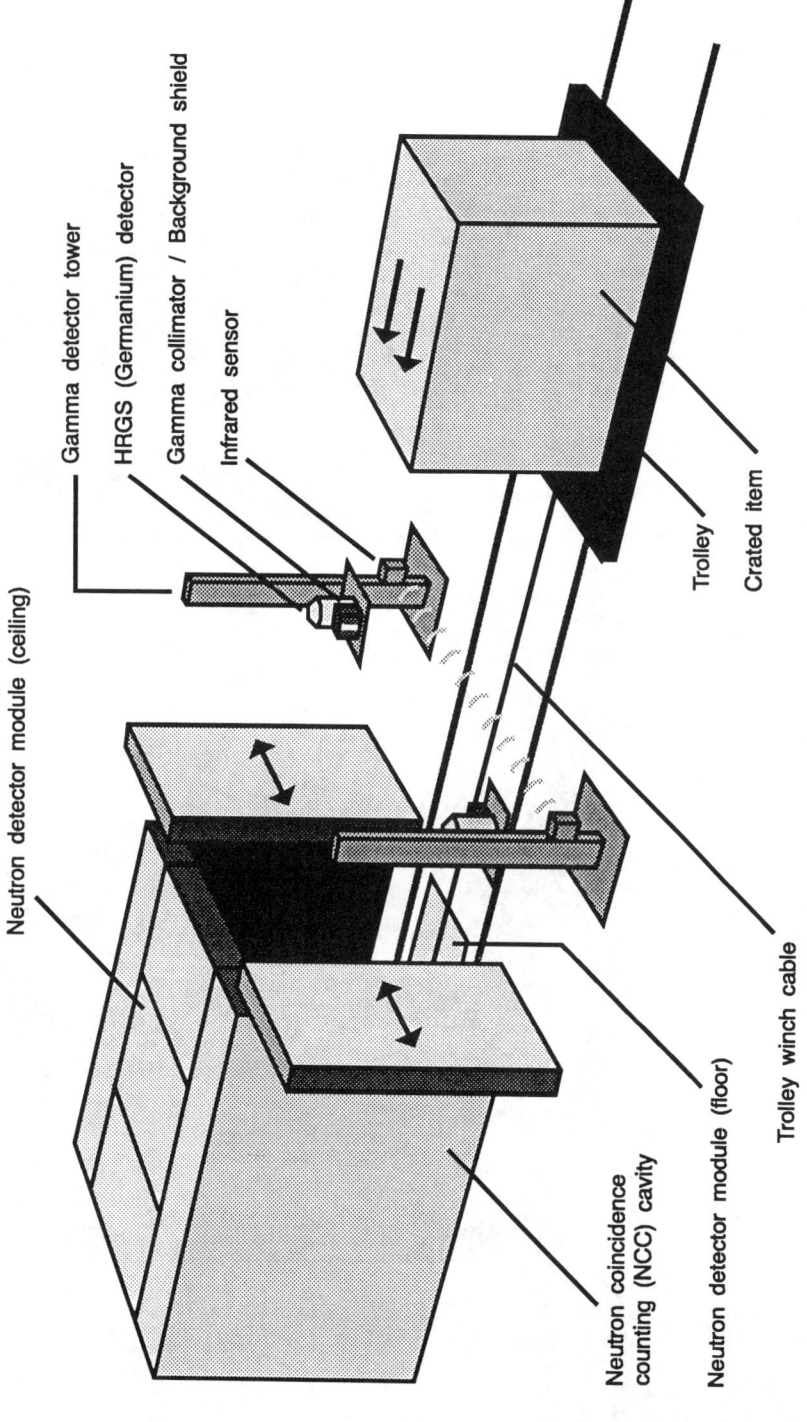

Gamma detector tower

HRGS (Germanium) detector

Gamma collimator / Background shield

Infrared sensor

Neutron detector module (ceiling)

Neutron coincidence
counting (NCC) cavity

Neutron detector module (floor)

Trolley winch cable

Trolley

Crated item

FIGURE 2   NEUTRON DETECTOR MODULE

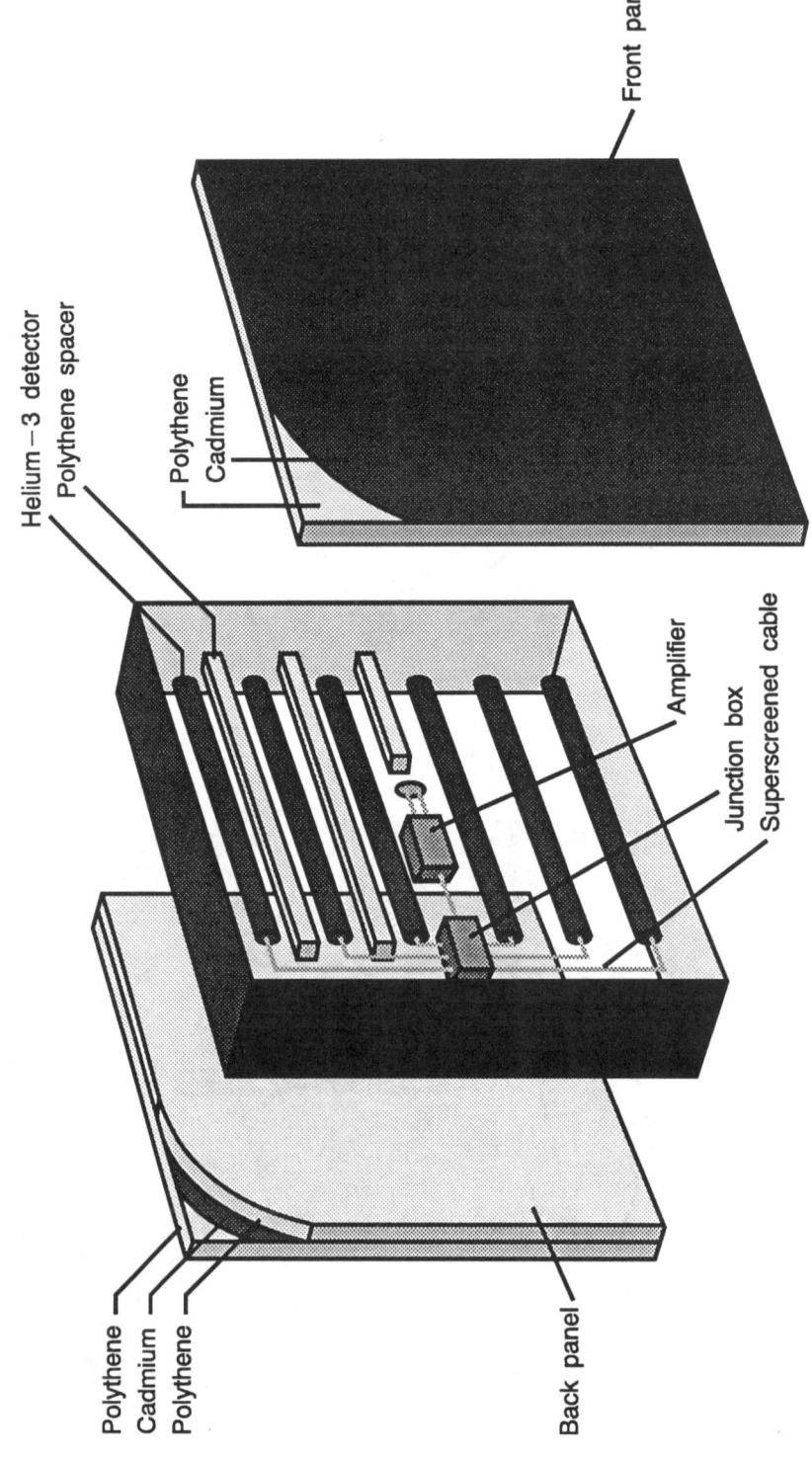

Front panel

Polythene
Cadmium

Helium−3 detector
Polythene spacer

Amplifier

Junction box
Superscreened cable

Polythene
Cadmium
Polythene

Back panel

FIGURE 3 TYPICAL GAMMA SPECTRUM FROM MAGNOX SWARF

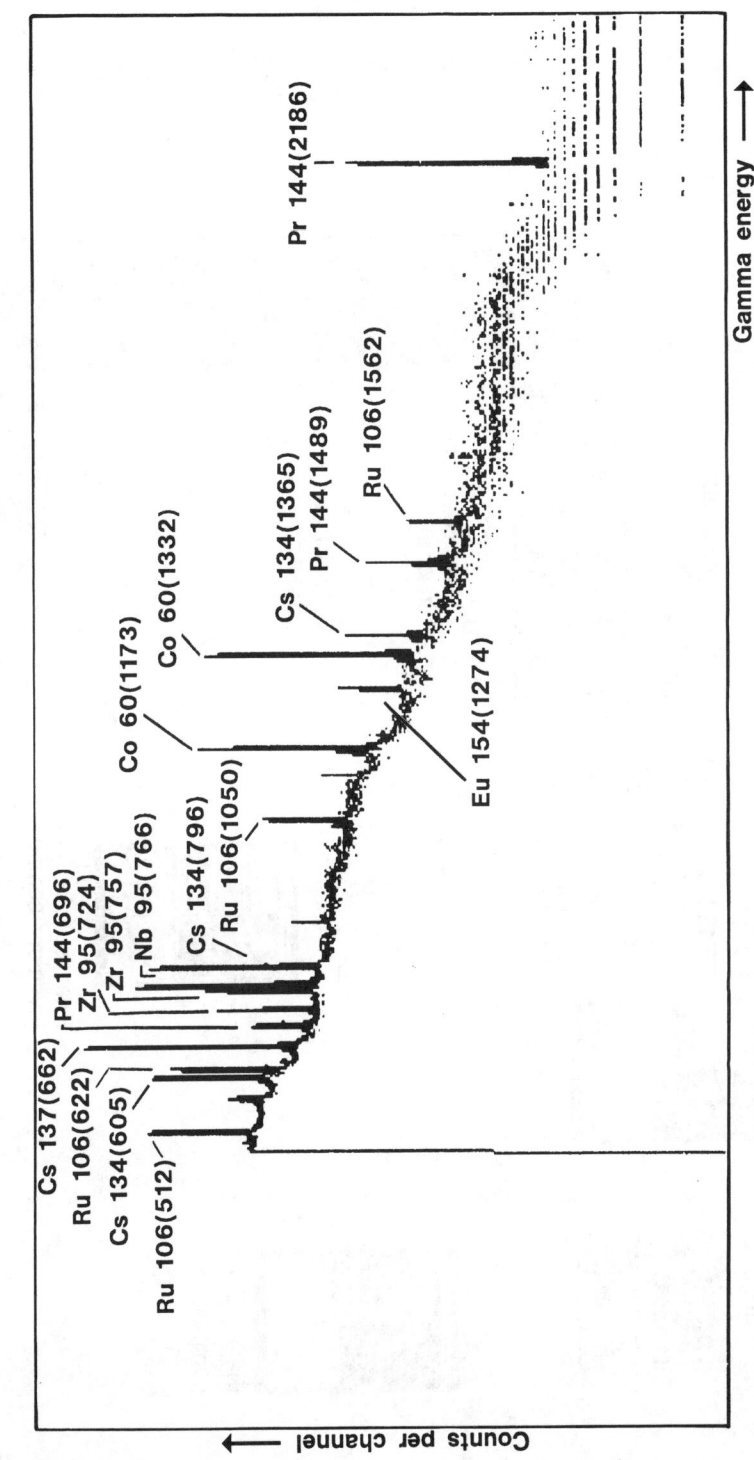

Numbers in brackets denote gamma energy of principal lines (keV)

FIGURE 4   SCHEMATIC LAYOUT OF MONITOR

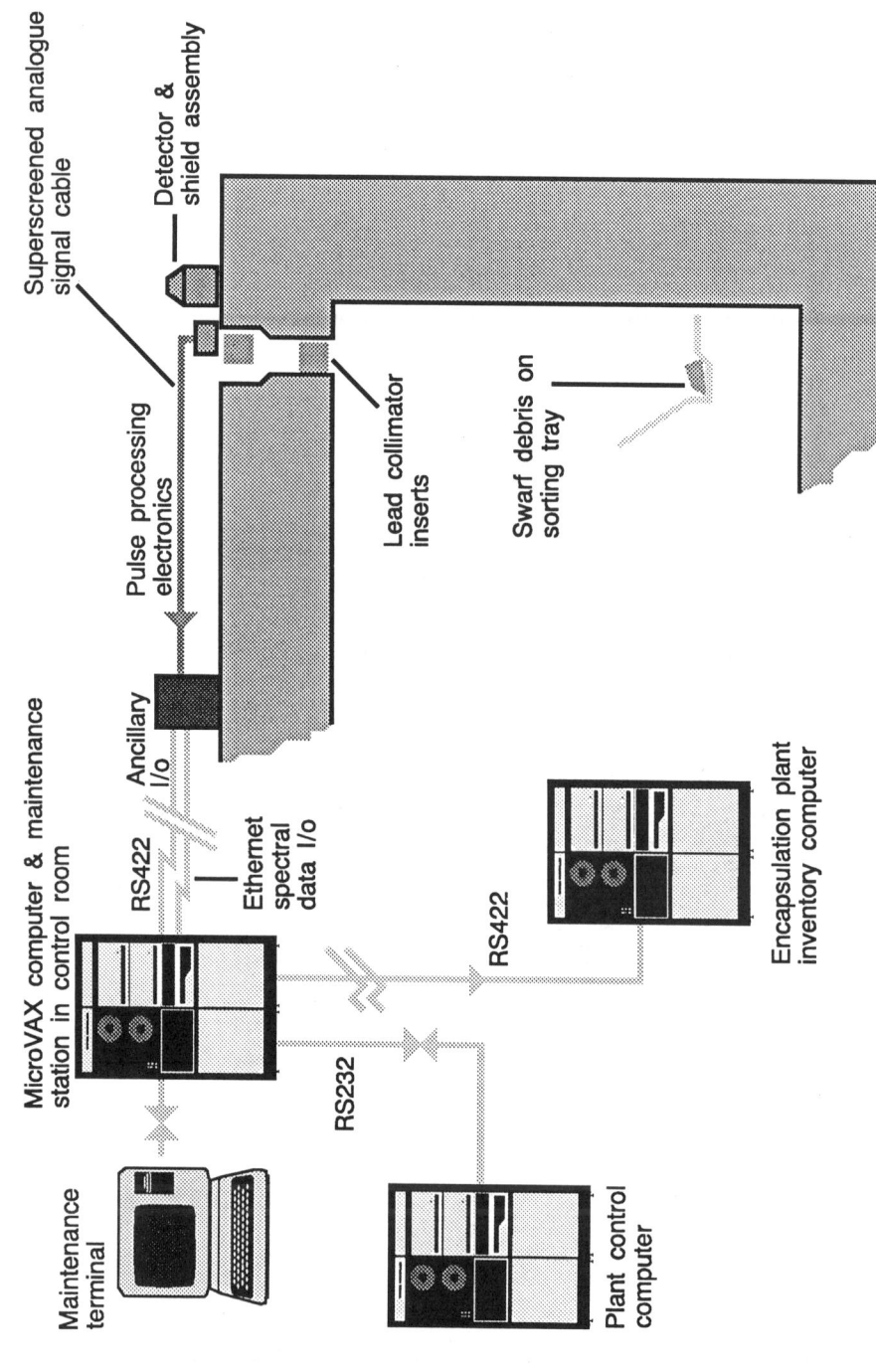

ON THE EFFECTS OF THE SPATIAL DISTRIBUTION OF PLUTONIUM IN WASTE MATRICES BY PASSIVE NEUTRON ASSAY.

B.H. PEDERSEN[1] and W. HAGE

Commission of the European Communities, Joint Research Centre,
Ispra Establishment, 21020 Ispra (Va), Italy.

Summary.
The purpose of this investigation is to examine the pair and triple correlation methods applied to the assay of Pu contaminated low β and γ radioactive waste. The paper gives the mathematical models, describes the experimental set up and reports the obtained assay results for two extreme waste matrices.

1. Introduction

PuO$_2$ present in radioactive waste generates neutrons coming from spontaneous fission events and (α,n) reactions. The neutron pair correlation technique [1-6] provides the necessary tools to determine the spontaneous fission rate F_s and the (α,n) reaction rate S_α. The additional parameters to be determined in the assay of radioactive waste is the neutron detection probability ε of the detection head due to the self shielding of the neutrons in a waste matrix. For this reason the triple correlation technique has been proposed [7,8]. In this paper both correlation methods are applied in order to examine the potential of both methods.

2. Theory

The Euratom Time Correlation Analyser measures the number of signals existing inside fixed time intervals τ. These intervals are opened either randomly or by each neutron signal of the pulse train. The quantities obtained are the frequency distributions $B_x(\tau)$ and $N_x(\tau)$ of the signals inside randomly and signal triggered observation intervals, respectively.

$B_x(\tau) \equiv$ number of events with x signals inside k randomly triggered intervals of duration τ.

$N_x(\tau) \equiv$ number of events with x signals inside intervals of duration τ in which each signal of the pulse train opens such an interval.

---

[1]On leave from
Imperial College of Science and Technology,
University of London, UK.

The frequencies $b_x^{\bullet}(\tau)$ and $n_x^{\bullet}(\tau)$ are obtained normalizing on the number of random interval triggers k and on the number of signals N_T collected during the measurement time T_M, respectively.

$$b_x^{\bullet}(\tau) = \frac{B_x(\tau)}{k} \tag{1}$$

$$n_x^{\bullet}(\tau) = \frac{N_x(\tau)}{N_T} \tag{2}$$

$b_x(\tau)$ and $n_x(\tau)$ are the probability counterparts to the measured frequencies $b_x^{\bullet}(\tau)$ and $n_x^{\bullet}(\tau)$. For the factorial moments of the order μ

$$m_{b(\mu)} = \sum_{x=\mu}^{\infty} \binom{x}{\mu} b_x(\tau) \approx \sum_{x=\mu}^{\infty} \binom{x}{\mu} b_x^{\bullet}(\tau) \tag{3}$$

$$m_{n(\mu)} = \sum_{x=\mu}^{\infty} \binom{x}{\mu} n_x(\tau) \approx \sum_{x=\mu}^{\infty} \binom{x}{\mu} n_x^{\bullet}(\tau) \tag{4}$$

rather simple recurrence formulas are obtained [8].

$$m_{b(\mu)} = \sum_{r=0}^{\mu-1} \frac{\mu-r}{\mu} \, m_{b(r)} \, m_{b(\mu-r)}^{\bullet} \tag{5}$$

$$m_{n(\mu)} = \sum_{q=0}^{\mu} m_{n(q)}^{\bullet} m_{b(\mu-q)} \tag{6}$$

with $m_{b(0)} = 1$ and $m_{n(0)} = m_{n(0)}^{\bullet} = 1$

The quantities $m_{b(\mu)}^{\bullet}$ and $m_{n(\mu)}^{\bullet}$ are simple analytical expressions of the neutron physical characteristics of the spontaneous fission sample, the moderator detector assembly and the settings of the Time Correlation Analyser (TCA).
For the mathematical description of $m_{b(\mu)}^{\bullet}$ and $m_{n(\mu)}^{\bullet}$ following assumptions are introduced.

1. The test sample has point geometry.

2. There is only fast neutron multiplication in the test sample.

3. Primary neutron emission and the following neutron multiplication chain occur contemporarily.

4. The time response function of the neutron detection is a pure exponential with decay time $1/\lambda$.

5. Primary neutron energy has no influence on the neutron detection probability.

6. There are no dead time losses of signals.

It is then:

$$m^{\bullet}_{b(\mu)} = w(\mu)\, \epsilon^{\mu}\, \tau \left[S_{\alpha} \left\{ (1-p)\delta_{1\mu} + \overline{v_{I(\mu)}(p)}p \right\} + F_s \overline{v_{s(\mu)}(p)} \right] \qquad (7)$$

$$m^{\bullet}_{n(\mu)} = f^{\mu}\, \frac{\epsilon^{\mu+1}}{m_{b(1)}} \left[S_{\alpha} \left\{ (1-p)\delta_{0\mu} + \overline{v_{I(\mu+1)}(p)}p \right\} + F_s \overline{v_{s(\mu+1)}(p)} \right] \qquad (8)$$

with $\displaystyle w(\mu) = \sum_{k=0}^{\mu-1} \binom{\mu-1}{k} (-1)^k\, \frac{1-e^{-\lambda\tau k}}{\lambda\tau k}$

and $f = e^{-\lambda T}(1-e^{-\lambda\tau})$

In many practical applications the isotopic composition of the fissile material is known. From this information the so called α-ratio is derived.

$$\alpha = \frac{S_{\alpha}}{v_{s(1)}\, F_s}$$

By means of this ratio the unknown S_{α} of eq. 7 and 8 can be eliminated.

the symbols used are:

S_{α} = (α,n) neutron emission rate of test item,
F_s = spontaneous fission rate of test item,
p = probability that a neutron generates an induced fission event,
ε = probability for detection of a neutron,
τ = observation interval,
T = delay between trigger signal and start of observation interval τ,
λ = fundamental mode decay constant of moderator-detector assembly,
$P_{jv}(p)$ = probability for the emission of v fast neutrons per prompt fission caused by a primary neutron generated by reaction j (j = I when induced fission, j = s for spontaneous fission),

$\overline{v_{j(\mu)}(p)}$ = μth factorial moment of the $P_{jv}(p)$ distribution.

$$\overline{v_{j(\mu)}(p)} = \sum_{k=1}^{\infty} \binom{v}{\mu} P_{jv}(p) \quad , \qquad \delta_{n,\mu} = \begin{cases} 1 & \text{for } n = \mu \\ 0 & \text{for } n \neq \mu \end{cases} \qquad (9)$$

The factorial moments of the $P_{jv}(p)$ distribution as function of neutron multiplication were obtained from [9] as follows:

$$\overline{v_{s(1)}(p)} = M\, v_{s(1)} \qquad (10)$$

$$\overline{v_{s(2)}(p)} = M^2 v_{s(2)} \left[1 + (M-1)\, \frac{v_{s(1)}\, v_{I(2)}}{v_{s(2)}\, (v_{I(1)}-1)} \right] \qquad (11)$$

$$\overline{v_{s(3)}(p)} = M^3 v_{s(3)} \left[1 + \frac{M-1}{(v_{I(1)}-1)v_{s(3)}} \left\{ v_{s(1)}v_{I(3)} + 2v_{s(2)}v_{I(2)} \right\} + 2 \frac{(M-1)^2}{v_{s(3)}} \frac{v_{s(1)}v_{I(2)}^2}{(v_{I(1)}-1)^2} \right] \tag{12}$$

$$\overline{v_{I(1)}(p)} = M \, v_{I(1)} \tag{13}$$

$$\overline{v_{I(2)}(p)} = M^2 (M-1) \frac{v_{I(2)}}{v_{I(1)}-1} \tag{14}$$

$$\overline{v_{I(3)}(p)} = M^3 (M-1) \left[\frac{v_{I(3)}}{v_{I(1)}-1} + 2(M-1) \frac{v_{I(2)}^2}{(v_{I(1)}-1)^2} \right] \tag{15}$$

and $M = \dfrac{1-p}{1-pv_{I(1)}}$

The experimentally determined factorial moments of the signal frequency distribution eq. 5 and 6 serve to obtain the factorial moments $m^{\bullet}_{b(\mu)}$ and $m^{\bullet}_{n(\mu)}$ which are functions of the sample, it's nuclear data [10,11], detector head parameters and the chosen observation intervals of the Time Correlation Analyser. Eq. 7 and 8 can be expressed by a common quantity $R_{(\mu)}$

$$R_\mu = \frac{m^{\bullet}_{b(\mu)} T_M}{w(\mu)\,\tau} = \frac{m^{\bullet}_{n(\mu-1)}}{f^{\mu-1}} N_T \qquad \text{for } \mu \geq 1 \tag{16}$$

Using eq. 16 to express $\overline{v_{s(\mu)}(p)}$ and $\overline{v_{I(\mu)}(p)}$ by means of eq. 10 through 15 the following expressions are obtained.

$$R_1 = \varepsilon \, F_S \, T_M \, M \, v_{s(1)} \, (1+\alpha) \tag{17}$$

$$R_2 = \varepsilon^2 \, F_S \, T_M \, M^2 \left[v_{s(2)} + (M-1) \frac{v_{s(1)} \, v_{I(2)}}{v_{I(1)}-1} \, (1+\alpha) \right] \tag{18}$$

$$R_3 = \varepsilon^3 \, F_S \, T_M \, M^3 \tag{19}$$

$$\left[v_{s(3)} + 2\,(M-1) \frac{v_{s(2)} \, v_{I(2)}}{v_{I(1)}-1} + (1+\alpha)(M-1) \frac{v_{s(1)}}{v_{I(1)}-1} \left\{ v_{I(3)} + 2(M-1) \frac{v_{I(2)}^2}{v_{I(1)}-1} \right\} \right]$$

Eq. 17 to 19 are the analytical expressions for the correlated singlets, doublets and triplets obtained with the two methods opening the observation intervals either randomly or by each signal registered.

3. The Waste Barrel Monitor and Samples

The Waste Barrel Monitor consists of three different elements the detector head, the electronic chains and the time correlation analyser for the data acquisition. The detector head has a cylindrical cavity housing a waste barrel up to 220 litres. The 60 He^3

neutron detectors are arranged in a 4π geometry inside 20 Polyethylene and Cadmium lined modules. A set of 4 He^3 neutron detectors is connected via an amplifier and discriminator directly to the TCA. The TCA is fed with 15 such signal inputs. The TCA processes the signal pulse train contemporarily for the two observation interval opening methods i.e. the randomly opened and the signal triggered intervals. The pulse train is automatically analysed in both methods for 16 observation intervals each of different duration in contrast to the present existing shift registers having only one observation interval. Thus each method gives 16 analysis results one for each interval size. This feature permits a check of the reliability of the obtained results. The adjustable interval range is such that the dead time and decay constant can be measured as well with high precision.

The stored frequency distributions of the two methods each with 16 time intervals are transferred after each measurement to a personal computer for analysis with the theory outlined above. The analysis itself can be performed according to 5 procedures [12] depending on the quantities required. In the case of radioactive waste with limited information on the waste matrix following two procedures are used:

Analysis Method 1.   Analysis Method 2.

Known   : α-ratio   Known   : M
Unknown: S_F, ε, M   Unkonwn: S_F, ε, S_α

The present set of experiments explore the potential of the multiplet analysis applied to the assay of fuel fabrication waste incorporated in various matrices. The parameters of investigation are:

1. Waste barrel size (20 1, 32 1, 60 1, 220 1).
2. Waste matrices.
3. Source arrangements.
4. Performance of the two statistical methods.

The waste barrels have axial re-entrant tubes for the accommodation of the PuO_2 samples. These tubes are arranged in two perpendicular axises. This paper reports only results obtained with a 60 litres barrel with either a concrete or a void matrix.

The PuO_2 samples have a diameter of about 10 mm. Only the metallic sample with its 33 mm diameter does not fit into the re-entrant tubes. The weight and isotopic composition of the samples are collected in table 1 [13].

SAMPLE No.	COMP.	Pu-wgt. [g]	Pu-238 [% wt]	Pu-239 [% wt]	Pu-240 [% wt]	Pu-241 [% wt]	Pu-242 [% wt]	Am-241* [ppm]
2	PuO_2	1.997	0.072	85.151	13.157	1.353	0.267	4830
20	PuO_2	4.983	0.072	85.151	13.157	1.353	0.267	4830
21	PuO_2	9.967	0.072	85.151	13.157	1.353	0.267	4830
22	PuO_2	19.919	0.072	85.151	13.157	1.353	0.267	4830
30	PuO_2	20.466	0.104	69.286	26.303	2.879	1.428	27510
40	metal	8.449	0.063	88.954	10.169	0.7	0.114	5880

* Refers to 30.07.82

TABLE 1. Weight and Isotopic Composition of Pu Standard Samples.

4. Influence of Detector Head Parameters

In the statistical model of the factorial moments of the signal pulse train it is assumed that the decay time $1/\lambda$ of the detector head and its waste barrel is a constant. It can be typical for a waste item but must be independent of the source position in the waste matrix.

| | C O N C R E T E | | |
Source Position R [cm]	Cf 1/λ [μs]	PuO$_2$ 1/λ [μs]	Pu-metal 1/λ [μs]
0.0	44.31 ± 0.20	44.17 ± 0.23	
5.0	44.83 ± 0.21	44.64 ± 0.40	
10.0	44.42 ± 0.32	44.28 ± 0.24	
15.0	44.30 ± 0.29	43.55 ± 0.34	
	V O I D		
0.0	43.83 ± 0.21	43.40 ± 0.47	43.42 ± 0.24
15.0	43.76 ± 0.18	43.72 ± 0.38	43.35 ± 0.15

TABLE 2. Decay time $1/\lambda$ of the Detector Head Response Function.

Table 2 gives for various neutron sources and radial positions R the decay time $1/\lambda$. There exists a small radial variation of $1/\lambda$ especially in a concrete matrix.

R [cm]	$M_1(1/\lambda \equiv true)$	$M_2(1/\lambda \equiv void)$	$(M_2-M_1)/M_1$ [%]
0.0	2.963 ± 0.152	2.991 ± 0.103	0.94
5.0	3.003 ± 0.186	3.015 ± 0.187	0.33
10.0	2.914 ± 0.166	2.923 ± 0.116	0.31
15.0	2.768 ± 0.177	2.769 ± 0.178	0.04

TABLE 3. Influence of $1/\lambda$ on Assay Result.

Table 3 gives the result obtained by placing a sample in various radial positions of a concrete barrel but analysing it first with the correct (M_1) then with the decay constant measured in void (M_2). The results show that it is not necessary to measure the decay constant for each waste barrel. Prefixed tables for specific waste types are sufficient.

The interpretation model permits the determination of the Pu^{240} mass equivalent with an unknown detection probability ε. Table 4 shows that ε must therefore be considered as an unknown quantity due to the influence of the source position and the waste matrix.

Care must be taken if the detector head is calibrated with a Cf^{252} source. Its detection probability is smaller than those measured with neutrons originating from Pu samples.

| | C O N C R E T E | | |
| Source Position | Cf | PuO$_2$ | Pu-metal |
R [cm]	ε [%]	ε [%]	ε [%]
0.0	6.589 ± 0.038	7.042 ± 0.022	
5.0	6.847 ± 0.042	7.400 ± 0.035	
10.0	7.970 ± 0.072	8.488 ± 0.048	
15.0	9.877 ± 0.086	10.182 ± 0.065	
	V O I D		
0.0	12.040 ± 0.077	12.079 ± 0.053	12.925 ± 0.094
15.0	12.274 ± 0.068	12.418 ± 0.053	13.380 ± 0.072

TABLE 4. Neutron Detection Probability.

5. Influence of Source Parameters

It has been shown that apart from the determination of the spontaneous fission rate F_s it is necessary to consider for NDA waste measurements the detection probability ε as a second unknown. Two other parameters directly related to the source characteristics are the α-ratio ((α,n) reaction rate divided by the spontaneous fission neutron emission rate) and its neutron multiplication factor M(p). In many cases the isotopic composition of the waste is well known such that a reasonable α-value can be estimated. In this case the model gives MPu240$_{eq}$, ε and M(p).

	C O N C R E T E			
	Signal Trigger		Random Trigger	
		Dev		Dev
α	MPu240$_{eq}$ [g]	[%]	MPu240$_{eq}$ [g]	[%]
0.500	3.482 ± 0.120	26.9	3.081 ± 0.130	12.3
0.525	3.337 ± 0.110	21.6	2.956 ± 0.130	7.7
0.550	3.199 ± 0.110	16.6	2.838 ± 0.120	3.4
0.575	3.069 ± 0.100	11.8	2.725 ± 0.120	-0.7
0.592	2.963 ± 0.150	8.0	2.627 ± 0.170	-4.3
0.625	2.828 ± 0.098	3.1	2.519 ± 0.100	-8.2
0.650	2.722 ± 0.092	-0.8	2.428 ± 0.100	-11.5
	V O I D			
0.500	3.079 ± 0.062	12.2	3.125 ± 0.086	13.9
0.525	2.951 ± 0.061	7.5	2.994 ± 0.081	9.1
0.550	2.829 ± 0.058	3.1	2.871 ± 0.078	4.6
0.575	2.713 ± 0.056	-1.1	2.753 ± 0.074	-0.3
0.592	2.638 ± 0.054	-3.9	2.676 ± 0.071	-2.5
0.625	2.500 ± 0.050	-8.9	2.537 ± 0.067	-7.6
0.650	2.408 ± 0.048	-12.2	2.442 ± 0.048	-11.0

Ref. mass: 2.744 g.

TABLE 5. Assay result MPu240$_{eq}$ as function of α-ratio, α(true) = 0.592

In Table 5 the sensitivity of the assay results is shown as function of the α-ratio used as input in the interpretation model. The results suggest for α and a void matrix a slightly lower α-value as 0.592. In concrete this is true as well for the Random Trigger technique. However for the signal triggered interval technique the α-ratio should be about 10 % higher in a concrete waste matrix. This discrepancy can not be explained by measurement errors.

Table 5 shows that the assay results are not very sensitive to the uncertainty of the α-ratio suggesting to use the analysis method 1.

The sensitivity of the assay results as function of the neutron multiplication factor M(p) is shown in table 6.

		$MPu240_{eq}$ [g]	
p [%]	M(p)	Sample 2.	Sample 22.
0.0	1.00000	0.221 ± 0.058	1.404 ± 0.066
0.5	1.01098	0.260 ± 0.069	1.634 ± 0.079
1.0	1.02232	0.392 ± 0.082	1.695 ± 0.094
1.5	1.03404	0.348 ± 0.096	2.154 ± 0.110
2.0	1.04615	0.398 ± 0.111	2.450 ± 0.127
2.5	1.05868	0.453 ± 0.127	2.772 ± 0.147
Ref. mass		0.275	2.744

TABLE 6. Assay result as function of M(p).

As expected the true assay mass for the smaller sample 2 is found for a smaller M(p) than is the case for the larger sample 22.
A wrong or neglected multiplication used as input produces for $MPu240_{eq}$ significantly wrong values. For this reason it is most convenient to use analysis method 1 in which the multiplication factor M, the detection probability ε and the spontaneous fission rate F_s is determined using the α-ratio as input.

6. Assay Results

The assay results were tested for 3 different cases important for NDA Pu waste measurements. These are:

1. Pu mass dependence in a void and concrete matrix (table 7).

2. Radial positioning of the Pu sample in a void and concrete matrix (table 8).

3. Two spatially separated Pu samples in a void and concrete matrix (table 9).

Each table gives assay results for 4 different techniques:

A. The TCA technique with signal triggered observation intervals (ST).

B. The TCA technique with randomly triggered observation intervals (RT).

C. The Shift Register technique (SR).

D. The Reduced Variance technique (RV).

In the TCA analysis A and B it was assumed that the α-ratio of the waste is known. Unknowns are the neutron multiplication M, the detection probability ϵ and the spontaneous fission rate. In the techniques C and D the neutron multiplication factor M is set equal to 1 leading to following simple expression:

$$F_s = \frac{R_1^2}{R_2 T_M} \frac{V_{s(2)}}{V_{s(1)}^2 (1+\alpha)^2} \tag{20}$$

Regarding table 7 the mass dependence is very well treated by the triple correlation methods in the case of a voided matrix (technique A and B), the pair correlation techniques leads to better results in a concrete matrix.

CONCRETE
$MPu240_{eq}$ [g]

Sample	Ref.	Triple Corr.		Pair Corr.	
		ST	RT	SR	RV
2	0.275	0.277 ± 0.023*	0.277 ± 0.021*	0.271	0.281
20	0.687	0.675 ± 0.064	0.692 ± 0.090*	0.687	0.689
21	1.373	1.474 ± 0.050	1.200 ± 0.042	1.336	1.392
22	2.744	2.959 ± 0.146	2.627 ± 0.166	2.563	2.700
30	5.951	6.599 ± 0.222	6.394 ± 0.214	5.792	6.061

VOID

Sample	Ref.	ST	RT	SR	RV
2	0.275	0.253 ± 0.020	0.235 ± 0.024	0.242	0.232
20	0.687	0.648 ± 0.035	0.617 ± 0.051	0.572	0.549
21	1.373	1.334 ± 0.034	1.333 ± 0.027	1.086	1.048
22	2.744	2.638 ± 0.054	2.676 ± 0.071	2.244	2.249
30	5.591	5.929 ± 0.135	6.018 ± 0.144	5.223	5.154

TABLE 7. Dependence of assay result on sample mass in the centre of a voided and concrete 60 l waste barrel.

In case of the concrete matrix and triple correlation the $MPu240_{eq}$ is generally overestimated with respect to the known reference values of the samples. At small sample weights (see measurement results of table 7 marked by a "*") the assay results showed large instabilities for the correlated triplets R_3 due to low count rates and discrepancies. For this reason these values were obtained only from selected values with $0.95 \leq M \leq 1.05$ taking the assay result only for the interpolated value M = 1. The error of the result was derived from the known fast multiplication of the sample (Sample 5: M = 1.0239 and Sample 20: M = 1.0384).
On the other hand the pair correlation (technique C and D) does not treat well the case of a void matrix. These observed discrepancies are subject of further investigations.
The detection limit of the triple correlation is in the present set up in the order of 20 mg and is lower for the pair correlation. Measures will be taken to reduce this limit further using the TCA both in pair and triple correlation.

Table 8 shows the assay result of sample 22 as function of the radial position R in both a void and a concrete matrix. In a void matrix the assay results are independent of the radial position. The triple correlation ST and RT give in the worst case an underestimation of about 4 %, in the pair correlation an underestimation of less than 21 %.

	CONCRETE MPu240$_{eq}$ [g]			
Position	Triple Corr.		Pair Corr.	
R [cm]	ST	RT	SR	RV
0.0	2.963 ± 0.152	2.627 ± 0.166	2.572	2.663
5.0	3.003 ± 0.185	2.766 ± 0.236	2.554	2.664
10.0	2.914 ± 0.164	2.822 ± 0.144	2.443	2.557
15.0	2.768 ± 0.177	2.890 ± 0.254	2.358	2.440
	VOID			
0.0	2.638 ± 0.054	2.676 ± 0.071	2.245	2.249
5.0	2.626 ± 0.060	2.671 ± 0.160	2.244	2.255
10.0	2.642 ± 0.081	2.651 ± 0.142	2.256	2.278
15.0	2.627 ± 0.082	2.638 ± 0.111	2.249	2.275

Ref. mass: 2.744 g

TABLE 8. Assay result as function of radial position of the PuO$_2$ sample 22.

The assay result of experiments placing PuO$_2$ sample 21 in the centre of the matrix and sample 22 in various radial positions are summarised in table 9. Again the assay results are independent of the radial position of the second sample in a void matrix. All assay results are lower than the reference mass in the worst case 4.4 % for triple correlation and 17 % for pair correlation.

	CONCRETE MPu240$_{eq}$ [g]			
Position	Triple Corr.		Pair Corr.	
R [cm]	ST	RT	SR	RV
0.0	4.539 ± 0.153	4.389 ± 0.317	3.836	4.011
5.0	4.406 ± 0.065	4.442 ± 0.267	3.840	4.013
10.0	4.296 ± 0.243	4.363 ± 0.193	3.744	3.868
15.0	4.205 ± 0.264	4.346 ± 0.241	3.511	3.654
	VOID			
0.0	3.997 ± 0.163	4.003 ± 0.145	3.391	3.418
5.0	3.983 ± 0.054	4.022 ± 0.054	3.396	3.441
10.0	3.937 ± 0.062	3.944 ± 0.105	3.405	3.443
15.0	3.943 ± 0.098	3.959 ± 0.103	3.406	3.452

Ref. mass Sample 21 and 22: 4.118 g

TABLE 9. Sample 21 in the centre position and sample 22 in various radial positions of a 60 litre concrete and void barrel.

In a concrete matrix exists for the signal triggered triple correlation method a pronounced underestimation with both sources at R = 0 and decreases with increasing radius of the second source position. In the case of the random trigger triple correlation method exists a systematic but constant overestimation. With pair correlation the $MPu240_{eq}$ is underestimated in both the concrete and void matrix.

7. Conclusions

The investigations reported in this work suggests the following conclusions:

1. There exists a radial dependence of the decay constant λ as function of the radial source position in a hydrogenous matrix. However the assay results are not very sensitive to this variation.

2. The radial dependence of the detection probability ε is quite substantial in a hydrogenous matrix and is significantly lower in a quasi void matrix. For this reason ε must be treated as an unknown in any interpretation model.

3. The assay results are not reliably if the (α,n) reaction rate S_α is part of the unknowns along with neutron multiplication.

4. Best results are obtained inserting the α-ratio i.e. the ratio of the (α,n) reaction rate to the spontaneous fission neutron emission rate.

5. The assay results are independent of the source intensity and their radial position.

7. Additional work on the models has to be done in order to reduce systematic deviations in the order of less than 20 % applying the pair correlation and of less than 10 % the triple correlation method.

The time correlation analyser proved to be a reliable instrument capable of applying both pair and triple correlation analysis methods.

8. Acknowledgements

Many thanks are due to K. Caruso for doing many of the calculations and to D. D'Adamo for confirmation of the isotopic composition of the used samples [14]. The authors acknowledge fruitful discussions with L. Bondar, R. Dierckx (JRC Ispra) and J.A. Mason (Imperial College of Science and Technology, London). Special thanks are due to Prof. W. Ameling, K. Kleinekorte and A. Karavas (Rogovski Institut, RWTH Aachen) for building the Time Correlation Analyser and to J.F. Gueugnon and K. Richter (TUI, Karlsruhe) for providing the samples.

References

1. G.BIRKHOFF, L. BONDAR, N. COPPO, J. LEY and A. NOTEA.
   SM-133/32, IAEA, Vienna (1970).

2. K. BÖHNEL.
   KfK 2203, Kernforschungszentrum Karlsruhe (1975).

3. N. ENSSLIN, J.E. STEWART and J. SAPIR.
   Nucl. Mater. Manage., 8, 60 (1979).

4. K.P. LAMBERT, J.W. LEAKE, A.I. WEBB and F.J.G. ROGERS.
   AERE-R-9936, United Kingdom Atomic Energy Authority (1982).

5. M. NEUILLY, B. THAUREL and J. MONNIER.
   CEA-CONF-6718, Commissariat à l'Energie Atomique (1983).

6. E.J. DOWDY, C.N. HENRY, A.A. ROBBA and J.R.PRATT.
   SM-231/69, IAEA, Vienna (1978).

7. L. BONDAR.
   SM-260/54, IAEA, Vienna (1982).

8. W. HAGE and D.M. CIFARELLI.
   Nucl. Sci. Eng. 89, 159 (1985).

9. W. HAGE and D.M CIFARELLI.
   Nucl. Instr. and Meth. A236,165 (1986).

10. M.S. ZUCKER and N.E.HOLDEN.
    6TH ESARDA SYMP., Venice (1984).

11. N.E. HOLDEN and M.S.ZUCKER.
    7TH ESARDA SYMP., Liege (1985).

12. D.M.CIFARELLI and W. HAGE.
    Nucl. Instr. and Meth. A251, 550 (1986).

13. J.F.GUEUGNON and K.RICHTER.
    T.U.I KARLSRUHE, Fabrication Report K0282049 (1982).

14. D. D'ADAMO.
    Private Communication (1989)

SESSION A3 — INTEGRATED AND/OR COMBINED SYSTEMS

Chairman : P. Bernard — CEA-CEN Cadarache

Seven papers were presented during this session :

- 4 papers presented devices combining active methods (using neutron generators or sources) with passive methods.

- 1 paper studied the use of a neutron generator.

- 2 other papers concerned passive measurement devices (gamma-ray spectrometry, neutron measurement).

Mr J. ROMEYER-DHERBEY's paper described 5 measuring systems : 3 systems using a Californium source (including a device for measuring small packages based on a "measurement at source" concept) and two systems using a 14 MeV neutron generator.

Yves BEROUD presented a number of devices applicable to different types of waste (size, gamma irradiation level), based on passive and/or active methods (using neutron generators or sources), one of which was transportable, for waste classification.

Mr VIDALIE described several measuring devices based on gamma-ray spectrometry and passive neutron counting, together with the network installed between different measuring stations for the automation, management and cross-validation of measurements, thereby enhancing the quality of results.

J.T. CALDWELL presented an integrated group of waste control stations for decontamination and decommissioning. The group is composed of 5 stations : imaging and passive neutron counting of glove boxes, checking decontaminated glove boxes using a neutron generator, checking that plutonium is not accumulated using passive couting methods, checking drums and loads in a neutron generator cell and measuring waste with a significant plutonium content by gamma-ray spectrometry and passive neutron counting.

R.J. ESTEP's paper concerned the performance of a small measuring cell, developed by LOS ALAMOS for highly radioactive waste. The DDT method is used with lead-protected detectors and a matrix-effect correction approach. Errors related to the position of the contaminant are greater when the matrix has a strong moderating or absorbent effect. The authors recommend sorting waste into several categories according to the granulometry of the contaminant.

K.P. LAMBERT's paper presented passive devices (gamma-ray spectrometry and passive neutron counting) developed at HARWELL and designed to determine quantities of fissile material and the alpha content of Pu waste.
The devices made were portable (small volumes), and designed to be modular and/or configurable.
Gamma-ray spectrometry methods on rotating 210-l drums were applied.

F.J. SCHULTZ, in his paper, described the principles of the non-destructive measurement and control devices used in the OAK RIDGE waste processing and handling project. This program is backed by the DOE.
Measurements are based on the use of a LINAC electron accelerator used for : gammagraphic imaging (non-destructive examination), active neutron interrogation, active photofission interrogation. The installation of a 6 MeV accelerator is planned.

Particular points of discussion were :

- characteristics and performances of neutron generator tubes applied to waste management.

- advantages of active-passive methods in directly measuring the mass of fissile material and determining alpha activity in a representative way while disregarding the burnup of the spent fuel at the origin of the waste.

- the good results obtained using neutron generator methods for taking measurements on hulls coming from spent fuel reprocessing.

- depending on methods used, the effects and significance of data available on waste (isotopy, chemical form, granulometry of the contaminant, characteristics of the matrix) as regards the creation of a "guaranteed" value.

DETERMINATION OF ALPHA ACTIVITY AND FISSILE MASS CONTENT IN SOLID WASTE BY SYSTEMS USING NEUTRON INTERROGATION

J.ROMEYER DHERBEY,G.LACRUCHE,R.BERNE,J.AUGE
L.MARTIN DEIDIER,M.BUTEZ

COMMISSARIAT A L'ENERGIE ATOMIQUE
IRDI/DEDR
DEPARTEMENT DE RECHERCHE PHYSIQUE
Service de Systèmes d'Aide à l'Exploitation
Centre d'Etudes Nucléaires de CADARACHE , FRANCE

Summary

The Quantitative control (determination of heavy nuclides and alpha activity) of alpha radioactive wastes is necessary,particularly to determine if the waste is in accordance with the surface storage limits.
In order to reduce the uncertainty on the alpha activity resulting from unknown isotopic composition, inhomogeneity of heavy nuclides in the matrix,combination of several methods is necessary.
In the paper we present the Cadarache development work in the NDA of solid waste using the Californium shuffler,14 Mev neutron generator,and also passive techniques such as neutron emission measurement and gamma spectrometry.
Experimental systems combining active and passive methods are presented (COSAC,BANCO,DANAIDE,PROMETHEE).

1. INTRODUCTION

The DEPARTEMENT OF RESEARCH IN PHYSICS (DRP) at CEA/CADARACHE is actively engaged in the development of PASSIVE and ACTIVE neutron and gamma techniques,for the measurement of liquid and solid waste .
For solid waste,the aims of the measurement systems are the determination of the fissile mass content (safety and mass balance aspects)and the determination of the alpha activity (surface storage).
Their application field concerns particularly the fuel cycle(enrichment,fuel manufacturing,reprocessing,dismantling).
The focus of effort has been on the neutron interrogation with a californium shuffler source (with delayed neutron detection and more recently with delayed gamma detection),and the differential die-away technique using the SODERN GNT-02 neutron generator (prompt neutron detection between the 14 Mev pulses) .

The development of active interrogation techniques has been influenced by the following facts :
-in active interrogation,the determination of the 239 Pu mass is very much less sensitive to an uncertainty on the isotopic composition than in passive neutron counting,
-For Uranium contaminated wastes the active method is the only way to measure accurately the 235 U content.
-The low level of the surface storage criteria (0.1 CIα/T for acceptance,0.01CIα/T mean value on the storage site), and the great number of drums to be measured necessitate low detection limit systems,with a short counting time.

In the following,the detection limit is defined as being equal to twice the mass that gives a relative statistical error of 100 %,at a confidence level of 95 % (2 σ).

The detection limit ,that is necessary to obtain,for surface storage, depends on the Pu origin and measurement errors.For example,if we consider 100 liter drums (representative in France of the most part of drums before conditionning),and a PWR PuO2 (45000 MWD/T,4 years cooling),it is necessary to certify that the plutonium mass is lower than 35 mg,for acceptance criteria,and the detection limit has to be lower (typically 10 to 15 mg),because it is necessary to take into account the several possible errors that could underestimate the alpha activity).

The main causes of errors came from:
-The counting statistic
-The non-homogeneous repartition of the plutonium inside the drum
-The difference between the real and the calibration matrix
-The uncertainty of the isotopic composition
-The uncertainty of the geometry of plutonium (self absorption and self shielding).

In order to minimize the repartition and the matrix effects,a device (COSAC),designed for the measurement of 20 liters,non gamma irradiating bags has been developped.(252 cf source,delayed gamma and delayed neutron).

It is possible to sort the bags,constitute homogeneous drums,and to optimize the number of drums having an alpha activity lower than the acceptance criteria.

For larger volumes,the neutron emission of the source has to be greater (the fission rate being proportional to approximatively the inverse of the cavity volume),and the number of detectors extended .

The BANCO (252 cf,delayed neutrons 220 l drums) and DANAIDE facilities (252 cf,delayed neutrons ,800 l drums with high gamma dose rate) are presented.

Last is presented the experimental device PROMETHEE,using a 14 Mev neutron generator,and recent results concerning high gamma dose rate drums.

For all the active systems,the passive neutron counting is used when important masses are to be measured ,in order to minimize the self shielding effect (especially for matrix with a high hydrogeneous content ,for which correction based on epithermal interrogation doesn't apply).

2. MEASUREMENTS SYSTEMS WITH 252 CF SOURCE

2.1 PHYSICAL PRINCIPLE

The Californium source is alternatively introduced inside the cavity (about 10 s per cycle),to induce fissions,and removed in an isolated storage position (about 10 s per cycle).

During irradiation,fission fragments are produced and some of them decrease by emitting gammas and, more rarely, neutrons.In "thermal" interrogation,the cavity is composed of light elements in order to slow down neutrons from Cf and to maximise the fission rate.

When the source is in the storage position,the delayed gammas and/or the delayed neutrons are counted.

The delayed neutron signal may be considered as the sum of six decreasing exponentials ,with a half life between 0.2 and 60 s,and energies between 250 and 500 kev.

The most part of the delayed neutrons are produced by the first three exponentials.

The delayed gamma (for time between 1 to 100 s after fission) may be considered as one exponential with a half life about 5 s.,with a mean energy of 1 Mev.

The signal processing consists of searching for an amplitude term and a background term ,by fitting a sum of exponentials to the signal,the amplitude term being proportional to the fissile mass content.

The detection of delayed gammas is interesting because several hundred more delayed gammas are produced than delayed neutrons.

The disadvantage is that this method is sensitive to "parasitical" gamma rays , and doesn't apply to gamma irradiating drums,or large drums having a matrix that could activate (for example stainless steel or aluminium).

For important masses where self shielding may occur ,the "epithermal" interrogation (absorbing sheets inside the cavity to harden the neutron spectrum) reduces this effect for non hydrogeneous matrix.

2.2 THE COSAC DEVICE FOR 20 L BAGS

From earlier measurements ,on several matrix,with a non-optimal experimental system ,we have recently built a system designed for fissile mass and alpha emitter content analysis at the origin of waste production (20 liters bags).

As shown in figure 1,the bag is positioned inside an irradiation and counting chamber,and rotates during measurements.

The chamber is comprised of:

-A plastic jacket receiving the waste bag,surrounded by a lead sheet to cut the 60 kev gamma ray emission of 241 Am.

-Four scintillators panels,with photomultipliers for delayed gamma counting

-A polyethylene bloc housing 12 He3 counters for delayed neutron and passive neutron counting (total and real coincidences).

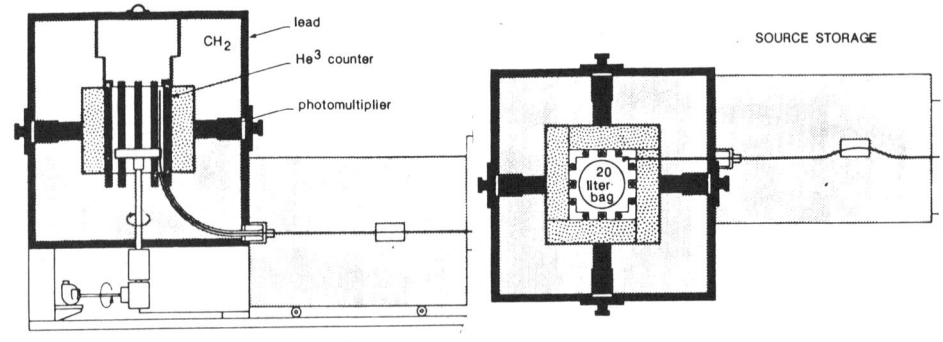

FIG 1 COSAC FACILITY AT CEN CADARACHE

A mechanical system conveys the Cf source $(4.10^8$ n/s) inside a guide-tube between a storage cask and its active position inside the irradiation and counting chamber.

The delayed gamma counting is used to obtain a low detection limit (1 mg of Pu from PWR for a 600 s counting time) .

The fig 2 shows a delayed gamma signal and the curve fitting with theorical time constant value.

In delayed neutrons the detection limit is 13 mg of PU from PWR ,for a measuring time of 600 s.(The delayed neutrons and the delayed gamma are measured simultaneously).
The delayed neutron information is used for redundancy.

The passive neutron counting (8% efficiency,1c/s background) is used when a wide measuring range is necessary and is combined with the active methode result to correct self shielding effect for important masses .

The detection limit is 4 mg of Pu from PWR,in total counting and 14 mg for real coincidences for a measuring time of 600 s.

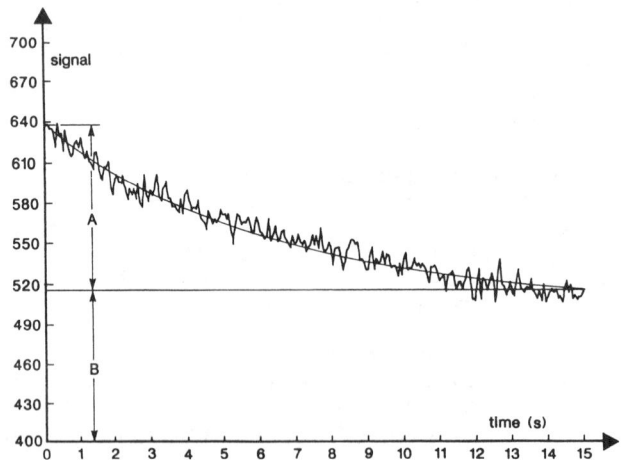

FIG 2 DELAYED GAMMAS SIGNAL AND CURVE FITTING

2.3 THE BANCO DEVICE FOR 220 L DRUMS

The system is represented on the fig 3.

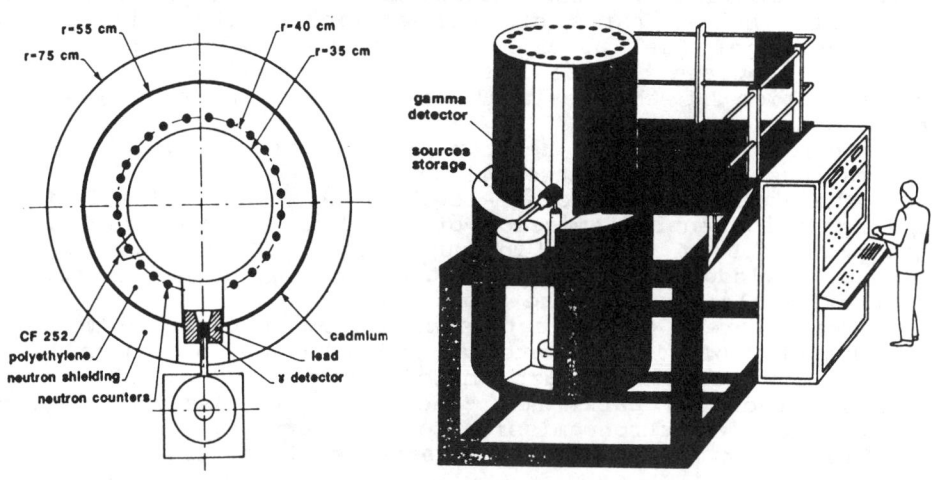

FIG 3   BANCO FACILITY AT CEN/CARARACHE

In passive mode , neutron counting (21 He counters ,1 meter length) and gamma spectrometry (HPGe diode) are performed simultaneously,the diode scanning along the vertical axis of the rotating drum.

For light matrix density (≈0.2), the detection limit is ≈10 mg of Pu from PWR,in total counting,and ≈40 mg with real coincidence measurement (shift register).

When no corrections are made the repartition effect may be more than 40 %.

In gamma spectrometry,special collimators have been designed to reduce the repartition effect (≈10%) at the expense of the detection limit.

The detection limit is 20 mg of Pu from PWR for the 208 Kev ray,and ≈ 1g for the 414 Kev ray of the 239 Pu.

When the gamma counting is sufficient ,the results of the gamma scanning are used to correct the passive neutron counting from repartition effects.

In active mode,the storage of the 252 Cf source is under the counting cell.

The delayed neutrons are detected by the same He3 counters as those used in the passive counting ,the signal being counted on multichannel scalers.

The detection limit has been measured at 30 mg of 235U, and 200 mg of Pu from PWR,with a 1,1 mg of 252Cf (2,6 10^9n/s),and a 1000s measuring time.

2.4 THE DANAIDE EXPERIMENT FOR 800 L DRUMS CONTAINING HULLS

The delayed neutron method, with 252Cf source has been recently applied to the measurement of 800 liter drums containing hulls (parts of fuel elements after cutting and dissolution of fuel assemblies in a reprocessing plant) in order to complete the passive neutron counting method for the determination of the residual fissile mass content and the alpha activity.

In passive mode, the neutron emission is mainly due to the spontaneous fissions of the 244 Cm.

The statistical error is low, but the main problem is that the specific neutron emission of the fuel is proportional to the power 4,5 of the fuel burnup, and an imprecision on the burnup knowledge causes important errors on the determination of the fissile mass and the alpha activity.

The interest of the active method is to directly measure the quantity of fissile material independently of burnup (for example low irradiated parts mixed with very well dissolved highly irradiated parts not "seen" by the passive counting because of low neutron emission for low burnups).

The experimental system is described in figure 4.

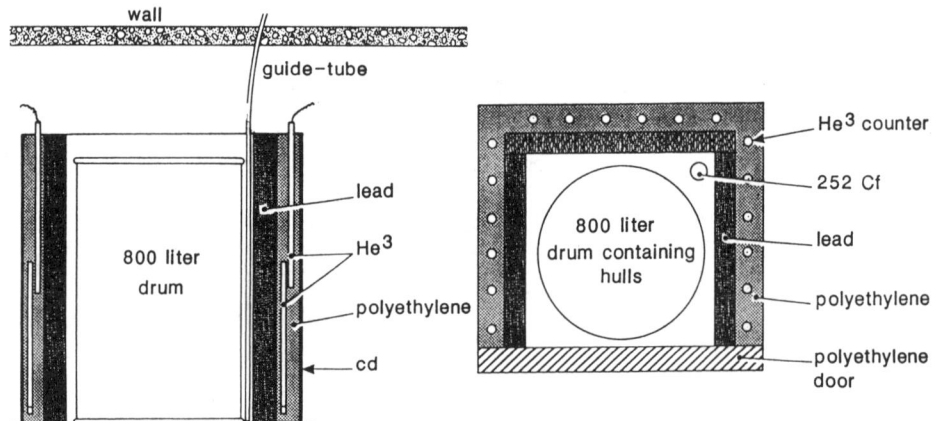

FIG 4 DANAIDE FACILITY AT CEN CADARACHE

In this configuration, 10 cm of lead protect He3 counters from the high gamma dose rate, in the case of a real drum.

The cavity is 2m high and 18 He3 counters (1 m long) are on the upper part, 18 on the lower part.

A modulated 252 Cf source (7.10^8 n/s) moved by a mobile shuffler unit was positioned on the top of the counting cell. Measurements have been made by positioning fuel pellets at several axial locations. On the horizontal plane, pellets were installed at 19 different locations to simulate a homogeneous repartition, the drum being filled with inactive hulls.

From these experimental results we obtain a detection limit, for a real drum, of about 6 g of fissile mass, for a 3 hours counting time and a source of $3 \cdot 10^9$ n/s. (and a neutron background of about 10^4 c/s)

3. MEASUREMENTS SYSTEMS WITH 14 Mev NEUTRON GENERATOR
3.1 PHYSICAL PRINCIPAL

In The differential die away technique,the 14 Mev neutrons produced in a short time (≈ 15 microseconds) by the neutron generator are slowed down inside the cavity and induced fissions.(after ≈ 0.5 ms the fast neutrons produced by the generator are in the low energy region).

The low energy ("thermal") neutrons flux decrease in time with a half life depending on the cavity and the matrix (near 1 ms in the PROMETHEE cell,in a empty cavity).

The he3 detectors are surrounded by a low energy neutron absorbing sheet ,to detect essentially the fast neutrons (≈ 2 Mev) emitted when a fission event occurs.

The number of neutrons emitted in coincidence with the fission process (prompts neutrons) is several hundred times greater than the number of delayed neutrons and consequently this method has a lower detection limit than the detection of delayed neutrons with Cf source, for a same interrogating neutron source emission .

For gamma irradiating drums ,and for He3 detectors (1 m lengthg) working in pulse current detection, a protection (usually lead) is necessary to have a gamma dose rate , at the location of the detector lower than about 2 rads par hour .

Interaction of energetic neutrons with lead increase the neutron background (partially by the inelastic diffusion process) and deteriorate the detection limit.

The pulsed neutron generator method has several potentialities :
-The time decrease of the signals may be correlated to the matrix composition
-In passive mode ,the combination of signals from detectors surrounded or not by absorbing sheet may be used to check if the matrix corresponds to those used for calibrations.
-For important masses,where self shielding may occur,the combination of delayed/prompt neutrons or thermal/epithermal interrogation may reduce this effect ,especially for matrix with a small hydrogeneous content.

3.2 PROMETHEE 1 FOR 220 L DRUMS

The PROMETHEE assay is a modular system for experimenting pulsed neutron generator techniques.

The cell (fig 5) accomodating standard 220 l drums is set up with panels mounted on rails,allowing easy structure modifications.The drum rotates during measurements (17t/mn).

The detection system includes twelve packages of three counters (1 m of utile length) on the lateral walls.Four of them are used for prompt neutron detection.

Three counters (45 cm of utile length) are positioned on the top and three others on the bottom of the cell.

A flux monitor is located at the corner of the cavity.

The neutron generator is the SODERN/GNTO2 ,working in pulsated mode (duration of pulse = 15 microseconds,time between pulses = 8 ms for the following results).

The amplifiers are outside the cavity,the pulses are counted on 8 MCS cards associated to a PC/AT computer.

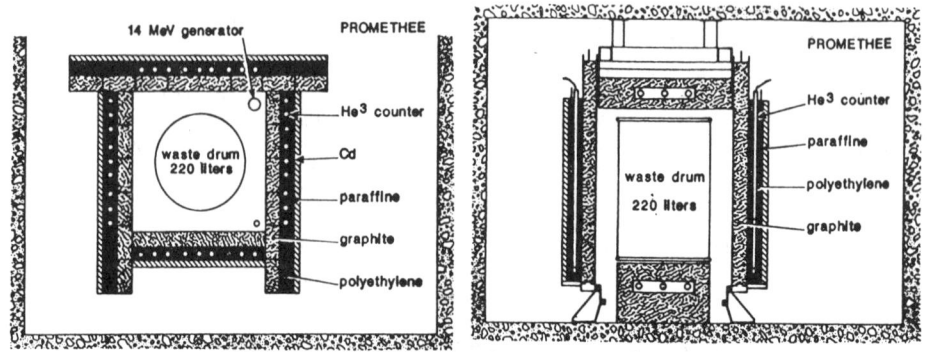

FIG 5 PROMETHEE FACILITY AT CEN/CADARACHE

The system could measure drums with a gamma dose rate up to 25 R/h without any modification on the signal.

This enables us ,for example,the measurement of 220 l asphalt ,gamma irradiating drums.In this case the detection limit is 25 mg of 239 Pu,for a measuring time of 15 mn.

For a higher gamma dose rate a shielding is necessary in order to have a neutron counting rate independant from the gamma dose rate level.

Recent experiments have been made by surrounding a 100 l drum by 4 cm of lead,in order to determine the detection limit for a industrial system ,having to measure drums with a gamma dose rate up to 500 R/h.

Experiments have shown that the absorption of low energy neutrons by the lead was approximatively compensated by the inelastic reactions (n,2n) and (n,3n) on lead.

The neutron flux is increased during the first milliseconds and decreases faster (absorption by lead);The "thermal" flux half time being reduced by 13 % (fig 6) (0.68 ms against 0.78 ms for an empty drum).

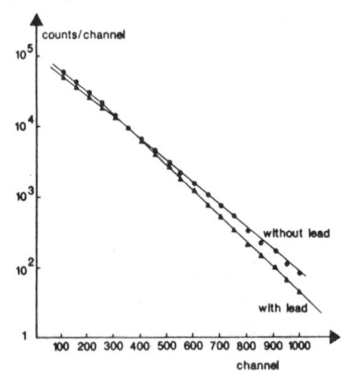

FIG 6 VARIATION OF FLUX MONITOR SIGNAL WITH 4 CM OF LEAD AROUND A 100 L DRUM.

If in passive counting the neutron background remains the same,in active mode the background is more important especially with inox steel matrix (fig 7).

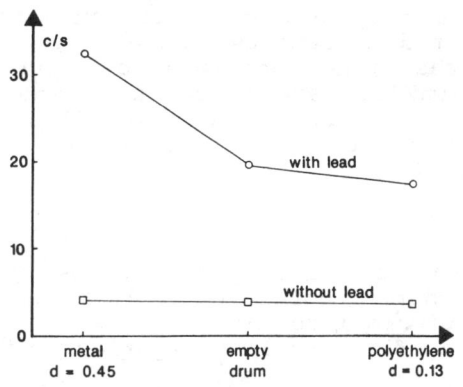

FIG 7 VARIATION OF BACKGROUND IN ACTIVE MODE DUE TO LEAD

In the case of gamma irradiating drums the detection limit,for a 15 mn measuring time varies between 4.6 mg of 239 Pu (13 kg polyethylene matrix) to 17.4 mg (45 kg stainless steel matrix).

Without lead the limits are respectively 2.8 and 5 mg.

3.3 PROMETHEE 2 FOR 800 L DRUMS CONTAINING HULLS

The PROMETHEE cell has been modified to experiment the DDT technique to the measurement of 800 l drums containing hulls.Preliminary calculations have shown the interest of the method for UO2-PUO2 Mixed OXyde(MOX) fuels ,having a neutron emission about 20 times greater than for Uranium OXyde fuels,the aim being to detect some grammes of residual fissile mass .

The size of the cavity (fig 8) is the same as for the DANAIDE experiment.

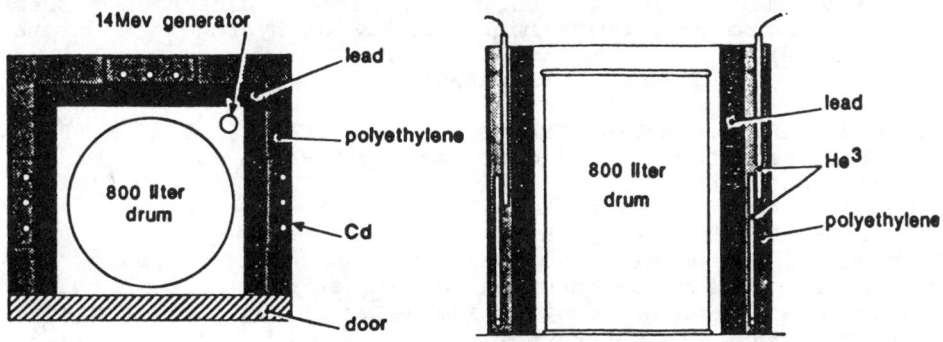

FIG 8 MODIFICATION OF PROMETHEE FOR THE MEASUREMENT OF 800 L DRUMS CONTAINING HULLS

Six blocks of three counters (1 m length) are on three lateral walls,Three of them on the upper part and three of them on the lower part.

The graphite walls are replaced by 10 cm of lead.

As experiments are only beginning all results are not yet available but first measurements have shown that the neutron background and the time decrease of the neutron flux monitor,with lead walls, are of the same order than for the measurements with lead surrounding a 100 l drum(cf 3.2) ,and thus the change of the graphite wall by lead will not be an important penalization.

4.CONCLUSIONS

The measurement of solid wastes necessitate following several rules in order to obtain reliable results.

-measurement of small containers

-constitution of several categories with homogeneous matrix

-combination of several methods,especially when a wide measurement range is necessary.

Measurements at the origin of waste production,on small bags, enable us to sort the bags to constitute homogeneous drums,and in this way to minimize the error due to a non homogeneous repartition.

The low detection limit for a short counting time,enables us the measurement of 10000 twenty liter bags (2000 equivalent 100 l drums) a year.

This type of process,with a low cost device,permits us to have a sensitivity below the 0,1 CIα/T acceptance criteria for surface storage and to warrant the obtained results.

For gamma irradiating wastes,or large size drums,the neutron generator method that gives also information about matrix characteristics is well adapted.

Our development work on the solid waste field is presently focused on the qualification of multimethod interpretation software and the determination of the application field of the combined thermal/epithermal methods,to reduce the self shielding effect, for wastes with important plutonium masses and to measure drums with high density of hydrogen (concrete for example)

REFERENCES

/1/ P.BERNARD,J.ROMEYER DHERBEY,J.PINEL,J.CLOUE,A.GIACOMETTI
"Méthodes de détection radioactives applicables aux usines de retraitement"
RECOD 87 .PARIS. 23-27 aout 1987
/2/ Y.BEROUD,J.ROMEYER DHERBEY
"A panel of non destructive assay systems for transuranic wastes using the pulsed neutron techniques"
Topical meeting on non destructive assay of radioactive wastes
20-22 nov 1989 CEN CADARACHE.
/3/ P.BERNARD,M.CANCE,Y.BEROUD
"Experimental systems using an active method for the measurement of low alpha emitter grade wastes"
ENC'86 GENEVE JUIN 1986
/4/ R.BERNE,G.SIMON,H.SOULIER
"Installation de comptage des fûts de dechets alpha"
RECOD 87 PARIS

A RANGE OF NON-DESTRUCTIVE ASSAY SYSTEMS
FOR TRANSURANIC WASTE USING PULSED NEUTRON TECHNIQUES

Yves BEROUD, (SGN, France)
Jacques ROMEYER-DHERBEY, (CEA, Cadarache, France)
Roger BERNE, (CEA, Cadarache, France)

ABSTRACT

Several active assay techniques are available but differ by the particles detected and by the energy level of interrogation neutrons. The detected particles may be prompt fission neutrons, delayed fission neutrons or gamma radiation.

The neutrons can be supplied by a Cf 252 or Am-Be source with a mechanical displacement system or by a 14 MeV pulsed neutron generator.

A system designed primarily for alpha-emitter analysis of small waste bags at the waste source is presented.
Interrogation neutrons are supplied by a Cf 252 source and the system detects delayed gamma photons. The sensitivity limit of this system is 1 mg Pu PWR for light matrix materials with a counting time of 10 mn.

A general-purpose assay system for 100-liter 200 liter drums is presented. Interrogation neutrons are supplied by a pulsed generator and the detected particles are prompt and delayed fission neutrons. The sensitivity limit of the system is 1 mg Pu PWR for low density drums with a counting time of 20 mn.

Results obtained with the systems described, considerations for large waste packages and gamma radiation emitting drums, specific industrialization factors and quality assurance are discussed.

INTRODUCTION

Together with the CEA and Cogema, SGN took an early interest in developing techniques that would enable radioactive waste generators to perform effective processing, volume reduction and packaging of their waste. SGN has continued this effort and now offers high-performance systems for measuring the alpha activity of packaged or unpackaged waste, which is mandatory for proper management of stored fissile materials.

The systems presented mainly employ active techniques that induce fission events using an external neutron source and detect the neutrons and gamma radiation produced by these fission reactions.
Passive methods such as gamma spectrometry and neutron counting are however not rejected because they are still complementary to and often

combined with active techniques in order to cover a wider range of applications and to ensure more representative measurements.

Many parameters affect the quality of measurements, including package size, chemical and isotopic form of the contaminant, matrix type and density. The importance of these factors means that no all-purpose method exists, and the technique must be adapted to the type of waste involved and to the requirements of the waste generator.

Three families of assay systems are presented for 20-liter bags, 100-liter and 200-liter drums, which may or may not contain radiation-emitting materials, and for large waste packages produced during equipment dismantling. Special attention is given to the 20-liter bag assay systems, which employ a novel technique for monitoring waste at the source.

The theoretical approach and results obtained on pilot facilities have been described in other publications [1] [2] . This paper will focus on the engineering, operation, safety and maintenance of the assay systems.

ACTIVE METHODS

Active methods, also called neutron interrogation methods, involve cyclically irradiating the waste with neutrons and detecting particles emitted by the resulting fission events in the contained fissile material. Several methods are available and differ by the energy level of the interrogation neutrons.

The detected particles may be prompt fission neutrons, prompt gamma radiation, delayed neutrons and/or delayed gamma radiation. The interrogation neutrons may be supplied by a Cf 252 neutron source or by a pulsed neutron generator.The systems described below use either a mechanically pulsed Cf 252 source or a 14 MeV pulsed neutron generator made by Sodern.

COSAC BAG COUNTING SYSTEM

This system is designed to measure the waste at the source in bags of approximately 20 liters. The bags are often filled with waste from glove boxes, which therefore has a light matrix that emits no radiation.

Principle

The unique feature of the system is that the fissile material is activated by a mechanically pulsed Cf 252 source and the delayed gamma radiation is detected. More precisely, the neutron source is alternately placed in the irradiation position during the activation phase and in the withdrawn position during the counting phase. The irradiation sequence for the fissile material lasts for 10 seconds, the removal of the source lasts about 1 second and the counting lasts 10 seconds.

In addition to other products, the induced fission events produce delayed gamma radiation and delayed neutrons. The delayed gamma radiation is detected by scintillator units and the delayed neutrons may be detected by optional He3 counters. With adequate calibration, the counting of these two types of emitters enables determination of the quantity of fissile material in the bag.

The system is designed to also detect delayed neutrons and spontaneous fission neutrons when a broad assay range is required.

The data processing software takes into account delayed gamma radiation, delayed neutrons, total passive counting and passive counting of spontaneous fission events. These results are compared with the data on the packages to ensure the validity of the measurements and determined the alpha mass and activity level.

Description

The COSAC system includes an irradiation and counting enclosure and a removal cask for the source (Fig. 1).

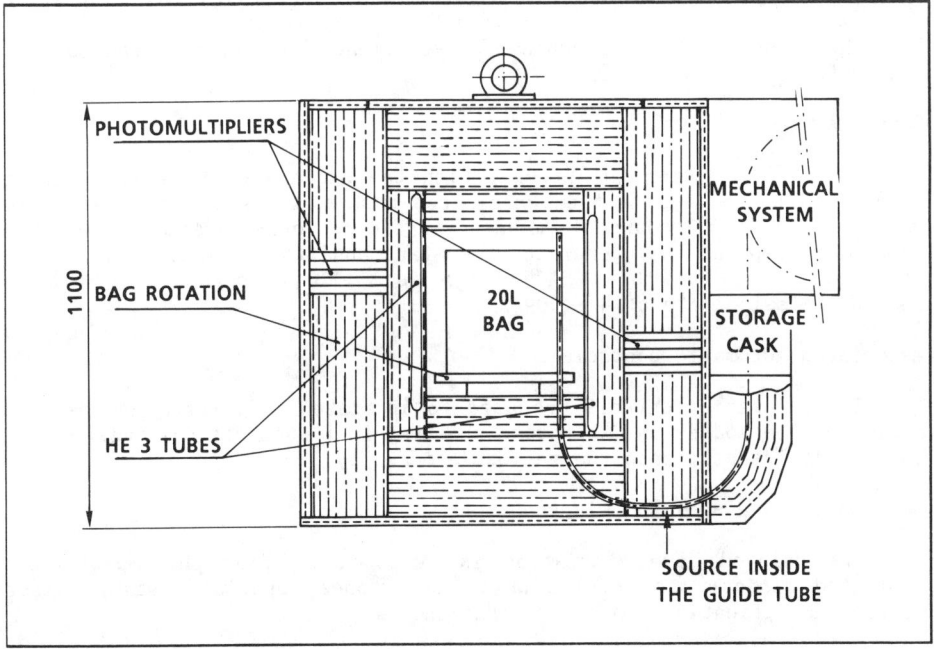

Figure 1 : System for small waste bags

The irradiation and counting enclosure includes a rotating disk-shaped tray on which a bag is placed with the gamma detectors and the optional neutron detectors. A guide tube brings the source from the storage position to the irradiation position near the bag. The enclosure is approximately in the shape of a cube measuring 120 cm on a side.

From the center, consists of:

- a cavity that receives the bag and contains the detectors;

- scintillator panels composed of a low-activation material with good efficiency to help moderate the fast neutrons. One photomultiplier per side panel is connected to the scintillator for gamma detection.

- a layer of polyethylene that provides biological shielding for operators when the source is in the irradiation position and a minimum background noise for the neutron counters when the source is in the withdrawn position.

- a layer of lead that provides biological shielding against gamma radiation for operators when the source is in the irradiation position and a minimum background noise for the gamma counters when the source is in the withdrawn position.

The removal cask is a cubical polyethylene block with a lead core at the center.

Source system

The source system is integral with the storage cask. It includes a motor used to drive a toothed wheel that pushes or pulls a helical cable on which the Cf 252 source is mounted. The cable drive system is operated by a microcomputer that generates the start and stop commands. The Cf 252 source selected for the COSAC system is a 170 microgram unit with a neutron emission rate of 4×10^8 n/s.

Associated automatic controls

The automatic controls are used to initiate the forward and reverse motion of the source and to rotate the tray that supports the bag.
The bag is inserted into the enclosure manually.

Detectors and associated electronics

The delayed gamma radiation is detected by four photomultipliers connected directly to modules which have three functions: high voltage supply, amplification and discrimination.

The delayed neutrons are detected by 12 He_3 counters connected to modules with the same functions as above. Several of these counters are however grouped for the same HV-amplifier-discriminator channel (Fig. 2). Data acquisition is performed by a microcomputer using multi-scale (MCS)

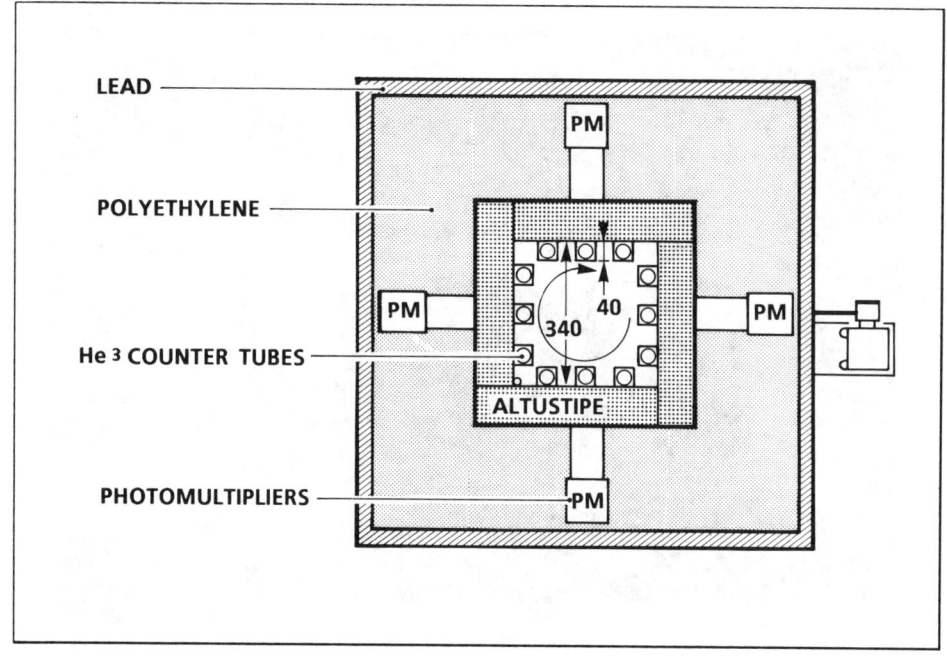

Figure 2 : System for small waste bags

printed circuit boards. The computer is also for data processing and operation of automatic control and protection devices.

Detection limit

The detection limit is 1 mg Pu 239 for 20-liter, light matrix, non-irradiating bags.The counting time is 10 mn and the activity of the source is 4×10^8 n/s (Fig. 3).The system is designed with optional delayed neutron detection capability that enables determination of the ratio of Pu 239 mass to the total fissile material mass.

Operation, safety and maintenance

The bag is inserted through the top of the enclosure after removing a polyethylene-lead plug by means of a mechanical system.

The protection devices provide assurance against radiation exposure hazards, prevent rotation of the waste package and restrict the source displacement limit switches. The detectors and electronic equipment are easily replaced. The source can be used for three years without significant deterioration of system performance. It is replaced by separating the removal cask from the enclosure.

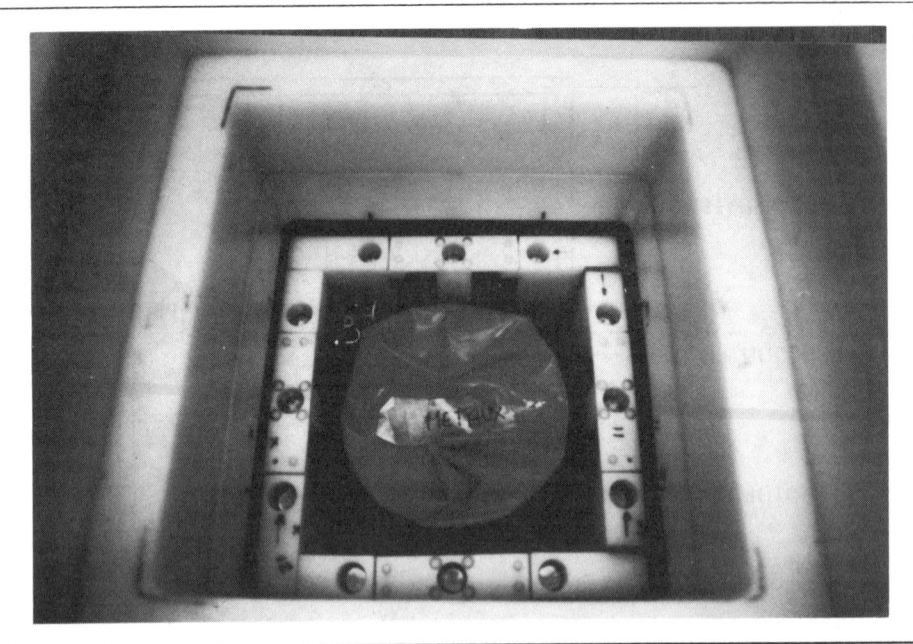

Figure 3 : COSAC

The source displacement system has been tested for over 1.5×10^6 cycles without observation of any incidents (1 cycle = 1 forward-reverse motion over a distance of 4 m). This test represents 3×10^6 reversals of direction. The source is sealed inside a double-shell casing. Its packaging and cable mounting device comply with the requirements of competent safety authorities. The system is designed to enable easy replacement of the source without risk of personnel exposure to radiation.

Radiation protection

The equivalent dose rate in contact with the COSAC system ranges between 65 µSv/h (6.5 mrem/h) and 400 µSv/h (40 mrem/h) depending on the location of the measurement points. To reduce the dose rate to a level of 25 µSv/h (2.5 mrem/h) prescribed by regulations, an exclusion area was set up at a distance of 1 m from the irradiation/counting unit and at a distance of 1.4 m from the storage cask.

SYSTEM FOR 100-LITER DRUMS WITH LIGHT MATRICES

The principle is the same as in the system described above for small waste bags: neutron interrogation with a modulated Cf 252 source followed by gamma and neutron counting.

The differences with respect to the system for small waste bags are the size of the chamber designed to accommodate 100-liter drums, the number of gamma and neutron detectors and the associated electronics [4].

The sensitivity limit is 5 mg Pu from PWR fuel in the case of light matrices with no spontaneous gamma radiation, a Cf 252 source activity of 4×10^8 n/s and a counting time of 20 mn.

GENERAL-PURPOSE SYSTEM FOR 100-LITER AND 200-LITER DRUMS

This system is designed for alpha emitter content analysis of highly diverse waste about which little prior information is available, as well as for gamma irradiating (hot) drums.

Principle

The system uses the Differential Die-away Technique (DDT) described in several publications [4]. The fissile material is activated by neutron interrogation and the system detects prompt fission neutrons. The interrogation neutrons are supplied by a pulsed neutron generator providing very short pulses of high amplitude. The neutrons in each pulse are thermalized in a graphite and polyethylene irradiation/measuring chamber and then trigger fission events in the fissile material contained in the analyzed waste drum.

Neutron generator

Neutron energy 14 MeV
Neutron emission (average) 10^8 n/s Neutron emission (peak) 10^{11} n/s
Target voltage 125 kV
Max. repetition rate 10 kHz
Min. width pulse 10 µs
Manufacturer Sodern

Description

The walls, ceiling and floor of the chamber have a multilayer construction with an inner layer of 10-cm thick graphite panels and an outer layer of 40-cm polyethylene panels.

The system combines passive neutron counting and prompt neutron detection. Operations involving the chamber are highly automated. This concerns for example opening and closing the chamber, drum insertion and removal, drum positioning at the center of the chamber and drum rotation (Fig. 4).

Detection and processing for spontaneous fission neutrons use 36 He3 counter tubes, HV-amplifier-discriminators, shift registers, coincidence analyzers and computer interface modules. Detection and processing for prompt neutrons employ 18 He3 counter tubes, plus two additional He3 tubes for neutron flux and drum monitoring, HV-amplifier-discriminators and a microcomputer with multi-scale inputs.

Automatic control devices and associated peripherals

The latest two assay systems designed use a neutron interrogation unit with a generator.

They also have a number of peripherals controlled by the central processing unit for automatic control monitoring, data acquisition and waste package routing. This includes :

- a ground-level or overhead conveyor,

- a drum gripper,

- weighing apparatus,

- a tagging unit,

- routing to other assay units,

- classification.

Software

The active/passive neutron counting unit includes data acquisition and processing software as well as an overall interpretation and expert evaluation software package.The interpretation software takes into account the available waste data and the information provided by the other assay units in order to generate the final result.

Gamma-irradiating drums

The gamma-irradiating drums with a contact dose rate of 2 rad/h or less * can be measured in the cell without alteration or significant deterioration of performance. Above this dose rate the He3 counters must be protected. This shielding is provided by a lead cylinder surrounding the target drum with a thickness determined by the dose rate of the waste package. Interaction of the neutrons with the lead deteriorates system performance, which dictated a trade off for specification of the shielding thickness.

*at the level of the counter tubes.

Detection limits

In routine operation, the detection limit is less than 3 mg Pu 239 for light matrices (paper, cotton) with low gamma radiation and a counting time of 15 mn.

Below are some of the detection limits achieved on Cadarache pilot facility.

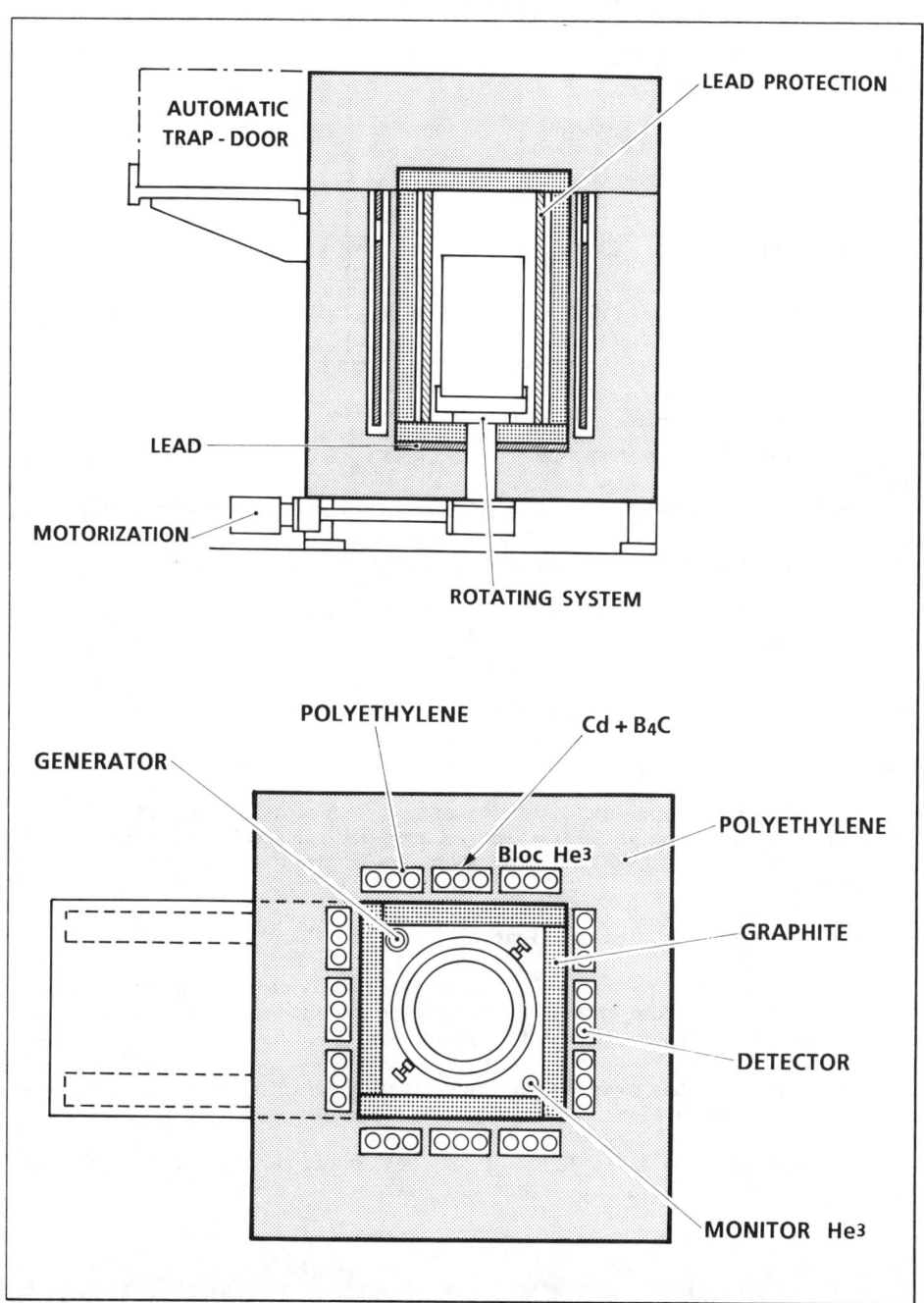

Figure 4 : General purple for 1001 / 2001 drums

MATRIX	MASS	VOLUME	MAX DOSE RATE	DETECTION LIMIT
POLYETHYLENE	13 Kg	100 l	25 Rad/h	2.8 mg
STAINLESS STEEL	45 Kg	100 l	25 Rad/h	5.0 mg
POLYETHYLENE	13 Kg	100 l	500 Rad/h	4.6 mg
STAINLESS STEEL	45 Kg	100 l	500 Rad/h	17.4 mg
ASPHALT	210 Kg	220 l	25 Rad/h	25 mg

Table 1 : Examples of detection limits

Safety and maintenance

The protection equipment and alarms are used at locations such as the drum entrance, drum centering, drum rotation, opening and closing of the enclosure door and operation and shutdown of the generator.

The detectors are accessible for maintenance and replacement. The electronic equipment is of the standard factory inspection type and the computers are PC/ATs or IBM compatibles. The system also enables performance of periodic recalibration checks and verification for deterioration using an Am-Be source.

Radiation protection

The equivalent dose rate in contact with the cell when the neutron generator is operating is 230 μSv/h (23 mrem/h). An exclusion area is set up around the cell to reduce this dose rate to a level prescribed by regulations.

Portable unit for waste classification

The last system presented is a portable unit employing the neutron interrogation technique with a neutron generator to monitor large waste packages resulting from equipment dismantling [5].

The method is the same as in the system described above for general-purpose 100-liter/200-liter drums.

The portable unit includes a 14 MeV neutron generator with two panels of polyethylene and He3 counter tubes inside (Fig. 5).

The sensitivity limit of the portable unit is estimated as a function of the waste package, the distance "d" and the angle "θ" between the two polyethylene panels (Table 2).

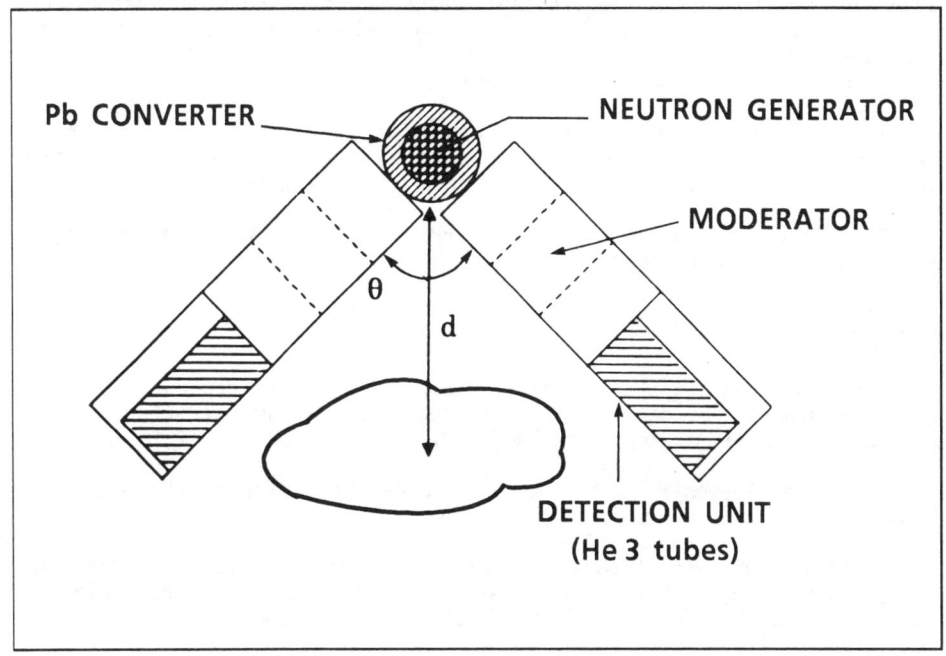

Figure 5 : Portable unit for waste classification

	U235			U235 IN 5 cm CH2		
COUNTING TIME(S)	500	2000	2000	2000	2000	2000
DISTANCE d (cm)	70	70	70	70	120	120
ANGLE Θ (°)	45	45	120	120	45	120
SENSITIVITY LIMIT (g)	3	2	10	14	22	57

Table 2 : Portable unit for waste classification sensitivity limits.

CONCLUSION

The alpha-emitter assay systems described in this paper have now advanced from the pilot development stage at the CEA's Cadarache and Bruyères-le-Châtel centers to the industrial stage. This family of systems meets most storage criteria and can be adapted to the financial and human resources of waste generators.

Emphasis was placed on the measurement of waste at the source, which significantly minimizes uncertainties and system cost thanks to the small dimensions of the package and characterization of the waste.

Further development work is required for some heavy waste packages and for analysis of large volumes.

REFERENCES

1) Determination of alpha activity and fissile mass content in solid waste by systems using neutron interrogation. J. Romeyer-Dherbey, J. Lacruche, R. Berne, J. Auger, M. Butez, L. Martin Deidier

2) Present developments of the waste measurement device at CEA - BIII - M. Ruffet

3) J. Romeyer-Dherbey. "Assay of fissile nuclei and alpha emitters in solid waste using active and passive methods. Physical principle of assays and description of equipment." CEA/DRP/SSAE report 88-R-001, Sept. 1988.

4) J. T. Caldwell et al. "Test and Evaluation of a High Sensitivity Assay System for Bulk Transuranic Waste." Los Alamos Laboratory, LA UR 83-2084.

5) G. Grenier, "Fissile Material Detection System." CEA BIII, Oct. 1985.

UNFOLDING OF AN INTEGRATED INSTRUMENTAL AND CONTROL SYSTEM
FOR MEASUREMENTS CARRIED OUT ON WASTE DRUMS

M. Vidalie
Centre d'Etudes de Bruyères-le-Chatel

Summary

Radiological measurements carried out on waste drums at B.3 Centre deal with gamma spectrometrical facilities and active questionning or passive counting by neutronic methods.

Those measurements must be interpreted in terms of quality insurance for the establishing of the contentious balance of activities.

We describe shortly the computerized net being outsettled and its use for the control-command of elementary processes in automatized measurements and for validation and management of results by a subsidiary computer.

1. QUICK INVENTORY OF MEASUREMENT FACILITIES

Five measurement devices are in use at Bruyères-le-Chatel for analysis of waste activities either inside global packages (bags or drums) or under special conditioning (normalised flasks or specific containers).

1.1. Galion

A low activity lead cell including a high purity and P-type coaxial Germanium detector for gamma spectrometry and enough enclosed volume for a 200 liters drum over a rotating support with 8 predetermined positions in height.

Measurement conditions are precised below :
Distance 50 cm between detector and drum or package axis;
Drum rotating at 8 rpm during a measurement time of 1800 s;
Measurement in three phases (lower, mean and upper parts);
Relative efficiency of detector is 25 %, Pic/Compton ratio 50, FWHM is 1.90 keV at 1.33 MeV, Detection limit 5 mg for 239Pu.

1.2. Cassandra

A double facility, performing both gamma spectrometry in 1800 s with a P-type coaxial detector and passive neutronic counting in 800 seconds by 24 tubes filled with ^{3}He and placed in a polyethylen cell around the rotating drum or package.

Measurement conditions are precised below :
Distance 1.5 m between detector drum or package axis;
Facultative interposition of lead or steel shield (1 to 4 cm);
Relative efficiency of detector is 22 %, Pic/Compton ratio 52,
FWHM is 1.85 keV at 1.33 MeV, Detection limit 60 mg for 239Pu;
Neutron counters are 24 tubes (150 NH 100) gathered 4 by 4,
Tube diameters are 25 mm, usefull length 1000 mm, under 4 bar,
Sensitivity to thermal neutrons 150 ips per $n/cm^2/s$, or 5 mg for 240Pu.

1.3. Celine

That device uses the method of active neutronic interrogation and Mr Ruffet details all characteristic elements in his presentation. We only have to remember here that the detection limit may decrease until 1 mg of 239Pu for 900 s counting duration with a cellulosic matrice.

1.4. Penelope

An other gamma spectrometrical facility with N-type high purity Germanium detector in a light lead cell and rather used for repackaging operations for obtaining of withdrawable drums.

Measurement conditions are precised below :
Distance 1.5 m between detector drum or package axis;
Relative efficiency of detector is 10 %, Pic/Compton ratio 40, FWHM 1.85 keV at 1.33 MeV, Detection limit 100 mg for 239Pu.

1.5. Amandine

A little lead cell with a high purity P-type Germanium detector whose purpose is very precise measurements on particular samples with volume less than 10 liters.

Measurement conditions are precised below :
Distance from 0 upto 50 cm between detector and sample;
Relative efficiency of detector is 15 % Pic/Compton ratio 460, FWHM is 1.75 keV at 1.33 MeV, Detection limit < 1 mg for 239Pu;

2. REFERENCE LEVELS FOR WASTE DISPOSAL

Three destinations are possible for nuclear waste packages of B.3 Center, depending mainly upon the mass and isotopic composition of transuranic elements included.

2.1. Definitive burying by ANDRA authority

In cases when activity of transuranians is low, about 30 mCi of 239Pu, in a 200 liters drums including two years old waste, and for which the isotopic quality in 241Pu does not risk to induce at 300 years an activity 241Am over 40 mCi, the detriment is considered very low and norms allow the sending to ANDRA for definitive storage at one of its Centres.

2.2. Temporary storage at Cadarache Centre

If this first threshold is overlapped, and until about 100 mCi inside a 100 liters drum, B.3 centre keeps responsibility of packages storage, but gives keeping to Cadarache Centre, in the expecting for outstanding new process economically valuable for retreatment at those concentrations.

2.3. Direct retreating at Valduc Centre

Direct retreatment is performed when it is already economically valuable, say over 1.25 Ci by 100 liters drum. That process is made in Valduc Centre and Plutonium can so be reused as such.

It is obvious that no intermediary packages must be produced (activities over highest limits for CAD but under lower levels for CVA); so reconditioning operations have to be carried out at B.3 centre for return to normal situations of removable drums.

3. UNFOLDING OF A COMPUTERIZED NET FOR WASTE MEASUREMENTS

3.1. Measurement level

Close to each one of the five measurement facilities, the measurement control-command is performed by IBM compatible micro-computers. Those computers are also in charge of automatic local recording of results and, although some further manual calculous are still necessary now, we are formalizing them more and more. This is obvious for gamma spectrometry cells where a professional software works under control of a database with which a series of complementary programmes have been written and tested :
autoabsorption corrections by the infinite energy method; waste Plutonium masses and activities at 2 and 300 years; acknowledge of isotopic composition to ANDRA spectral type.

3.2. Coherency expertizing of gamma and neutronic results

A central microcomputer, connected by RS232 link to each one of the computers of the measurement stations, is being programmed for checking that the estimations of plutonium masses obtained by the various methods are compatible together in their overall average and variation. About the validation of spectrometrical results, another software, working with algorithms different from those of the first one for deconvolution of peaks found into histograms, will be used as a judge in case of difference with results obtained by passive and active neutronic methods. At the level of the main database, the work of diary will be taken into account and the consistency with the range of isotopic compositions in relationship with the various origins of waste will also be checked.

3.3. Quality insurance in terms of contentious balance

Since the end of September 1989, transfers of analysis orderings and measurement results are formalized and applied systematically under the means of floppy exchanges after direct recording by the computers of the measurement net. That first automatization step, strictly consistent with informatic security rules excluding all physical links with outside, gives nevertheless opportunity for eliminate quite all rewriting errors for references and results between Group "Déchets" by whom waste are collected in nuclear operational facilities, Group "Laboratoire" in charge of analysis and Group "Transport" which performs the final removal to wayout.

4. FUTURE WORKS FOR COMPUTERISED NET UNFOLDING

4.1. In gamma spectrometry

Although the double computation of spectrometrical histograms be more of value for reaching reliable results than any other single analysis, it stays that existing software are not perfect and for that reason follow permanent changes. Mainly we do expect to be associated to future developments and qualification of spectrometrical expert-systems.

4.2. In neutronic methods

Passive counting neutronic method is quite fixed and we do not plan to have any major changes completed in near future. But, dealing with active interrogation method, as you could acknowledge after hearing from Mr Ruffet communication, quite any algorithms for composite matrices recognising have to be carried out further and rewritten then, what will be one of our main projects for next year periods.

Reseau Mesures Dechets CEB3/SPR/SMC/LDC
Synoptique Installation Informatisee STC

Centraliseur Mesures
type XT Amstrad 1640
DDHDEGA avec Journal

CASSANDRA gammaneutron
— avec Goupil G_5 AT 286
logiciel DBGAMMA + STC

GALION gamma bas bruit
— avec Goupil G_5 AT 286
logiciel DBGAMMA + STC

PENELOPE gamma simple
— avec Goupil G_5 AT 286
logiciel DBGAMMA + STC

AMANDINE gamma residus
— avec Goupil G_3 + FLEX
logiciel propre de STC

CELINE methode active
— avec Goupil G_5 AT 286
logiciel propre de STC

Interconnection of measurement stations

INTEGRATED FIVE STATION NONDESTRUCTIVE ASSAY SYSTEM FOR THE SUPPORT OF DECONTAMINATION AND DECOMMISSIONING OF A FORMER PLUTONIUM MIXED OXIDE FUEL FABRICATION FACILITY

J.T. CALDWELL, J.M. BIERI, R.D. HASTINGS, W.S. HORTON,
T.H. KUCKERTZ, W.E. KUNZ, K. PLETTENBERG, AND L.D. SMITH

Pajarito Scientific Corporation
Los Alamos, New Mexico, USA

Summary

The goal of a safe, efficient and economic decontamination and decommissioning of plutonium facilities can be greatly enhanced through the intelligent use of an integrated system of nondestructive assay equipment. We have designed and fabricated such a system utilizing five separate NDA stations integrated through a single data acquisition and management personal computer-based controller. The initial station utilizes a passive neutron measurement to determine item Pu inventory to the 0.1 gm level prior to insertion into the decontamination cell. A large active neutron station integrated into the cell is used to measure decontamination effectiveness at the 10 nci/gm level. Cell Pu buildup at critical points is monitored with passive neutron detectors. An active neutron station having better than 1 mg Pu assay sensitivity is used to quantify final compacted waste pucks outside the cell. Bulk Pu in various forms and isotopic enrichments is quantified in a combined passive neutron coincidence and high resolution gamma ray spectrometer station outside the cell. Item control and Pu inventory are managed with bar code labeling and a station integrating algorithm. Overall economy is achieved by multiple station use of the same expensive hardware such as the neutron generator.

1. INTRODUCTION, FACILITY HISTORY AND OVERALL D&D PROJECT PLANS

1.1 Nuclear Fuel Services (NFS) MOX Facility

The plutonium facilities at NFS-Erwin were constructed in 1964-1965. Between 1965 and 1972, NFS processed 812 kilograms of plutonium for four primary customers, 92% of the material being associated with one customer, the Southwest Experimental Fast Oxide Reactor (SEFOR). Some 2,000 PuO_2-UO_2, MOX, fuel rods were fabricated. The ^{240}Pu content of the SEFOR material was 8.3% with the material processed for other customers ranging to 16.3% ^{240}Pu. The weighted average ^{240}Pu content of all material processed is about 10.0%. The dominant chemical species of the residual Pu is believed to be oxide. High resolution gamma ray measurements support this conclusion, based on the **absence** of measureable light element (alpha, neutron) and (alpha, proton) reaction gamma rays-- other than those associated with oxygen-alpha reactions.

In 1988, Pajarito Scientific Corporation, as a sub- contractor, performed a comprehensive set of holdup measurements in the two NFS buildings involved in the MOX operations. The holdup in the two buildings was determined using a large-scale neutron imaging technique. A high sensitivity passive neutron detector was used with measurements being made throughout the volume of each building. A proprietary imaging analysis was used to convert the measurements into quantitative Pu holdup distributions. A three-dimensional display of a portion of these data is shown in Fig. 1. As can be seen, the distribution of Pu within the building consists of both discrete sources (example: peaks at x=2, y=30 and x=20, y=50 planes) and distributed sources. The information obtained from these measurements has played a major role in the planning of decontamination and decommissioning operations. For example, the indicated discrete sources will be removed before glove boxes are taken down, thereby reducing considerably item inventories prior to decontamination and decommissioning (D&D) activities.

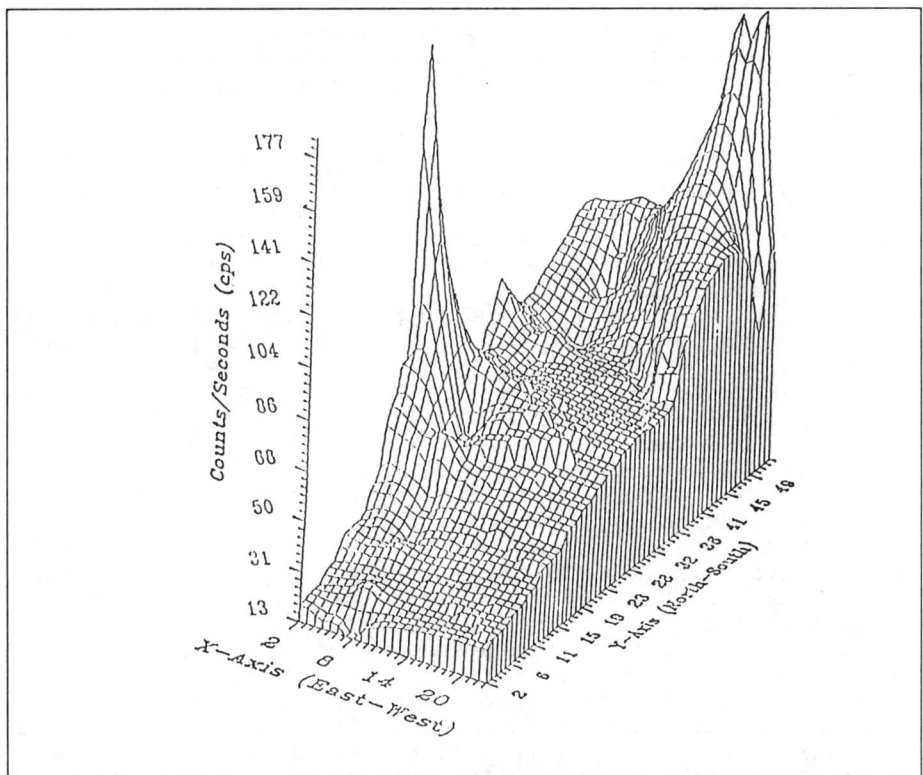

Figure 1. Typical neutron imaging data obtained during in situ holdup measurements. Scale for x and y axes is ft.

1.2 Overall D&D Facilities and Material Flows

EcoTek, Inc., of Erwin, Tennessee, was contracted in July, 1987, to manage the D&D activities of the NFS Pu facilities. Two buildings encompassing approximately 1000 m^2 of floor space are involved. Waste processing strategy centers around decontamination and sectioning with an ultrahigh pressure water jetting system incorporating a recirculated medium and subsequent volume reduction in a high-capacity shear/baler. Material control and decontamination effectiveness monitoring will be accomplished with the five station NDA system described in the present paper.

The primary waste stream is 136 gloveboxes containing process equipment. Additional sources are ventilation ductwork, process piping, conduit, scabbled concrete and contaminated soil. The decontamination and volume reduction activities are housed within a sealed, negative pressure stainless steel containment cell, the "Decontamination and Volume Reduction Facility" or DVRF. The DVRF consists of interchangeable, modular stainless steel panels and has overall dimension of 10 x 3 x 5 m. This is divided into two separate rooms, one of which functions as a material receipt airlock while the other houses the processing equipment. A completely communicating "bubble" of dimensions 3.0 x 1.5 x 1.8 m is attached to one side of the DVRF. This will house NDA Station 2, the "whole glovebox" active neutron system, sized to accommodate 80% of the unsectioned gloveboxes. This will monitor decontamination effectiveness after decon by the high-pressure jetting unit.

This is a unique design in that the partially (or totally) decontaminated glovebox is inserted into the Station 2 assay chamber, which is inside the sealed-but-contaminated DVRF--while all NDA hardware (neutron generator, detectors, electronics) remains outside in the radiologically clean area. This arrangement permits sensitive (10 nci/gm) monitoring to be accomplished without contamination of expensive equipment. An overall view of the DVRF and supporting structures is shown in Fig. 2. The DVRF bubble is located in the middle of Fig. 2--labeled as NDA Station 2.

The DVRF material flow is readily understood by reference to Fig. 2. Following removal of all discrete/visible source material, the intact gloveboxes are brought to the DVRF. The initial processing occurs at NDA Station 1, where passive neutron imaging measurements of the type (see Fig. 1) utilized for the in situ holdup campaign are made. This serves to establish the item inventory prior to decontamination. If the item inventory is small enough for safe DVRF operations, the item is inserted through the airlock into the decon cell. Following one or more decon cycles, the intact item is then placed into NDA station 2 to determine decon effectiveness and residual hotspot location.

When an acceptable level of decontamination has been achieved, the item is introduced into the shear/baler for size reduction. The output of this stage consists of waste "pucks" having dimensions approximately 400 x 400 x 100 mm. These are radsafe (double bagged) packaged and removed from the DVRF. Pucks are bar coded and stored until enough have accumulated to warrant an NDA Station 4 campaign. At this time, accumulated pucks are assayed and placed into 208-liter drums according to measured nci/gm levels. The aim of this final puck sorting is to maximize the number of filled drums in the <10 nci/gm category. It should be noted that Station 4 can be configured in two different assay chamber volumes--one optimized for waste puck measurements and a second for 208-liter drum measurements. Following the generation of a sufficient number of filled drums, Station 4 will be used to verify filled drum contents. In addition, drums of contaminated soil and other miscellaneous waste will be assayed in Station 4.

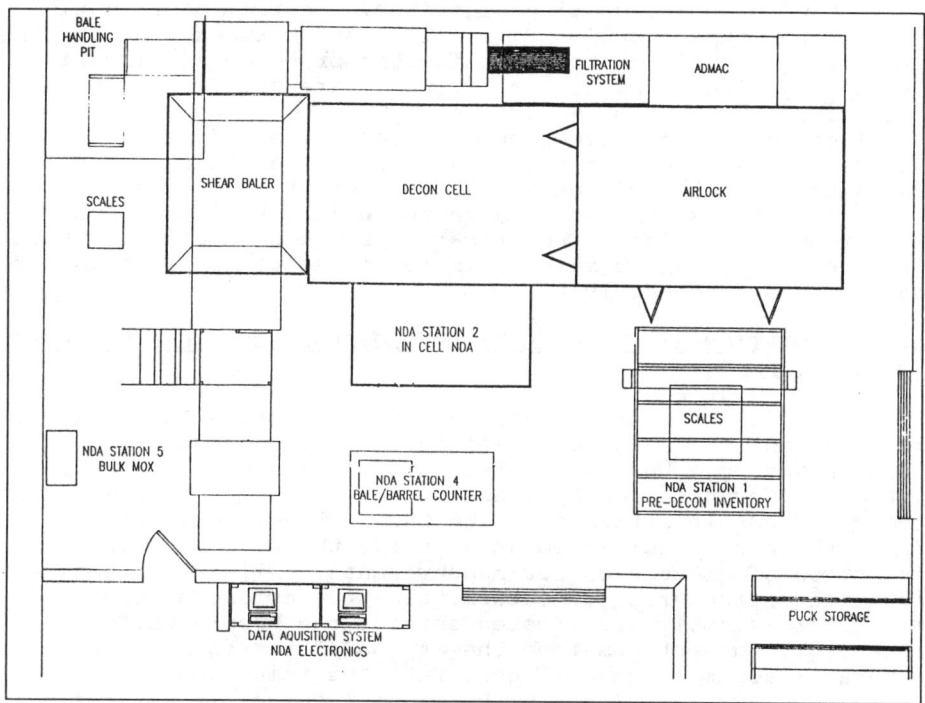

Figure 2. DVRF layout showing supporting hardware. Four of the five NDA stations are labeled. The Pu buildup monitors are located by the Filtration System and the Shear/Baler

All assay measurements and barcode readings are processed in the Data Acquisition/Management personal computer (PC) based controller shown at the bottom left of Fig. 2. Buildup of Pu in the filtration system and other regions (Station 3 passive neutron counter network) is also monitored through this controller. Operations personnel are alerted when buildup levels require action--for instance, removal of filters. The bulk Pu measurement (Station 5) determines the final inventory values of all significant Pu quantities, as well as the average ^{240}Pu percentages in such packages. This data is acquired and processed with the PC-based controller.

2. NDA STATION 1: PRE-DECON ITEM INVENTORY DETERMINATION

The 136 glove boxes to be decontaminated are quite large on the average. Twenty percent of these glove boxes have at least one dimension exceeding 3 m and 60% have at least one dimension exceeding 2 m. Station 1 utilizes a scanning/imaging passive neutron method to quantify the Pu in these large items. This station consists of two high sensitivity moderated ^{3}He detectors (dimensions: 200 x 100 x 1000 mm, intrinsic efficiency 20%). Measurements are taken with these detectors along two opposite sides and at different heights. The resultant data set is analyzed for total Pu content with an integrating algorithm. Spatial distribution data of the type shown in Fig. 1 is also obtained. An imaging algorithm is utilized to determine the presence and location of any significant Pu hot spots, which are noted for special attention during decon activities.

These detectors have been fabricated, tested and delivered. A detection sensitivity of 0.1 gm Pu has been demonstrated with hot spot location resolution on the order of 10 cm, which was demonstrated in the prior in situ holdup measurement campaign. The primary purpose of Station 1 is to provide an initial Pu mass estimate to determine if it is safe to commence decon activities.

3. NDA STATION 2: IN CELL WHOLE GLOVEBOX DECON EFFECTIVENESS MONITOR

Following an initial decon of an item within the cell, it is important to evaluate how effective the decon has been. This is accomplished in a large size (dimensions: 3.0 x 1.5 x 1.8 m) differential dieaway active neutron unit built into the indicated bubble attached to the DVRF. A vertical door (polyethylene walls) is located inside the DVRF to permit insertion of gloveboxes for measurements. All other Station 2 neutronic structures, detectors, electronics and neutron generator assembly are located outside the bubble walls. Interrogating neutrons from the MA-165C/Zetatron neutron generator assembly readily penetrate the thin bubble containment walls (aluminum) to cause fissions in any residual Pu materials in the gloveboxes. Correspondingly, the prompt fission neutrons (PFN) produced in these fission events readily penetrate the bubble walls to enter detection packages located in the external moderating walls.

Station 2 has been designed to achieve at least a 10 mg ^{239}Pu assay sensitivity for gloveboxes up to 3.0 x 1.5 x 1.8 m size. Some of its key features are: low background imaging PFN detectors, high efficiency neutron interrogation assembly and efficient interrogation geometry. At the date of this writing, Station 2 was being fabricated. The detector packages and interrogation assembly have been tested experimentally and meet design criteria. The Station 2 unit is scheduled for completion, testing and delivery in mid December 1989.

4. NDA STATION 3:   PU BUILDUP MONITORS

To further assure safe decon operations, moderated ^{3}He passive neutron detectors of the same design as used for Station 1 will be placed at points around the DVRF where significant Pu holdup can occur. Notably, the particulate holdup filters, solution ion exchange filters and the shear/baler assembly will be monitored. These detectors are calibrated (singles neutron rates) in terms of gm Pu per observed cps in the worst case geometry. Outputs will be monitored continuously and an alarm will be sounded whenever buildup exceeds a predetermined safe operating limit. These neutron-based buildup monitors are quite sensitive, having a demonstrated 0.1 gm Pu detection capability in typical geometries. The safety philosophy for this D&D operation is to provide a warning at a quantitative buildup level much lower than could reasonably be expected to cause problems. Station 3 functions as an independent safety indicator to the DVRF insertion Pu mass control described briefly above (Station 1). Station 3 has been fabricated, tested and delivered.

5. NDA STATION 4:   WASTE PUCK AND FILLED DRUM ASSAY UNIT

After an item has been deconned to as low a level as possible, it is volume reduced in the shear/baler. The output from this operation is a compressed puck having dimensions of 400 x 400 x 100 mm and weighing about 25 kg. These are the unit waste packages. They are double bagged at removal from the DVRF to assure a radiologically clean external surface. For final disposal purposes, it is most important to measure these pucks with a sensitivity below the 10 nci/gm level. The final disposal package is a filled 208-liter drum, which will contain up to 8 pucks. If the alpha contamination level of the drum contents can be certified to be below 10 nci/gm, the drum can be disposed of at a cost 10 times less than if the contamination level exceeds 10 nci/gm. This leads to a strategy in which individual pucks can be selected in a highly cost-effective manner for placement in a drum--if each puck is measured accurately at a level of sensitivity well below 10 nci/gm.

Utilizing an optimized design, we have achieved a demonstrated assay sensitivity well below 1 mg ^{239}Pu in a recent series of Station 4, differential dieaway, active neutron mockup measurements performed in the Pajarito Scientific test facility at Los Alamos. We project from these tests that we will achieve a routine 5 nci/gm or better assay sensitivity for typical size pucks in the completed Station 4 assembly. Station 4 is now being fabricated, with completion, testing and delivery scheduled for early January 1990. As stated above, Station 4 may also be configured as a 208-liter drum assay unit, utilizing the same detectors, electronics and neutron generator assembly. We project a 2 mg Pu sensitivity in this configuration.

It should be noted that a single MA-165C/Zetatron neutron generator assembly will service Station 2 and both configurations of Station 4. This will be accomplished with a

cart-mounted apparatus that can easily be wheeled to either
station location and inserted into the designed neutron
generator assembly slots provided for each station. The
adjacent locations of Stations 2 and 4 (see Fig. 2) minimize
the amount of assembly movement. We estimate that a time of
less than 5 minutes will be required to move and set up the
neutron generator assembly, starting with either location and
going to the other.

6. STATION 5: BULK PU ASSAY AND ISOTOPICS MEASUREMENT

All packages containing significant Pu amounts (roughly:
gm quantities or greater) will be measured in a combined
passive neutron coincidence (PNCC) and high purity germanium
(HPGe) counting station. This station has been fabricated,
tested and delivered. The PNCC portion has a novel, moveable
detector arrangement designed to accommodate differing size
packages and achieve the optimum assay sensitivity possible
for a given size package. The nominal, minimum-size
configuration occurs when the two slab geometry detector
packages (each slab package contains 4--50 mm OD x 610 mm
length x 2 atm fill pressure moderated ^{3}He counters--a slide
rail mounting facilitates configuration changing) are
positioned 250 mm apart. In this configuration, the measured
4 pi detection efficiency is 10%. In recent tests with Oak
Ridge National Laboratory Pu standards, we demonstrated a
detection sensitivity of 0.1 gm Pu (nominal 8.3% ^{240}Pu).
The other half of Station 5 consists of a Canberra
7229X-2020 extended-range portable HPGe system with suitable
peripherals including an S-20 MCA, portable cryostat and PC
interface. This has been purchased, tested and delivered.
The primary use of this instrument will be to acquire high-
resolution gamma ray spectra of the various U, Pu and Am
isotopes in the bulk MOX packages. These spectra will be
analyzed for Pu isotopic composition (required to quantify the
PNCC data) as well as Am and U composition information to
provide supporting information for all measurements.

7. DATA ACQUISITION, ANALYSIS AND MANAGEMENT SYSTEM

All counting electronics are interfaced through a standard
camac crate to an IBM/PC/AT clone-type controller. The
operating and hardware control software is an original
Pajarito Scientific C language program based on the passive-
active neutron system software developed in groups N-2 and
MEE-3 at the Los Alamos National Laboratory. Our program has
been greatly generalized to accommodate multiple independent
measurement stations and gamma ray spectral data as well as
all types of passive and active neutron measurement data.
Station 3 data will be acquired continuously; the other
stations will generally operate one at a time. All data
records will be archived in a spread sheet format with item
identification tags. The item inventory records will be
processed to determine how well the Pu in a glovebox entering
the DVRF is accounted for in the decon process. Barcode
labeling of waste pucks and bulk Pu packages will be tracked
through this system to facilitate final disposal and to

provide a certifiable drum inventory. This system will also be
interfaced to the NFS data management system. All associated
hardware for this system has been delivered and the
operational software is currently undergoing checkout.

8. MA165C/ZETATRON NEUTRON GENERATOR SYSTEM

    This neutron generator system was developed by the Los
Alamos and Sandia National Laboratories several years ago for
use in the USDOE's transuranic waste assay program. Currently
there are 12 of these systems in routine operational status,
including 2 recently put into service by Pajarito Scientific
Corporation. In 1981, Sandia Labs performed an extensive set
of life time tests with 9 zetatron tubes operating in a 100-
Hz maximum output mode. These tests were conducted in an
essentially continuous fashion until each tube failed--over
several consecutive weekend periods. The tube output was
carefully monitored and a continuous record of each tube's
output as a function of time was recorded. The results of
these tests indicated a mean Zetatron tube lifetime of
53,000,000 pulses with a mean drop off to 50% of initial
output after 36,000,000 pulses.
    In the years since Los Alamos implemented several of these
systems at various USDOE sites, it has been possible to
acquire additional lifetime and performance data for the more
stressful "real world" of routine assay operations. The
original Sandia studies were done under continuous operating
conditions--the usual NDA use of these systems in the USA is a
series of 2000 pulse measurements (40 sec @ 50 Hz) performed
on a waste drum followed by a several minute down time until
the next drum is ready to be measured. Perhaps 30-35 such
assays are performed in a typical day and measurements are
performed 200 or more days in a year's time. The zetatron
tube performance/lifetime history shown in Fig. 3 illustrates
this type of operation. In the period between February 1984
and June 1989, this tube was in "real world" use in the waste
NDA mode described above. The data shown is actually the
SWEPP drum assay system flux monitor reading for a standard
2000-pulse run taken with a standards drum at least once each
day waste drums are assayed. During the period shown, more
than 25,000 assays (more than 50,000,000 pulses fired) were
performed with this system. As can be seen, this tube reached
its approximate 50% of original output point a bit more than 4
years after installation and service startup. The actual
number of pulses fired in each year since startup has been
reasonably constant--in the first 18 months most of the
operations were dedicated to extensive matrix calibration
studies and operator training.
    In the first two years, 15,000,000 pulses were fired and
an average of 12,000,000 pulses per year have been fired since
the SWEPP facility began routine waste certification
activities in mid 1985. We estimate about 40,000,000 pulses
were fired prior to the system reaching its 50% of initial
output point. As can be seen in Fig. 3, the drop off in
performance to that point was more or less linear. Actually,
the drop at the 4-year point was caused by an external power
supply failure--and was not directly associated with the

zetatron system. However, it is clear that in the 4-5 year period following installation, the tube began to suffer failure symptoms, ultimately dropping off to an unuseable output. Nonetheless, this is a remarkable performance for the type of use the tube was put to and for the length of time involved. In fact, only the zetatron tube assembly had to be replaced to reestablish operations at SWEPP following this failure. Although none of the other USDOE installations have had as extensive useage as SWEPP--all zetatron systems have recorded similar performance histories. The results of the original Sandia lifetime/performance studies appear to be applicable to the more stressful routine waste NDA operations mode of useage.

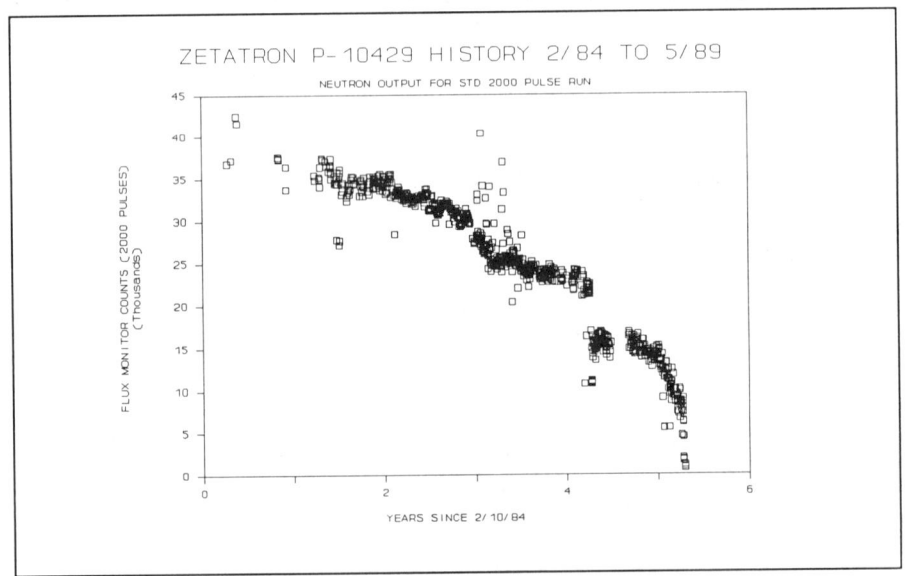

Figure 3. Peformance/Lifetime history of zetatron tube
P-10429, installed at SWEPP/Idaho in Feb. 1984.
It was in routine service for more than 5 years.

9. CONCLUSION

An integrated five-station NDA system designed to support and enhance D&D operations at a former MOX fuel fabrication facility has been described. Passive gamma, passive total neutron, passive coincidence neutron and differential dieaway pulsed active neutron detection systems are used for a variety of purposes in an integrated fashion. All data is acquired, analyzed, archived, and managed through a single IBM/PC/AT clone-type controller interfaced to a camac crate and operating under a comprehensive C language program control. We believe this integrated and comprehensive use of state-of-the-art NDA technology is an excellent means to achieve a safe, efficient and economic D&D operation.

REFERENCES:

1. KUNZ, W.E.; ATENCIO, J.D. and CALDWELL, J.T. (1980).
    J. Institue Nucl. Materials Management, IX, 131-137.
    A 1 nCi/g Sensitivity Transuranic Waste Assay System
    Using Pulsed Neutron Interrogation.

2. CALDWELL, J.T.; HASTINGS, R.D.; HERRERA, G.C.; KUNZ, W.E.
    and SHUNK, E.R. (Sept. 1986). Los Alamos National
    Laboratory Report LA-10774-MS. The Los Alamos Second-
    Generation System for Passive and Active Neutron Assays of
    Drum-Size Containers.

3. KUCKERTZ, T.H.; CALDWELL, J.T.; MEDVICK, P.A.; KUNZ, W.E.
    and HASTINGS, R.D. (May 1987). Proceedings 9th ESARDA
    Symposium on Safeguards and Nuclear Material Management,
    London, U.K., 389-393. Making Transuranic Assay
    Measurements Using Modern Controllers.

A PASSIVE-ACTIVE NEUTRON DEVICE FOR ASSAYING REMOTE-HANDLED TRANSURANIC WASTE

R.J. ESTEP, K.L. COOP, T.M. DEANE,* and J.E. LUJAN
Los Alamos National Laboratory, U.S.A.

Summary

A combined passive-active neutron assay device was constructed for assaying remote-handled transuranic waste. A study of matrix and source position effects in active assays showed that a knowledge of the source position alone is not sufficient to correct for position-related errors in highly moderating or absorbing matrices. An alternate function for the active assay of solid fuel pellets was derived, although the efficacy of this approach remains to be established.

1. INTRODUCTION

We have developed and constructed a passive-active neutron (PAN) assay device for certifying remote-handled transuranic (RH-TRU) waste for eventual implacement at the Waste Isolation Pilot Plant (WIPP) near Carlsbad, New Mexico. This device is essentially a smaller, detector-shielded version of earlier PAN assay devices built at Los Alamos National Laboratory [1], combining the active differential die-away technique (DDT) with passive neutron coincidence counting to yield two separate but complimentary fissile mass measurements. Because of design differences and the difficult nature of the RH-TRU wastes to be assayed, the so-called "second-generation" matrix effects correction algorithm [2] used in the earlier Los Alamos PAN devices was not used with this device.

According to WIPP criteria, RH-TRU waste is defined as transuranic waste with an exposure rate of more than 0.2 R/h at the surface of the container. As the name implies, such waste must be handled remotely to ensure that workers are not exposed to excessive levels of radiation. The RH-TRU waste generated at Los Alamos consists mainly of irradiated breeder-reactor fuel pellets containing mixtures of uranium and plutonium of varying isotopic composition and degree of burn-up. These have been reduced in metallurgical studies to sections of fuel pellets and to grindings and cuttings from the pellets. The composition of this waste, therefore, varies from absorbed solutions to gram-size lumps of fissile material, in various matrices. Before assaying, the 3.8-L cans of waste are sealed inside 21-cm-diam by 30-cm-high steel cans. The surface gamma-ray exposure rates of these cans can be as high as 1000 R/h. Because of these high exposure rates, the assay device was designed to be operated inside a hot cell.

The principal design modification required for assaying RH-TRU waste was the addition of 15 cm of lead shielding in front of the detectors to attenuate the intense gamma-ray fields from the waste. The ^{3}He proportional counters used in PAN assay devices detect low-energy

---

*Service Academy Research Associate, U.S. Navy.

neutrons via the ^{3}He$(n,p)^3$H reaction, which produces a charge pulse significantly larger than those produced by gamma rays. When assaying contact-handled TRU waste, this allows nearly complete discrimination against gamma radiation. Unfortunately, the gamma-ray rates from the RH-TRU waste are so high that without heavy shielding of the detectors, pulse pileup can defeat the discrimination.

Because it is necessary to place the sample as close as possible to the "on-can" flux monitor (see Sec. 2) to get a useful matrix effects correction, we were concerned that the cadmium shielding around that detector would locally depress the thermal flux and that a uniform interrogation could only be obtained by raising the sample up some distance from the on-can detector. Therefore, in addition to developing a (uniform) matrix effects correction for the DDT assays, we have tested whether taking DDT measurements with the waste container in two different positions can significantly improve the assay accuracy.

2. DEVICE DESIGN

Figures 1a and 1b show the schematic design of our RH-PAN device. Its dimensions are 142 by 142 cm by 86 cm high, and it is constructed mostly of lead, polyethylene, and graphite, with steel and aluminum supports (not shown). The neutron source for DDT assays is a sealed 14-MeV neutron generator (zetatron). The upper and lower parts of the moderating cavity are packed with additional graphite not shown in the figures, but indicated by the dashed lines in Fig. 1a.

Imbedded in each of the four vertical walls is a pair of ^{3}He detectors: one bare and one shielded (against thermal neutrons). The shielded detectors are surrounded inside-to-outside with 1.3 cm of

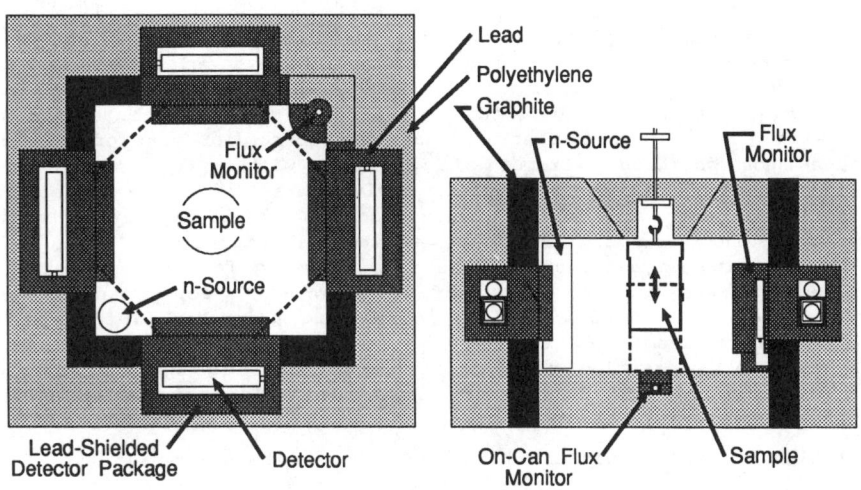

FIGURE 1. (a) Top cut-away view of the RH assay device. The dashed lines indicate the location of additional graphite at the top and bottom of the unit. (b) Side cut-away view of the device indicating the up and down measurement positions for the sample by solid and dashed lines.

polyethylene, 0.16 cm of cadmium, and 0.32 cm of borated rubber. The shielded detectors provide the basic signal for the DDT assay. The bare and shielded detector counts are summed for passive neutron coincidence counting. The total system efficiency (bare plus shielded) for counting neutrons from ^{240}Pu is approximately 1.1%.

A ^{3}He detector mounted in the corner of the cavity opposite the neutron generator monitors the flux for DDT assays. To first order, this detector gives a count rate proportional to the thermal neutron flux inside the cavity, and is used for normalization. Imbedded in the floor of the device is a partially cadmium-shielded, boron-lined proportional counter, referred to as the "on-can" flux monitor. The cadmium shielding, which was carefully designed to minimize the amount of cadmium used, blocks thermalized neutrons entering from the sides but is open to neutrons entering from above. To first order, this detector measures the thermal neutron flux inside the waste container and is used to correct for matrix effects.

The polyethylene lid to the chamber is suspended from above (not shown) by the chain of a remotely-operated crane motor that is mounted on a sliding track. To allow for reproducible dual-positioning of the sample can, the lid is attached to a sliding rod with adjustable stops. Figure 1b shows the two measurement positions for the sample, referred to here as "up" (solid lines) and "down" (dashed lines). The sample can attaches to the end of the sliding rod by a small hook.

3. EXPERIMENTAL METHOD

We made a series of measurements using real and mockup waste cans to characterize the active and passive systems. To derive a matrix effect correction for the active assay and to study the use of two measurement positions, over forty (4000-pulse) DDT assays were performed in both the up and down positions on an 8.67-g, 0.005-cm-thick enriched ^{235}U foil (93% ^{235}U). This was a fairly large foil, having a surface area of ~90 cm^2. The foil was placed at various positions and orientations within the container, in matrices ranging from no-matrix to closely-packed high density polyethylene blocks (highly moderating matrix) to Borax mixed with vermiculite (highly absorbing matrix). Some of the measurements were made with the ^{235}U foil flat against the bottom or top of the can. Because such position extremes are unlikely with real waste, we have excluded those data points when deriving a function to estimate correction factors and in our evaluation of the dual-position method. They are useful, however, in making qualitative comparisons.

To determine an alternative mass formula valid for fuel pellets, DDT measurements were made on a set of 13 stainless steel cylinders (each 0.9 cm in diameter by 5.1 cm long) containing unirradiated fuel pellets with diameters from 0.67 cm to 0.82 cm and lengths from 0.64 cm to 0.88 cm. The composition of the fissile material in the cylinders was ~20% plutonium and 75% uranium of varying isotopic ratios. The cylinder contents fell into three categories:

1) high burn-up plutonium (12% ^{240}Pu) with depleted uranium;

2) low burn-up plutonium (6% ^{240}Pu) with moderately enriched uranium (40% ^{235}U); and

3) low burn-up plutonium (6% ^{240}Pu) with highly enriched uranium (93% ^{235}U).

(The isotopic percentages are relative to the element.) Active measurements were made on single cylinders and on combinations of cylinders, with fissile masses ranging from 2 to 180 g. (Unless otherwise stated, fissile masses are expressed as the equivalent mass of low burn-up plutonium, i.e., as 94% ^{239}Pu and 6% ^{240}Pu. For this reason, what we are calling the "true" masses of the mockup samples will generally be different for passive and active assays.) The measurements were made in matrices varying from no-matrix, to high-density polyethylene blocks (highly moderating), to steel blocks (highly absorbing).

Following the DDT assays, the non-irradiated fuel pellet mockup samples were assayed using 900-s and 30000-s (overnight) passive coincidence counts. These data were used to characterize the passive system.

All data were collected in an IBM AT-based system using four Ortec ACE multichannel scaler cards under the control of computer software developed at Los Alamos National Laboratory [4].

4. RESULTS AND DISCUSSION

4.1 Active results
4.1.1 Correction for matrix and position effects
Figure 2 shows the measured shielded-detector response in the down position vs the on-can response in the down position for DDT assays of the ^{235}U foil. The foil positions are indicated in the figure by different symbols. The background-corrected shielded and on-can responses were normalized to the corresponding flux monitor counts, and the shielded response was further normalized so that a value of 1.0 was obtained for the case in which the foil was centered in the can with no matrix present. Because the same mass was measured in each case, with this normalization the inverse of the shielded response can be taken as the active correction factor. Figure 3 shows the shielded detector response in the up position vs the on-can response in the down position for the same mockup cans measured in Fig. 2.

Figure 2 shows that the correction factor required for the down measurement with a given matrix is smallest when the fissile material is near the top of the can, and largest when the material is near the bottom or at the center. By contrast, Fig. 3 shows that the correction factor required for the up measurement is smallest when the material is either near the top or the bottom of the can, and largest when the material is in the center. This demonstrates that the sample receives a more uniform interrogation when the can is in the up position, and confirms our hypothesis that the on-can flux monitor causes a significant local depression of the thermal flux.

Because the up and down measurements respond differently as a function of the foil position, one might suppose that position of the foil could be (approximately) determined by taking the ratio of the two measurements. This is verified in Fig. 4, which shows the ratio of the down-to-up shielded responses vs the approximate height of the foil within the canister. Although there is a great deal of scatter in the data, the general trend is that the down-shielded to up-shielded ratio increases more-or-less linearly with the height of the foil.

For the down measurement alone, the active mass is given by

$$MA = a_1 \cdot SHD \cdot CF , \tag{1}$$

where MA is the active fissile mass, SHD is the normalized down-shielded response, a_1 is a constant calibration factor, and CF is the matrix effect correction. In the simplest idealization, the correction factor, CF, should be proportional to the inverse of the down on-can response.

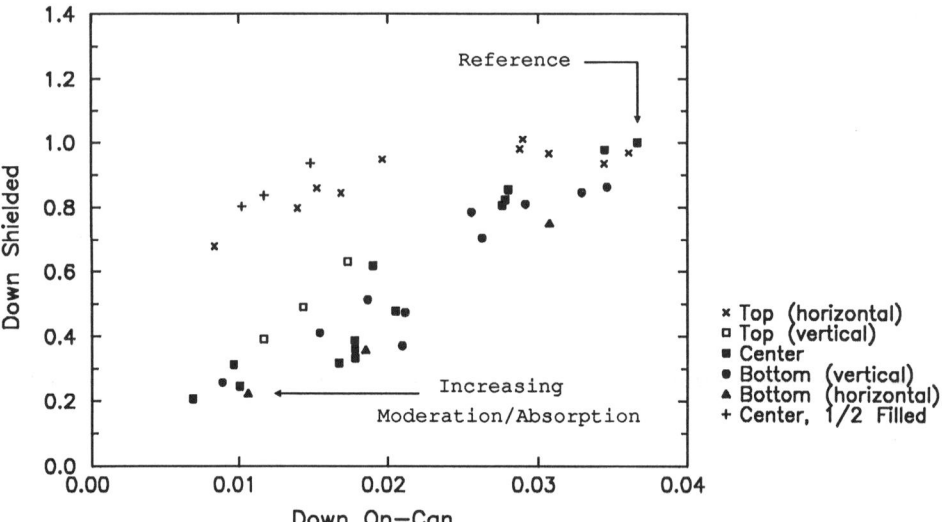

FIGURE 2. The down shielded detector response vs the down on-can response for active assays of the ^{235}U foil. The down shielded response is normalized to equal 1.0 for the indicated reference point, which represents the no-matrix case with the foil centered in the can.

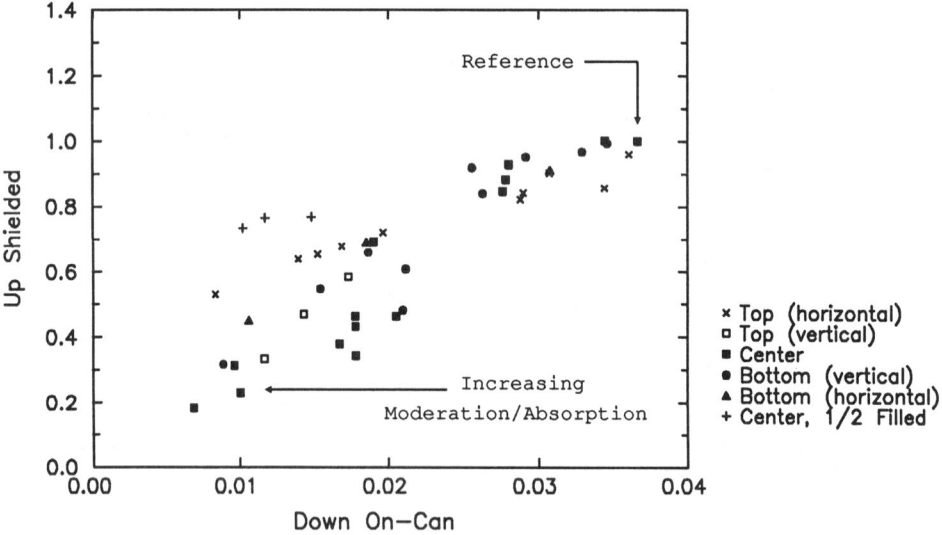

FIGURE 3. The up shielded detector response vs the down on-can response for active assays of the ^{235}U foil.

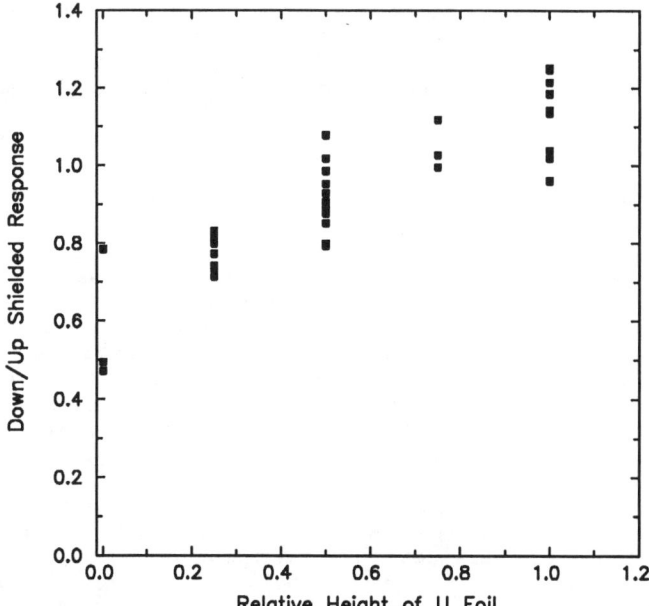

FIGURE 4. The ratio of the down shielded response to the up shielded response for active assays of the ^{235}U foil vs the approximate height of the foil in the waste can.

Caldwell, et al., [2] used a particular exponential function of the on-can (barrel) flux monitor response for an absorber correction in their "second-generation" approach. Because we are using the same variable to correct for both absorber and moderator effects, we have not restricted ourselves to their function. Instead, we have tested several functions. The one that seems to give the best results on the ^{235}U foil data in both Figs. 2 and 3 (excluding the extreme position cases) is

$$CF = constant \cdot ONCAN^{a_2} \quad , \qquad (2)$$

where ONCAN is the down on-can response, and a_2 is a constant to be determined.

To test whether an additional measurement in the up position can improve the results, we have compared results from Eq. 1 with those obtained using the more general function,

$$MA = (a_3 \cdot SHD + a_4 \cdot SHU) \cdot CF \quad , \qquad (3)$$

where SHU is the shielded response in the up position, and a_3 and a_4 are constants to be determined. Note that the weighted average in Eq. 3 represents the only way to combine SHU and SHD that is additive for multiple sources. That is, to the extent that Eq. 3 works as a position correction for point sources, it is also guaranteed to work on distributed sources.

Figure 5 shows the ratios of the calculated mass to the true mass as a function of the down on-can response, computed for both the single- and double-measurement approaches. The coefficients a_i were determined with an unweighted, iterative least-squares fitting procedure. Thus, the mass formula for the single-measurement approach is

$$MA = 2.65 \cdot SHD \cdot ONCAN^{-.93} \quad ; and \qquad (4)$$

the mass formula found for the double-measurement approach is

$$MA = 1.82 \cdot (0.124 \cdot SHD + 0.876 \cdot SHU) \cdot ONCAN^{-1.0} \quad . \qquad (5)$$

To evaluate the goodness-of-fit of Eqs. 4 and 5, we can compare the standard deviations in the final, calculated masses. With the single measurement, the standard deviation in the mass is 16.4%. This is much larger than the 1-3% error that would be expected based on counting statistics alone. With the double measurement, the standard deviation in the mass is 15.2%, only slightly better than we obtained for the single-measurement approach. Based on this, we have decided against routinely using the double-measurement approach with this device, because the benefit is too slight to justify the extra assay time.

That the double-measurement approach did not give significantly better results than the single-measurement approach suggests that the position-related errors are less important than originally supposed. This may be due to the small size of the our RH cans relative to the source-detector distances (compared with the larger, contact-handled PAN assay units) and to the relatively large size of the uranium foil. If so, then it would be unwise to generalize our results to the larger units. Another explanation is that it is not the absolute position of

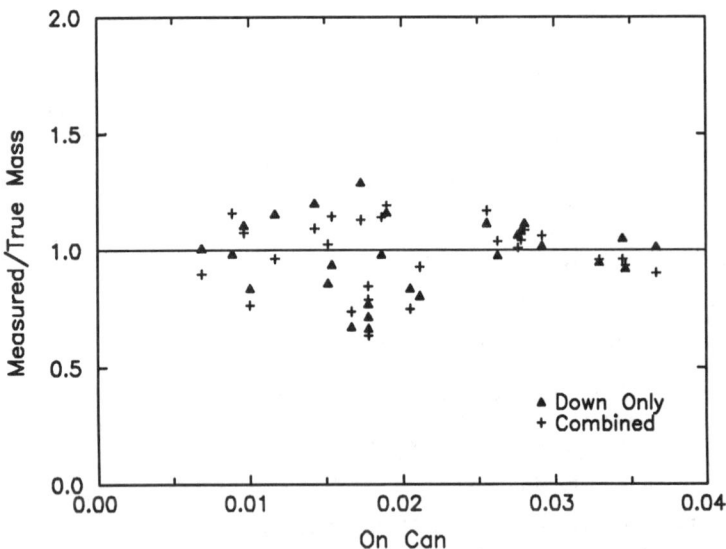

FIGURE 5. The ratio of the measured to true fissile mass for active assays of the ^{235}U foil vs the down on-can response.

the source that is significant, but the position of the source relative to whatever absorbing or moderating materials are present. This view is supported by DDT measurements made with the can filled only halfway with matrix material, and the ^{235}U foil placed flat on top of the matrix. Three such measurements are shown in Figs. 2 and 3. The assay values are approximately the same as when the can is completely filled with matrix and the foil is laid flat on the top. (These data points were not used in fitting Eqs. 4 and 5.)

It has been observed in other DDT systems that the DDT assay error generally increases with the correction factor [2]. In Fig. 5, our errors seem to be largest for intermediate values of the correction factor. This problem can be attributed to the higher uniformity of those matrices that required the largest correction factors (mixtures of vermiculite and Borax). The group of points with the highest scatter (between ONCAN = 0.012 and 0.02) had matrices of stacked blocks (2.5- to 5-cm edges) of polyethylene and iron. While these were stacked as tightly as possible, there were necessarily some gaps.

We estimate that the lower limit of detection (4 times sigma of background) for DDT assays with this device under ideal conditions is ~50 mg of enriched uranium for a 4000-pulse (80-s) irradiation.

4.1.2 Alternate mass function for fuel pellets

Some of the Los Alamos RH-TRU waste will contain intact or large segments of fuel pellets. Such waste is difficult to assay using the DDT method because the fissile material is in the form of lumps, and only the outer skin can be penetrated by the interrogating thermal neutron flux. We have derived an alternate function for estimating the fissile mass when it is known that only fuel pellets are present, based on the assumption that the geometry of the pellets is regular enough that approximately the same amount of self-shielding occurs with each pellet. To the extent that this approach works, it can be applied to wastes of unknown composition to estimate the maximum fissile content.

To determine the alternate mass function, DDT measurements were made on the mockup samples of non-irradiated fuel pellets in various matrices (see Sec. 3). Figure 6 shows the measured mass of the various fuel samples (calculated from Eq. 4) vs the true fissile mass. The formula used to fit this data (the curve in Fig. 6) was

$$MA = 0.47 \cdot M^{2/3} \quad ,$$

which gives the alternate function,

$$MPELLET = 3.1 \cdot MA^{1.5} \quad , \tag{6}$$

where MA is the mass calculated for the normal case, M is the true mass, and MPELLET is the estimated fuel pellet mass. Because of the large amount of scatter in Fig. 6, it is difficult to judge how well the function actually fits our fuel pellet data.

4.2 Passive results

The same mockup fuel pellet samples used above to obtain the alternate DDT calibration were also used to characterize the passive neutron coincidence system. Figure 7 shows the measured passive mass vs the true mass of the samples. The measured masses are calculated from the coincidence rate using the formula

$$MP = 408 \cdot RATE \quad , \tag{7}$$

where RATE is the net true coincidence rate (counts/second) and MP is the passive mass. Because of the low efficiency of the system, the statistical (counting) errors on the calculated masses are relatively

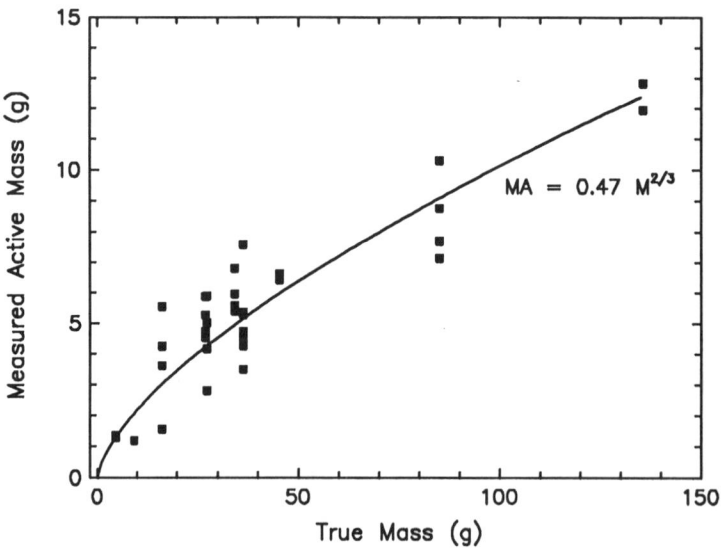

FIGURE 6. The measured active fissile mass of the mockup fuel pellet samples vs the true fissile mass.

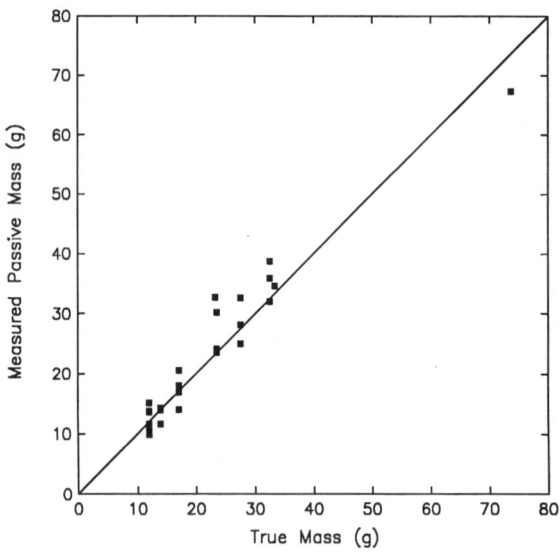

FIGURE 7. The measured passive fissile mass of the mockup fuel pellet samples vs the true fissile mass.

large. For the data in Fig. 7, the counting error averages 26% for the 900-s counts and 7% for the 30000-s counts. The scatter in the data in Fig. 7 can be mostly accounted for by this error. The sensitivity of the passive assay system is ~10 g for the 900-s counts and falls to approximately 2 g for the 30000-s counts (overnight counts).

4.3 Results with actual wastes

To date, only 10 cans of actual waste have been made available for assay. Table I summarizes the results from those measurements, expressed in equivalent grams of low burn-up plutonium. All that we know about the waste cans is that they contain grindings from irradiated fuel pellets (no intact pellets) and hot cell debris, and that the surface exposure rates are ~10 to 100 R/h. The fissile material present is presumed to be a mixture of uranium and plutonium of unknown isotopic composition. For purposes of accountability, a fissile mass of 0.5 g had been previously assigned to each of the cans. This was no more than a guess by the operator(s) who generated the waste and should not be taken as the true mass. In addition to the "normal" active assay result, Table I shows the fissile mass computed using the special fuel pellet function (Eq. 6).

For each waste can, the passive assay showed a fissile mass of <10 g, which is below our sensitivity limit for 900-s passive assays. Because we know that grindings (and not intact fuel pellets) are involved, the normal active assay should be reliable. As shown in Table I, the normal DDT assay gives a fissile mass range for the 10 cans of 0.15 to 1.64 g, with an average value of 0.91 g. This demonstrates that the DDT assay system is not impaired at this exposure rate and that, for these 10 cans, the nominal value assigned by the operators was not greatly in error.

Can Number	Active Mass		Passive Mass
	Normal (g)	Pellet (g)	(g)
88	0.25 ± 0.04	0.38 ± 0.19	<10
84	1.32 ± 0.21	4.7 ± 2.3	<10
85	0.22 ± 0.04	0.33 ± 0.16	<10
80	1.32 ± 0.21	4.7 ± 2.3	<10
83	1.05 ± 0.17	3.3 ± 1.7	<10
78	0.16 ± 0.02	0.19 ± 0.09	<10
87	0.50 ± 0.08	1.1 ± 0.55	<10
86	1.62 ± 0.26	6.4 ± 3.2	<10
81	1.63 ± 0.26	6.5 ± 3.2	<10
82	1.05 ± 0.17	3.4 ± 1.7	<10

TABLE I. Measured fissile mass for actual waste cans.

5. CONCLUSIONS

Our observation that making DDT measurements in two positions results in only a slight improvement in the assay accuracy implies that non-uniformity in the interrogating flux is not a primary source of error with this device. Position-related errors are largest when the matrix is highly moderating or absorbing and are probably due more to the amount of absorber or moderator immediately surrounding the source than to the absolute position of the source. Indications are that non-uniformity of the matrix is our chief source of assay error, and that the effects of source position and matrix distribution are tied together in such a way that merely knowing the position of the source is not sufficient to improve the assay significantly.

The alternate equation for DDT assays of fuel pellets seems to work moderately well up to 140 g, and it may be that this approach will prove useful for certifying wastes of unknown form as being below the 200-g limit set by WIPP. However, we feel that further testing of this approach, perhaps with real wastes, is required. We also note that when the error estimate is added to the assay result, as little as 12 g of finely divided fissile material may appear to exceed the 200-g limit with this approach. We intend to further evaluate this method by comparing it with the combined thermal-epithermal neutron [3] (CTEN) method.

Because of the difficult nature of the wastes to be assayed and because of the large assay errors that can be expected, we have tentatively recommended that the operators classify the wastes into three categories:

1. waste composed of finely divided fissile material;
2. waste containing fuel pellets; and
3. waste of unknown form.

With the first category, the combination of the (normal) DDT and passive assays should give the best estimate of the fissile mass in the cans. With waste known to contain fuel pellets, we recommend that the alternate fuel pellet function approach be used, but only to confirm the estimates of the mass made by the operators who generated the waste. For the third category, we can obtain upper and lower fissile limits using both approaches, but we're hopeful that measurements with the CTEN device will provide more definitive results.

ACKNOWLEDGMENTS

The authors wish to acknowledge the contribution of John Caldwell, who initiated this project. Krag Allander and Clarence Herrara assisted in the construction of the device. The neutron generator was assembled and maintained by Ray Hastings.

REFERENCES

1. KUNZ, W.E.; CALDWELL, J.T; ATENCIO, J.D.; BERNARD, W.; FRANCE, S.W.; HERRERA, G.C.; HSU, H.H.; KUCKERTZ, T.K., and PRATT, J.C., "Current Status of the Multi-Isotopic Waste Assay System," Proc. Am. Nucl. Soc. Topical Meeting on Treatment and Handling of Radioactive Wastes, Richland, Washington, USA, April 19-22, 1982 (Battele Press, April 1982).
2. CALDWELL, J.T.; HASTINGS, R.D.; HERRERA, G.C.; KUNZ, W.E., and SHUNK, E.R, "The Los Alamos Second-Generation System for Passive and Active

Neutron Assays of Drum-Size Containers," Los Alamos National Laboratory report LA-10774-MS (September 1986).
3. COOP, K.L.; CALDWELL, J.T., and GOULDING, C.A., "Assay of Fissile Materials Using a Combined Thermal/Epithermal Neutron Interrogation Technique," Proceedings of the Third International Conference on Facility Operations - Safeguards Interface, pp. 333-338, November 29 - December 4, 1987, San Diego, California, USA.
4. ESTEP, R.J., "GSHELL: A Data Acquisition Software Package for IBM-PC Compatible Computers," Los Alamos National Laboratory report, to be published.

THE DEVELOPMENT OF PASSIVE NEUTRON AND PASSIVE GAMMA MEASUREMENT TECHNIQUES FOR THE ASSAY OF PLUTONIUM IN SOLID WASTES

K.P. Lambert and J. Farren
Instrumentation Branch, Harwell Laboratory
AEA Technology,
Oxfordshire, OX11 ORA, England

Summary

Instrumentation Branch of the Harwell Laboratory is actively engaged in the development of passive neutron and passive gamma techniques for the measurement of the plutonium content of alpha bearing radioactive waste. The development areas are multidisciplinary and encompass physics, mechanical engineering, electronics hardware, computer software and gas counter technology; culminating in the design and production of plant-specific waste assay systems. The current range of instrumentation is suitable for the assay of waste drums of up to 210 litres volume, similar instrumentation is under development which will accommodate 500 litre waste drums.

1. INTRODUCTION

We examine the problems of the non-destructive assay of alpha bearing wastes and describe the Harwell family of engineered waste assay systems which have been developed to provide plant operators with the tools to quantify the plutonium present in waste streams. The quantitative assay of fissile isotopes by passive non-destructive methods is based on high-resolution gamma spectroscopy and neutron counting. These measurement techniques are implemented by segmented gamma scanners and passive neutron counters, the latter operating in the coincidence mode for plutonium accountancy purposes and in the total count mode for fail-safe upper limit estimation of plutonium mass when this is below the detection level for coincidence counting.

The main problems encountered are the non-uniform distribution of fissile material in the waste matrix, absorption of emitted radiation by the matrix which depends on both the matrix composition and density and the type of radiation being detected. For gamma ray assay, the self absorption of gammas within the the fissile materials is important. Typical wastes can vary from dense, highly active, head-end material to loosely packed soft tissues arising from laboratory operations. Consideration of these problems strongly supports the argument that the precursor to the choice of assay technique and its subsequent calibration is the categorisation and segregation of waste streams at source.

2. PASSIVE NEUTRON SYSTEMS

2.1 The passive neutron technique

The objective of these systems is to measure the passive neutron emission form both hard and soft waste and to relate it to the mass of the plutonium contaminant. The basis of the technique is to thermalise the fast neutrons emitted from the plutonium by a polyethlene moderator into which are embedded ^{3}He proportional counters for thermal neutron detection.

Accurate plutonium assay requires coincidence electronics to separate the time correlated spontaneous fission neutrons, which are the signature of ^{238}Pu, ^{240}Pu, ^{242}Pu, from the random (α,n) events which depend on the species and quantity of light elements present in both the plutonium compounds and in the waste matrix. The measured spontaneous fission neutron coincidence-rate is a function of the ^{238}Pu, ^{240}Pu, ^{242}Pu in the sample, and hence the total plutonium mass can be derived if the isotopic composition of the plutonium is known.

For plutonium masses below the detection limit for coincidence counting the total neutron count-rate can be used to give an upper limit on the plutonium mass; whereas the total neutron count-rate has components from both spontaneous fission neutrons and (α,n) events, the calculation considers all events as being from spontaneous fission which results in a fail-safe estimate of plutonium mass[1].

Measurement error due to non-uniformity of the plutonium distribution within the waste drum is reduced by attention to the design of the moderators and the disposition of the proportional counters within them to minimise the variation in counting efficiency.

We now consider the pertubation effect of the waste matrix on the emitted fast neutrons. In general, hard non-combustible matrices will scatter and absorb neutrons but will have little moderating effect resulting in a large proportion of source neutrons being detected. However soft combustible matrices, in particular the hydrogenous components, will moderate and subsequently capture a fraction of source neutrons. The usual approach to this problem is calibration with representative matrices containing well characterised sources. An alternative solution has been jointly developed at JRC Ispra and DNE Dounreay which applies a theoretical correction factor to the observed neutron count-rate to take into account the influence of the matrix on the plutonium's fast neutron emission rate [2].

2.2 Electronic development

Our electronic development has concentrated on two areas, ie. composite amplifier/discriminators and neutron coincidence units.

The charge amplifier/discriminator unit type 996006-1 comprises of an EHT filter, counter load resistor, hybrid thin film charge sensitive pre-amplifier/discriminator (Amptek-111), digital pulse width control circuit and a line driver, all housed in a single tubular enclosure as shown in Figure 1. The unit can either be directly coupled to a single proportional counter or connnected to a series of counters using superscreened cable which provides high immunity against electromagnetic interference, this being of particular importance for equipment operating in an electrically noisy plant environment [3]. The fast pulse shaping time constant (0.15 μs) enable the amplifier to discriminate efficiently against gamma pulse pile-up events making it ideal for use with proportional counters operating in high gamma fields. For example, we have operated a 25mm diameter, 4 atmosphere ^{3}He counter with an Ar/CH$_4$ quench gas at gamma dose-rates up to 400 mG/h and extended the dose-rate tolerance to 450 mG/h by use of CO_2 quench gas.

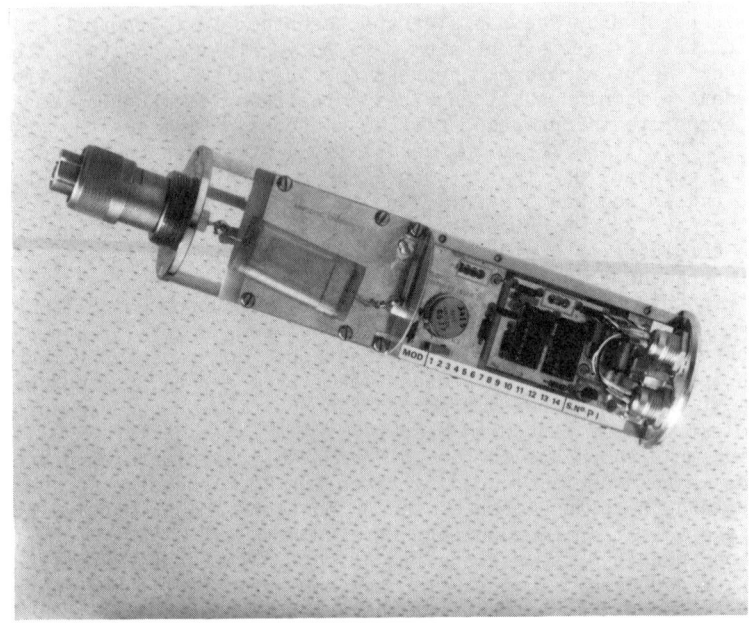

FIGURE 1.   Type 996006-1 charge amplifier/discriminator

The function of the neutron coincidence unit is to separate time correlated events from random (α,n) events in order to derive the spontaneous fission neutron coincidence-rate. Our coincidence unit is based on the generic design of Böhnel [4] but uses only a single coincidence gate whose contents are interrogated twice, once at time T = 0 to give the Real+Accidental count and again after a delay of 4096 μs to give the Accidental count. The coincidence gate, pre-delay and long delay are each generated using a RAM which is addressed sequentially with a counter to form a shift register. The data synchoniser which synchronises input data pulses with the 4MHz clock incorporates a derandomising buffer to minimise synchroniser dead time losses; experience has shown that input count-rates of up to 3.5MHz can be processed without significant losses.

Control logic is centred around an Intel 8031 micro controller which is used to manage the setting up of measurement parameters, eg. counting time, coincidence gate width and pre-delay length in addition to the aquisition of data. Operational control of the unit is achieved via a system controller, typically an IBM-PC or equivalent computer, capable of offering an RS232 serial communications facility. A major advantage of computer control of coincidence gate width, variable from 4μs to 1024μs in 4μs steps, and pre-delay length, variable from 0.25μs to 64μs in 0.25μs steps is that the detector dieaway time and hence corresponding gate width and the optimum predelay value can be measured automatically. The dieaway time is determined by measuring Real coincidence-rate as a function of pre-delay for a spontaneous fission source and the pre-delay for minimum bias by measuring $[(R+A)-A)]/A$ as a function of predelay for a random neutron source, $(R+A)$ and A being the Real+Accidental and Accidental coincidence-rates respectively.

2.3 Engineered passive neutron systems

Our system development programmes have addressed the specific areas of assay chamber design for both small transportable and large installed systems, focusing on techniques which provide simple loading/unloading methods for waste containers and the optimisation of moderator and detector configurations resulting in uniform counting efficiency throughout the volume of the assay chambers. Modular neutron detector systems, which can be configured in a variety of ways, have also been developed to facilitate the plutonium assay of irregular shaped packages arising, for example, from plant decommissioning.

2.3.1 Passive neutron monitor for packages up to 12 litre capacity

This transportable system, as shown in Figure 2, is designed for both the measurement of small waste packages for waste management purposes and cans of plutonium product for nuclear materials accountancy. The assay

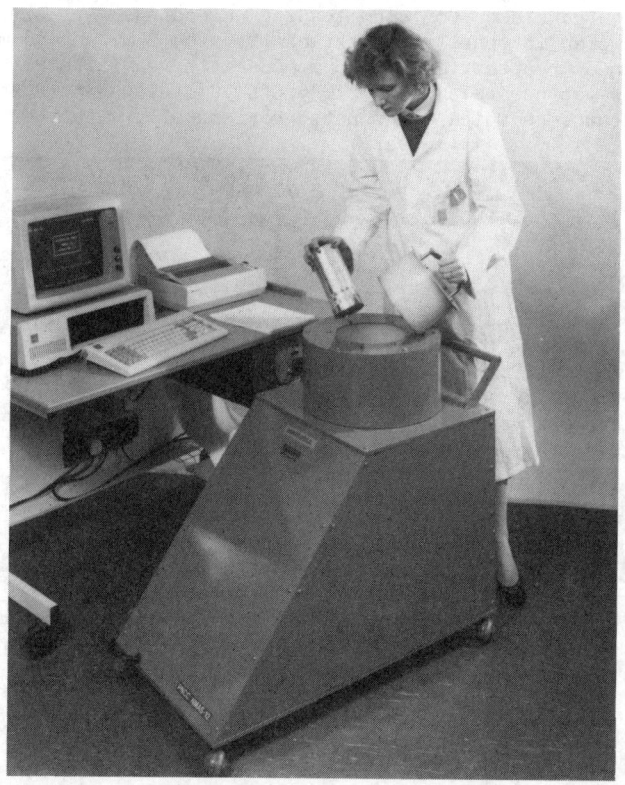

FIGURE 2.   Transportable passive neutron monitor

chamber consists of an annular polyethlene well counter fitted with 18 off 25mm diameter, 4 atmosphere ^{3}He proportional counters. As the polyethlene annulus forms an undermoderated assembly, the efficiency at the top and bottom of the chamber is increased preferentially with respect to the centre by the addition of outer polyethlene rings. In conjunction with polyethlene end plugs fitted with aluminium inserts, this results in a coincidence counting efficiency uniform to within ± 1.5% over 450mm of the total 460mm internal height. The inner surfaces of the sample cavity and the end plugs are covered with 1mm cadmium sheet to reduce thermal neutron leakage from the detector area back into the sample cavity thus minimising thermal neutron induced fission in the sample, whilst the outer surface of the detector body is also clad with 1mm thick cadmium sheet to reduce the system response to background neutrons. The system is provided with microprocessor controlled coincidence electronics, with operational control and data analysis performed by a personal computer.

The typical 1σ spread on set of 6 repetitive counting periods of 200s (1200s counting time) is ± 5% for 1g of ^{240}Pu. Precision improves as 1/ T where T is counting time. Precision degrades as a function of increasing random neutron count-rate. Typical measurement accuracy is ± 5% (1σ) for 1g ^{240}Pu.

2.3.2 Passive neutron monitor for 210 litre drums

The assay chamber, as shown in Figure 3, is a rectangular steel frame structure with front door for drum loading/unloading. All six sides of the structure are clad internally and externally with 25mm and 38mm thick polyethlene sheet respectively to form moderating cavities with a 75mm air gap. This air gap houses a total of 36 off 50mm diameter, 2 atmosphere ^{3}He proportional counters which are mounted horizontally with the inter-counter

FIGURE 3.   Passive neutron monitor for 210 litre drums

distance being adjusted to give a uniformity of coincidence counting efficiency within ± 5%. Extensive use is made of superscreened cable to connect the counters to the charge amplifier/discriminator units, we group the counters 4 per amplifier on the sides and 2 per amplifier in the roof and base of the chamber. The resulting 10 TTL signal channels are monitored by light emitting diodes prior to mixing and transmission to the coincidence electronics; this provides an efficient fault diagnostics tool and significantly reduces maintenance time.

The limits of detection are typically 10mg ^{240}Pu and 6mg ^{240}Pu respectively in the coincidence and total count modes for a total counting time of 1000s and a background count-rate of 20c/s. Precision of measurement in the coincidence mode is ± 5% at 1σ for 6 repetitive 300s counting periods, whilst typical accuracy is ± 20% at 1σ for 1g ^{240}Pu.

2.3.3 Reconfigurable passive neutron monitor

The economic requirement for a passive neutron assay system which could be moved to the assay site and reconfigured to accommodate a variety of shapes and volumes of waste package led to the development of our neutron slab monitor. The main components are a series of 8 polyethlene slabs 1050mm x 270mm x 100mm, each housing 3 off 50mm diameter, 2 atmosphere ^{3}He proportional counters; again superscreened cable is used to connect the counters within a unit to the composite charge amplifier/discriminator unit, whilst flexible coiled leads are employed for inter-slab power and signal connections to facilitate system reconfiguration.

The system employs both coincidence and total neutron counting electronics which are interfaced to a personal computer for system control and data processing purposes. Figure 4 illustrates the system in its

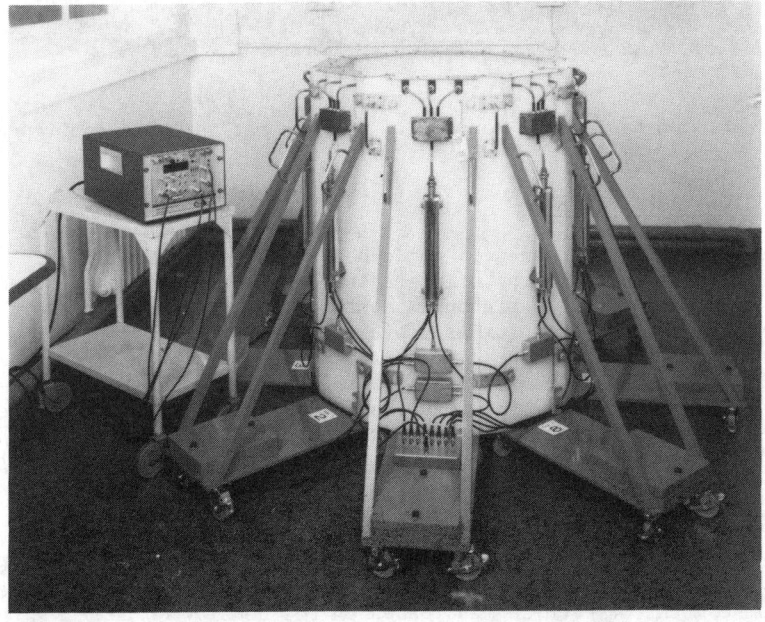

FIGURE 4.   Portable slab monitor in closed octagonal format

closed octagonal format surrounding a 210 litre drum, in this configuration its neutron counting efficiency is 15% for 1.95MeV neutrons; whilst in Figure 5 the system is shown reconfigured as 2 rows each of 4 slabs for the measurement of a rectangular box, in this format the counting efficiency is typically 13% for 1.95Mev neutrons.

FIGURE 5.   Portable slab monitor in an alternative, open configuration

3.   PASSIVE GAMMA SYSTEMS

Development in this area is concerned with the application of high resolution gamma spectroscopy, implemented on a segmented gamma scanner, for the measurement of the plutonium content of soft waste and the determination of the plutonium isotopic composition.

When measuring plutonium contaminated waste, the 414 keV photopeak from ^{239}Pu is measured simultaneously with the 400 keV photopeak from a ^{75}Se transmission source. This enables an estimate to be made of the density of the particular segment being measured, thereby allowing the area of the ^{239}Pu photopeak to be corrected for matrix absorption, but not for plutonium self absorption. In addition to the direct measurement of the ^{239}Pu mass from the 414 keV photopeak, measurement of ^{241}Pu from its ^{237}U daughter and determination of the ^{239}Pu/^{241}Pu ratio, enables an estimate of the ^{240}Pu equivalent isotopic composition to be given for material of known origin with its ^{241}Am content unmodified. A summation is made of the total plutonium content of each segment based on the density corrected ^{239}Pu mass and the estimated plutonium isotopic composition.

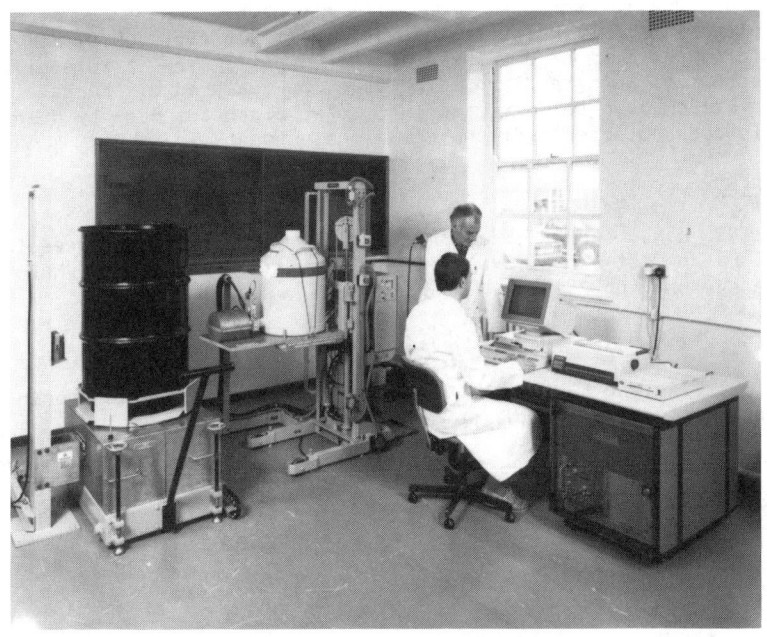

FIGURE 6.   Segmented gamma scanner for 210 litre drums

Figure 6 gives an overall view of our segmented gamma scanner.
Measurement error due to the non-uniform distribution of fissile material
is minimised by rotating the waste drum for radial inhomogeneity and
sequential measurement of vertical segments for vertical inhomogeneity.
The system consists of a turntable on to which the drum to be measured is
loaded. This rotates at 10rpm and can accept containers up to 1250kg
maximum weight and 210 litre capacity. To the side of the turntable a high
resolution intrinsic germanium detector views the drum through a lead
collimator. A hydraulically operated platform is used to raise the
detector enabling the drum measurement to be made in fixed vertical
increments of 50mm, ie. a measurement is made on each segment of the drum.
An external radiation source located on the far side of the drum from the
detector is used to measure the density of the waste matrix. Gamma spectra
are accumulated by a multichannel analyser, with complete system control
and data processing being performed by a dedicated personal computer.

5.   CONCLUSION

Our development programmes for passive neutron and passive gamma
counting techniques have resulted in the generation of a comprehensive
range of non-destructive, plant specific, plutonium assay systems which are
used for both nuclear waste management and nuclear materials safeguards
purposes.
The combination of a passive neutron system and a segmented gamma
scanner is an ideal solution to the requirement to measure the plutonium
content of drums of alpha bearing waste. The segmented gamma scanner is

capable of assaying soft waste and will report both the ^{239}Pu mass and total plutonium mass in addition to determining the plutonium isotopic composition of both hard and soft waste. The ^{240}Pu equivalent mass content of both hard and soft waste, as measured by the passive neutron system, can be converted into total plutonium mass by using the isotopic data obtained from the gamma measurements.

REFERENCES

1. LEAKE, J.W., LAMBERT, K.P. and WARNER, M.C. (1979). Proceedings of ESARDA 1st Annual Symposium on Safeguards and Nuclear Material Management, pp 284-288.

2. BREMNER, W.B., et al. (1982). EUR 8020 EN.

3. BEST, D.A. and FOWLER, E.P. (1983). The Radio and Electronic Engineer, Vol 53, No.4, pp 147-152.

4. BOHNEL, K. (1975). KFK 2203.

KPL/RMR(10563)-031089

NONDESTRUCTIVE ASSAY AND NONDESTRUCTIVE EXAMINATION OF REMOTE-HANDLED TRANSURANIC WASTE AT THE ORNL WASTE HANDLING AND PACKAGING PLANT*

F. J. Schultz
Oak Ridge National Laboratory

J. T. Caldwell
Pajarito Scientific Corporation

Summary

The purpose of this investigation is to examine the use of an electron linear accelerator (LINAC) in the performance of nondestructive assay (NDA) and nondestructive examination (NDE) measurements of remote-handled transuranic wastes. The system will be used to perform waste characterization and certification activities at the Oak Ridge National Laboratory's proposed Waste Handling and Packaging Plant. The NDA and NDE technologies which were developed for contact-handled wastes are inadequate to perform such measurements on high gamma and neutron dose-rate wastes. A single LINAC will provide the interrogating fluxes required for both NDA and NDE measurements of the wastes.

1. INTRODUCTION

The Oak Ridge National Laboratory's (ORNL) Waste Handling and Packaging Plant (WHPP) is a key element in the Department of Energy's (DOE) transuranic (TRU)[1] waste program for repackaging remote-handled (RH)[2] and special-case (SC) TRU wastes generated in the course of Defense Programs activities. WHPP (see Fig. 1) is proposed as a fiscal year (FY)

---

*Prepared by the Oak Ridge National Laboratory, Oak Ridge, Tennessee, operated by Martin Marietta Energy Systems, Inc., for the U.S. Department of Energy under contract DE-AC05-84OR21400.

[1]DOE Order 5820.2A defines TRU waste to be, without regard to source or form, waste that is contaminated with alpha-emitting TRU radionuclides with half-lives greater than 20 years and concentrations greater than 100 nCi/g at the time of assay. Heads of DOE Field Elements can determine that other alpha-contaminated wastes, peculiar to a specific site, must be managed as TRU waste.

[2]RH-TRU waste is defined by DOE Order 5820.2A as that packaged TRU waste whose external surface dose rate exceeds 200 mrem/h. Test specimens of fissionable material irradiated for research and development purposes only and not for the production of power or plutonium may be classified as RH-TRU waste.

1992 ORNL capital line-item project at a total estimated cost of $245 million. The projected operational start of the WHPP is FY 1996.

The mission of the WHPP is to receive, repackage, characterize, certify, and ship RH-TRU waste to the Waste Isolation Pilot Plant (WIPP), the DOE geologic repository for TRU wastes located near Carlsbad, New Mexico. The ORNL was chosen as the site for construction of the WHPP because approximately 90% of the RH-TRU waste stored in the entire DOE system is currently in storage at ORNL. The proposed WHPP would also receive RH-TRU solid waste from other DOE sites such as the Idaho National Engineering Laboratory, Argonne National Laboratory, and the DOE facilities located at Hanford, Washington.

An essential part of the WHPP operational flowsheet includes evaluation and characterization of the preliminary waste form and certification of the processed or product waste form. When the packaged RH-TRU waste first enters the Processing Cell of the WHPP and before the waste is removed from each container (see Fig. 2), an evaluation and characterization of the waste form will be performed by x-ray examination to determine its condition, for example, to ascertain whether free liquids or compressed gas cylinders are present in the waste package. While still present in the Processing Cell, the fissile content of the package will also be determined by neutron interrogation techniques. This will help ensure criticality control and help maintain the proper TRU concentration throughout the solid waste and, to a lesser degree, the sludge processes. Before final preparation of the waste for shipment to the WIPP facility (for example welding of the drums into the RH-TRU canister), the processed waste form will be assayed by pulsed-neutron interrogation techniques and examined by real-time radiography (RTR) to verify compliance with established WIPP waste acceptance criteria.

One of the major research areas supported by the WHPP Solids Technology Development Program is a LINAC-based RH-TRU waste NDA and NDE or RTR capability. NDA and RTR measurements of RH-TRU waste packages are complicated by typical surface dose rates up to 10,000-rad/h gamma and 1-rad/h neutron. Most NDA and NDE instrumentation and techniques [1] used to certify contact-handled (CH) TRU wastes cannot be operated in the high gamma and neutron fluxes anticipated to exist in the WHPP hot cells [2]. X-ray sources of 420 kV or less and imaging devices used routinely to inspect CH wastes are not suitable for examination of RH wastes because of these high radiation levels. The constraints of placing all NDA and RTR hardware into a limited portion of a multifunction facility present yet more design challenges.

One option which meets the WHPP NDA and RTR waste package characterization and certification requirements is provided by the proper application of a 6-MV electron LINAC. The LINAC, which must be properly shielded and directionally collimated for optimum interrogation performance, is the driving source for penetrating RTR measurements as well as for high-sensitivity NDA measurements based upon a combination of the pulsed-neutron differential dieaway technique and an independent pulsed-photofission method [3,4].

2. INTENDED APPLICATION OF THE LINAC BASIC INTERROGATION PHYSICS

The source of the interrogating fluxes for both the NDA and RTR measurements of the waste is produced by the acceleration of a high-energy beam of electrons into a heavy metal target. The most efficient conversion of the electron beam into x-ray radiation is obtained with materials of high atomic number (high-Z elements). Of these elements, tungsten (Z=74) offers the best combined efficiency and physical properties. To ensure maximum radiation output, the thickness of the target material should be slightly greater than the range of the electrons in that material. Targets should be designed to produce the minimum focal spot size consistent with their high radiation output (e.g., the Varian Linatron™ 1000 achieves focal spot sizes of less than 2 mm).

The sequence of interactions which occur during the RTR examination of the contents of RH-TRU waste packages is given below.

electron beam ---> internal tungsten converter ---> bremsstrahlung radiation ---> waste container ---> imaging system.

The LINAC bremsstrahlung radiation beam profile is shaped to waste interrogating requirements using internal lead or depleted uranium metal collimators. The internal collimators not only provide proper shaping of the interrogating beam for the NDA/NDE applications but also serve to alleviate building wall shielding requirements by reducing side- and back-angle secondary photon radiation. Approximately 2 ft of normal density concrete (2.35 g/cm^3) are required to reduce the relatively low-energy (compared to the primary beam) off-angle radiation dose rate from a 6-MV bremsstrahlung source by a factor of 100. Consequently, a relatively small amount of high atomic-number shielding around the bremsstrahlung target greatly reduces the secondary shielding requirements of building walls off the beam axis.

The NDA measurement utilizes two unique interrogating fluxes: pulsed neutron differential dieaway technique [5,6,7,8] and pulsed photofission interrogation. The sequence of interactions which occur during pulsed neutron differential dieaway NDA of RH-TRU waste packages is given below.

electron beam ---> internal tungsten converter ---> bremsstrahlung radiation ---> beryllium converter target ---> photoneutrons ---> waste container ---> prompt-fission neutrons ---> detectors

Additionally, the sequence of interactions which occur during the pulsed photofission neutron interrogation of the RH-TRU waste packages is given below.

electron beam ---> internal tungsten converter ---> bremsstrahlung radiation ---> waste container ---> prompt neutrons ---> detectors

The shielding requirements for the wall which intercepts the primary bremsstrahlung beam are considerable. Calculations [9] indicate that a 6-MV LINAC producing a usable NDA/NDE interrogating dose rate of about 1000 rad/min measured at 1 m from its internal conversion target would require a normal-density concrete shield wall that is 60 in. (1.5 m) thick to reduce the externally transmitted dose rate to 1 mrad/h, assuming the LINAC is operated 40 ft from the wall. The shielding requirements escalate rapidly with increasing energy and beam dose rate levels. For example, a 9-MV LINAC delivering 3000 rad/min at 1 m from the target would require a normal concrete shielding wall that is 100 in. (2.5 m) thick to reduce the external dose rate to 1 mrad/h at 40 ft. Practical considerations of this type lead one to the conclusion that the interrogating driving source of choice is a low-energy LINAC delivering as low a beam dose rate as can be tolerated.

A linear accelerator installation, such as that using a Varian Linatron™ 1000, includes an x-ray head, a control console, and modulator and pump stand cabinets. The modulator cabinet contains the main power supplies, pulse modulator, and power distribution electronics. The pump stand cabinet contains the heat exchanger, water pump, and closed-loop water tank. The x-ray head would be placed in the WHPP Packaging Cell (see Fig. 3), while the ancillary equipment would remain outside the hot cell.

3. RH-TRU WASTE NDA/NDE APPLICATION

Figure 3 shows the schematic layout of a LINAC-based NDA/NDE measurement system within the Process and Packaging Cells of the WHPP facility. This drawing illustrates the multifunctional approach required to accomplish the combined NDA/NDE measurements, that is, using a single LINAC driving source to perform NDA/NDE measurements in two adjacent hot cells. In this particular concept, the LINAC head assembly is constrained to move along a linear track parallel to one of the cell walls. The LINAC photon beam is produced in a direction which is perpendicular to this track and directed at the NDA and RTR waste inspection stations. Figure 4 is an artist's conceptual drawing of a RTR waste inspection station. Figure 5 is an engineering drawing of the RTR waste inspection station.

As many as four waste inspection stations may be accommodated with this concept: an RTR and NDA waste inspection station located in each cell. In this concept, two beam access ports are required to allow passage of the bremsstrahlung beam from the Packaging Cell into the Process Cell. In other proposed concepts, both the RTR and NDA waste inspection stations within the Process Cell are positioned in a staggered arrangement and would be serviced with the same beam access port [11].

3.1 NDA measurement techniques

A pulsed-LINAC driving source, producing an intense neutron interrogating flux, provides a high-sensitivity NDA technique suitable for examination of RH-TRU wastes. The pulsed neutrons are produced by the interaction of the incident bremsstrahlung radiation with an insertable beryllium converter (see Fig. 6) through a (gamma, neutron) reaction. Beryllium possesses a (gamma, neutron) threshold reaction energy of 1.67 MeV; therefore, the bremsstrahlung from a 6-MV electron beam produces large quantities of photoneutrons in such a converter layer. This interrogating technique would utilize the same differential dieaway neutron interrogation method which was developed in the early 1980's by the Los Alamos National Laboratory (LANL) for the assay of CH-TRU wastes [5,6,7,8].

The bremsstrahlung from high-energy electron can also be used to directly produce a photon-induced fission interrogation of TRU waste. A prototype NDA system based on simultaneous differential dieaway and photofission interrogations was constructed and tested in 1984 by personnel from LANL and EG&G Santa Barbara using a commercially available electron LINAC in an existing facility at the Lawrence Livermore National

Laboratory (LLNL) [4].   The availability of both pulsed-neutron and pulsed-photofission interrogations provides enhanced discriminatory measurement capability,

The two NDA station concepts shown in Fig. 3 are readily serviced with a single 6-MV LINAC. The LINAC x-ray head would be located in the Packaging Cell, with an appropriately designed waste inspection assay station in the adjoining Process Cell. Incoming waste packages would be examined and characterized in the Process Cell for screening purposes. The NDA waste inspection station located in the Packaging Cell would be used for final waste form certification of the processed and packaged wastes.

3.2 NDE(RTR) measurement technique

RTR measurements are routinely performed in industrial settings, using commercially available 2-20 MV electron LINACs of the type proposed for the WHPP. A survey of the RH waste examination technologies, which were in use at that time, was performed for the DOE in 1986 [10]. The authors of this work also performed several experimental evaluations of the various RTR hardware configurations required for examination of RH wastes with ^{60}Co and ^{192}Ir gamma-ray sources (approximately 1000 Ci) used to simulate typical RH waste package dose rates. Their conclusions were that high-energy electron accelerators, which include 6-MV LINACs, are suitable for RTR characterization and certification measurements performed on RH wastes packaged in various size containers.

The high energy of the pulsed LINAC bremsstrahlung is capable of achieving an instantaneous dose rate of 10^7 rad/min measured at 1 m (e.g., the Varian Linatron^{TM} 1000 is capable of 1000 rad/min dose rate at 1 m in a 4-μs pulsed mode). This interrogating flux is nearly 10^6 times the dose rate produced by a 1000-rad/h RH waste drum measured at 1 m. Interrogating fluxes of this magnitude have been shown to penetrate sludge waste in drums (a waste form which will be encountered in the WHPP) while preserving the required imaging sensitivity of low-Z objects such as free liquids [10].

One observation resulting from the aforementioned study was that when the level of interrogating bremsstrahlung radiation exceeded the external dose rate of the examined package by a factor of three or more, excellent quality RTR images were realized. This factor will determine the ultimate imaging limitations of the WHPP RTR measurements.

However, it should be noted that the basic LINAC bremsstrahlung radiation is produced in about 2- to 5-μs pulses, with a typical repetition rate of 50 to 200 Hz. Simple gating strategies in which one or more of the RTR components (in particular the imaging system) are operated in a similar pulsed fashion can be used to increase LINAC-signal-to-package-background ratios.

4. DEVELOPMENT TASKS

The FY 1990 funded LINAC-based NDA/NDE development tasks, which must be addressed to begin the effort required to achieve an operational system for use in examination of RH-TRU wastes in the WHPP, are given below.

1. Develop a detailed test plan.
2. Begin and complete LINAC demonstration site preparation. A test site located at another Oak Ridge Reservation facility has an available 6- to 8-MV LINAC for the WHPP development work.

3. Complete preparation of demonstration phase documentation (e.g., Quality Assurance Plan and Safety Assessment).
4. Define and assess available RH-TRU waste characterization data.
5. Perform Monte Carlo calculations to parameterize NDA system.
6. Prepare LINAC NDA cost estimate.
7. Begin design of RH-TRU NDA system.
8. Prepare interim report.

Some of the issues which require resolution before an operational system can be realized include reliability and maintainability of the LINAC and the NDA and RTR waste inspection stations while in a hot cell environment; waste package surface dose rates up to 10,000-rad/h gamma and 1-rad/h neutron; primary beam shielding requirements; the LINAC output requirement for NDA and RTR measurements; design of the NDA and RTR waste inspection stations; and lower limits of detection for the NDA measurement techniques.

5. CONCLUSIONS

The WHPP facility RH-TRU waste characterization and certification requirements can be served by a multifunction, multistation LINAC-based NDA/NDE system. A single 6MV electron linear accelerator can be used as the driving source for both RTR and active NDA measurements. A careful choice of LINAC size and intensity, which minimizes building shield wall requirements, coupled with efficient use of the inherent signal-to-background advantage of pulsed operation should make the integration of a 6-MV LINAC into the WHPP facility a successful and cost effective venture.

REFERENCES

1. SCHULTZ, F.J. and CALDWELL, J.T. DOE Assay Methods Used for Certification of CH-TRU Waste, ORNL-6485, in publication, December 1989.

2. CALDWELL, J., et al. "Experimental Evaluation of the Differential Die-Away Pulsed Neutron Technique for the Fissile Assay of Hot Irradiated Fuel Waste." Proceedings of the ANS Topical Meeting on Treatment and Handling of Radioactive Wastes," Richland, Washington, April 19-22, 1982.

3. FRANKS, L.A., et al. "High-Sensitivity Transuranic Waste Assay by Simultaneous Photon and Thermal-Neutron Interrogation Using an Electron Linear Accelerator," Nucl Inst. and Meth. 193 (1982) 571-576.

4. KOCINSKI, S.M., et al. "An Electron Accelerator-Based System for Assay of Transuranic Wastes,: LA-UR-84-200, 1984, paper presented at the International Linear Accelerator Conference, Dormstadt, Federal Republic of Germany, May 7-11, 1984.

5. KUNZ, W.E.; ATENCIO, J.D. and CALDWELL, J.T. LA-UR-80-497, "A 1-nCi/g Sensitivity Transuranic Waste Assay System Using Pulsed-Neutron Interrogation," Proceedings J. Institute Nucl. Mat. Mgt. IX, page 131-137, 1980.

6. CALDWELL, J.T., et al. "Test and Evaluation of a High-Sensitivity Assay System for Bulk Transuranic Waste," Proceedings INMM 24th Ann. Mtg., Vail, Colorado, July 10-13, 1983.

7. SCHULTZ, F.J., et al. "Neutron and Gamma-Ray Nondestructive Examination of Contact-handled Transuranic Waste at the ORNL TRU Waste Drum Assay Facility'" Oak Ridge National Laboratory Report ORNL-6103, March 1985.

8. CALDWELL, J., et al. "The Los Alamos Second-Generation System for Passive and Active Neutron Assays of Drum-Size Containers," Los Alamos National Laboratory Report, LA-10774-MS, September 1986.

9. 6 and 9 MV bremsstrahlung shielding and collimation calculations provided by Varian Associates, Inc., Radiation Division, Palo Alto, California.

10. BROWN, B.W. and MIKESELL, C. R. "Real-Time X-Radiography for Examination of Remotely Handled Radioactive Objects" INEL, Informal Report, EGG-SD-7399, September 1986.

11. CALDWELL, J., et al. "Nondestructive Assay and Real-Time Radiography Requirements for Remote-Handled Transuranic Wastes at the Proposed ORNL WHPP Facility," DOE Model Conference, October 3-7, 1988.

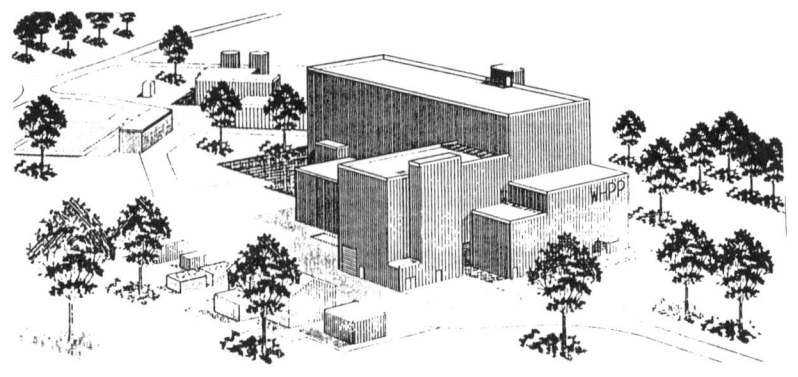

Figure 1. Architect's rendering of the Waste Handling and Packaging Plant.

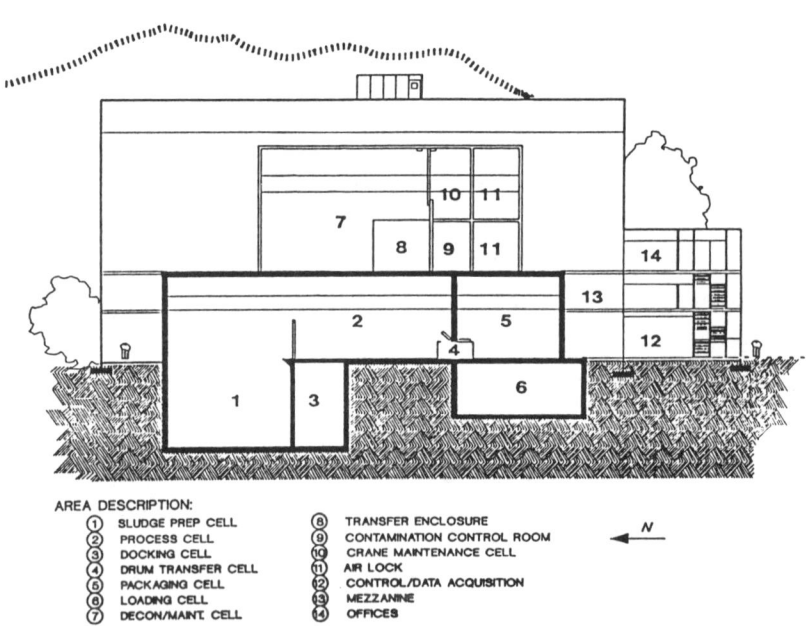

AREA DESCRIPTION:

① SLUDGE PREP CELL
② PROCESS CELL
③ DOCKING CELL
④ DRUM TRANSFER CELL
⑤ PACKAGING CELL
⑥ LOADING CELL
⑦ DECON/MAINT. CELL
⑧ TRANSFER ENCLOSURE
⑨ CONTAMINATION CONTROL ROOM
⑩ CRANE MAINTENANCE CELL
⑪ AIR LOCK
⑫ CONTROL/DATA ACQUISITION
⑬ MEZZANINE
⑭ OFFICES

N ←

Figure 2. Sectional view of the Waste Handling and Packaging Plant.

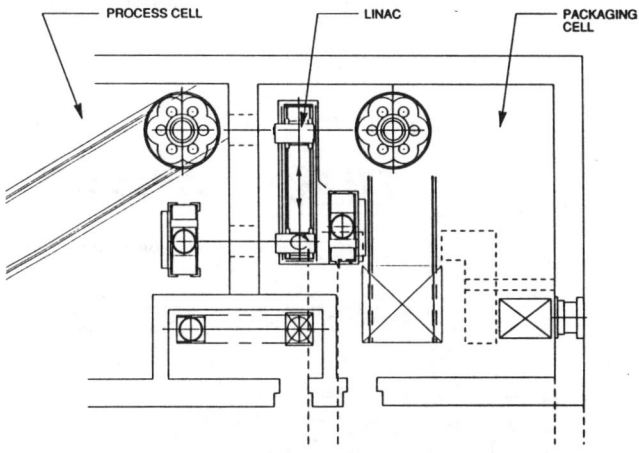

Figure 3.   Schematic view of the LINAC and NDA/NDE waste
            inspection stations.

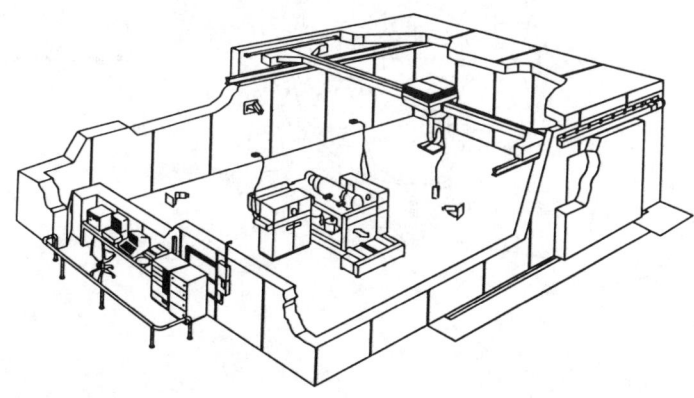

Figure 4.   Artist's conceptual drawing of an NDE waste
            inspection station located in a hot cell.

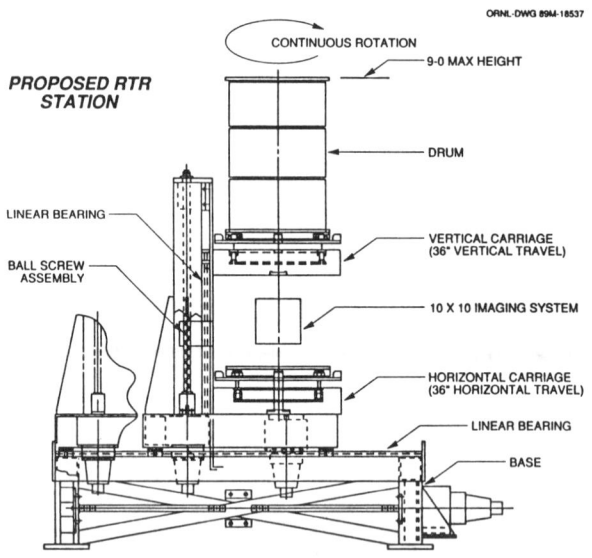

ORNL-DWG 89M-18537

CONTINUOUS ROTATION

9-0 MAX HEIGHT

PROPOSED RTR STATION

DRUM

LINEAR BEARING

VERTICAL CARRIAGE
(36" VERTICAL TRAVEL)

BALL SCREW
ASSEMBLY

10 X 10 IMAGING SYSTEM

HORIZONTAL CARRIAGE
(36" HORIZONTAL TRAVEL)

LINEAR BEARING

BASE

Figure 5.   Engineering drawing of an NDE waste inspection station.

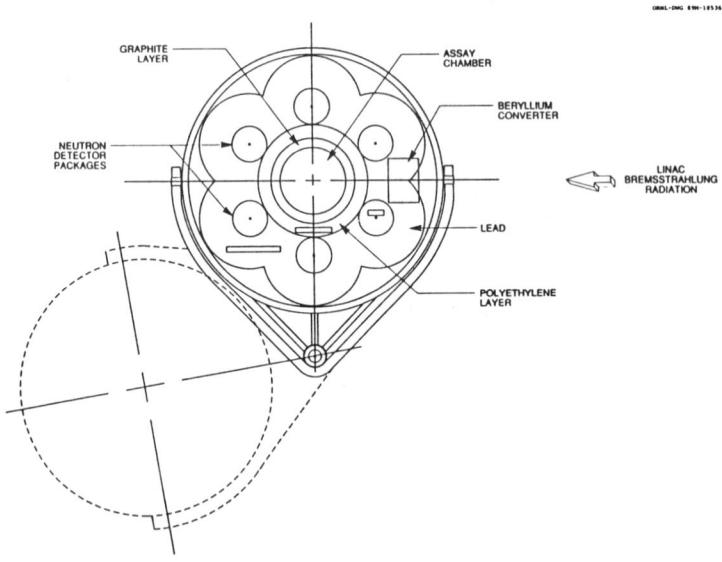

ORNL-DWG 89M-18536

GRAPHITE
LAYER

ASSAY
CHAMBER

BERYLLIUM
CONVERTER

NEUTRON
DETECTOR
PACKAGES

LINAC
BREMSSTRAHLUNG
RADIATION

LEAD

POLYETHYLENE
LAYER

Figure 6.   Artist's conceptual drawing of an NDA waste inspection
             station.

— 164 —

PART B

PERFORMANCE AND OPERATING

EXPERIENCE

SESSION B1 - PERFORMANCE

Chairman : J.C. Alder - Nagra, Baden

Opening statement

The subject-matter of the session is the presentation of some systems implemented for non-destructive assay of the radioactive content of the waste, a report on their performance and analysis of their response under the various conditions of real operations.

The aim of the session is to obtain an overview of the behaviour of such systems, which were conceived and developed for specific purposes, when they are confronted with real wastes or with more or less known dummy wastes in round-robin tests.

The conclusions of the session are as follows. A numerous waste and experimental set-up parameters have a bearing on the assay results, these have to be fully identified and understood in order to avoid the stated uncertainties in the results being too small. The range of waste parameters acceptable for measurement must be clearly defined. Any prior information on the waste to be assayed would improve the accuracy of the results and should be collected. Gaining confidence in the general performance of implemented assay systems appears consequently to be of importance.

PRESENT DEVELOPMENT ACTIVE NEUTRONIC INTERROGATION CELL AT CEB.III

P. RUFFET
P. CHOPLIN
CEA - CENTRE D'ETUDES DE BRUYERES-LE-CHATEL
BP 12
91680 BRUYERES-LE-CHATEL

Summary

The major characteristics of the active neutron interrogation cell (CELINE) for alpha waste measurements in Bruyères-le-Châtel Nuclear Research Center are briefly summarized. The effects of parasitic neutron sources up to 5.10^5 n.s^{-1} and of gamma irradiation up to 8 R.h^{-1} are taken into account. An experimental method for matrices identification inside 200 l. drums is presented.

1 - INTRODUCTION

Legal rules dealing with nuclear waste, and especially for low activity α waste having to be definitively stored overground, implies the use of high-quality measurement devices. So, at the difference of passive measurement methods, the active interrogation by 14 MeV neutrons had to progress in the domains of matrices identification and of associated software.

2 - ACTUAL CONFIGURATION OF THE MEASUREMENT CELL CELINE

The active device CELINE is now automatized and instrumented as shown in figures 1 :

- a 14 MeV pulsated neutrons generator (tube SODERN GTN02) with an equivalent fluence of $1,5.10^{+8}$ n.s^{-1} and frequency 40 Hz at lifetime 8.10^{-5}s. The tube is placed in the corner of a cubic chamber, covered with graphite and polyethylen, inside which are measured the 200 l. drums. An automatic system for introducing and inside rotating drum is under software full control.

- a set of neutron detection with four blocks of three counters with ^{3}He (model 150 NH 100 LCC) linked two by two (measurement lines 1 and 2). Those counters are into the walls of the cell. The ceiling and the botton are not instrumented.

- electronics counting and integrating with baseline fast recovery, associated with a micro-computer GOUPIL, gives facility to compute the nuclear fissile amount (^{239}Pu) in waste, depending upon the matrices inside the drum, and accordingly to the software carried out by CEB.3/SPR team (programs like "CHORUF" or "MATRICE", see figure 2).

- an overall checking device of the cell using an Am-Be (α-n) source is made by pneumatic transfert under control of the computer. The activity of that neutron source is 4.10^4 n.s^{-1}.

- some other peripheric devices as a plotter, video color or black/white video screens and a printer give easy record of all necessary data for normalized mass measurements.

- Three neutron fluence monitors
   . m_1 for counts normalisation versus pulsated active flux (an helium counter inside a polyethylen sphere placed outside the cell).
   . m_2 and m_3 for matrices identification, m_3 under cadmium shield.

The general fitting is shown also in figure 1.

3 - METHOD IN USE

3.1 General remarks

We deal with two regions of interest in a multiscale signature (MCS with $\delta T = 10$ ms). Those regions include all the events coming from the measurement lines 1 and 2, once digitized and integrated. We consider :

- region 1 between 0,45 and 5 ms after 14 MeV active neutronic pulse;

- region 2 between 5,45 and 10 ms after the pulse, the one of delayed neutrons; it is considered as the background level in "with drum" condition, but includes at the measure time :

   . outside background continuum and ambiance;
   . background of counters (cosmic γ rays...)
   . spontaneous fissions (for instance ^{240}Pu or ^{244}Cm);
   . delayed neutrons;
   . α,n reactions (for instance Pu-Be) or other.

Our measurement duration is 600 s and the count C (normalized as C* versus the reference active flux).

3.2 Choice of matrices

For evident reasons dealing with our own future use, we have selected seven specific matrices well represented in table I.

4 - STUDY OF THE RESULTS

Many parameters control the obtained results (see table I). We consider :

4.1 Dispersion bound to Geometry

The position of a point-source situated into any place inside the matrice has a very strong influence over the major count C*.

For the five first kinds of matrices selected, we have studied the dispersion to the measurement linked with isolated 32 mg masses of ^{239}Pu placed in nine points selected as representative of most extreme possible variations.

Let us recall that, during measurement, the drum is rotating at 6 r.p.m.

The recorded results furnish :

- standard mean values;
- standard maximum uncertainty bound to geometrical effect;
- forescast estimate of homogeneous contamination.

4.2 Pseudo-homogeneous contamination

In classical situations, we consider that in an α waste drum, the matrix is homogeneously contamined. This is generally true, but we shall see later the limits of that hypothesis.

Te mass calibration versus the normalized count C* is carried out between 0 and 300 mg by use 5 standard sources of ^{239}Pu :

- under filament shape;

- of well-known isotopic quality;

- of certified mass values;

situated in the definite points into each one of the 7 matrices.

Figure 3 shows the obtained results as well as the mass calibration curves in use : m = f (C*).

Then, the standard deviation is established from :

$$\sigma \% = \pm \, 300 \, \sqrt{(2 \mathrm{x} B^* + C^*)} \, / \, C^*$$

4.3 Other interferences on the measurement

4.3.1 Neutrons coming from :

- spontaneous fissions : isotopes with high specific emissions, like ^{244}Cm, ^{252}Cf or ^{240}Pu, or else...

- (α,n) reactions due to interaction of alpha particles with the matrix itself.

Although those neutrons do not disturb the measurement they will decrease the Count/Noise ratio and so increase the uncertainty on the result and the detection limit.

In a classical vinyl matrix, we have characterized the low influence of spontaneous fission neutrons from a ^{244}Cm source.

We remark that :

- neutronic background is increased in regions 1 and 2;
- normalized count C* varies slightly ($\delta \approx 20$ %).

In the same way, we have simulated the influence of (α,n) neutrons by the including of two Am-Be sources into the center of a drum made up with technological matrices.

With the first source (2.10^4 n.s^{-1}), no change can be observed on the multiscale drawing but, with the other (4.10^5 n.s^{-1}), the background level is increased and the obtained value for a standard mass of 269 mg (^{239}Pu) is only 134 mg. This indicates the startup of an electronic saturation at the actual of parasitive neutronic fluences (about 10^5 n.s^{-1})

4.3.2 Gamma irradiation level

The study of the limit has been carried out by two ways, by studying variations induced :

- on the background signaturs;
- on the measurements of a 269 mg (^{239}Pu) standard source.

Two matrices have been tested precisely : "technologique" and "Labo chaud" with interfering source of ^{137}Cs and ^{60}Co in the center of the drum.

The actual level limit is about 2 R.h^{-1} close to the counters, that means 4 R.h^{-1} close to the drum itself, and no difference can be shown for the measurement of a 269 mg (^{239}Pu) source.

But, for a level of 8 R.h^{-1}, an important change occurs on the background level, incompatible with a correct measurement of fissile material (see Figure 4).

5 - IDENTIFICATION OF MATRICES

It is based on the modification of the neutronic distribution in the cell, in absence of Plutonium, due to the matrix itself in the drum.

Our results are obtained for the 7 homogeneous known matrices, but this is not yet achieved for most practical cases, being taken into account the influence of hydrogeneous material.

A series of preliminary tests carried out with two proportional counters either protected or not by Cadmium and placed into the cell allow the automatic identification of some types of matrices. Those measurements qualify the homogeneous and heterogeneous matrix, and identify the fitting coeffects for the mass calculation.

They were carried out over standard drums whose matrix composition was well known and where it was possible to record the modification of the neutronic ambiant spectrum due to the matrix itself.

5.1 Characteriztion of homogeneus matrices

The results obtained at the end of that study are summarized in the curves on the figures 5 and 6.

Major facts :

The curves giving the count rate on monitors m_2 and m_3 versus matrix mass show various gradients for each kind of matrice.

The changes on the thermic neutronic spectrum of distribution depend mainly upon two parameters:

- the mass of the matrix in the measuring cell;
- the nature of that matrix.

The analysis of the response of the monitor m_3 shows that calibration curves are superposed for vinyl and cellulosic matrices, but that they differ much from those for graphite and metal matrices (iron, inox...).

At the contrary, the analysis of the monitor m_2 allows to discriminate between vinyl and cellulosic matrices, and, moreover to separate them from graphite and metal curves (those two last staying quite superposed).

So, we can conclude that the discuss of the counts given by m_2 and m_3 gives ability to determine the nature of matrices inside a 200 liters drum.

5.2 Characterization of heterogeneous matrices

That study was carried out on four matrices of two kinds with mass distribution being respectively :

- 1/3 iron and 2/3 vinyl;
- 2/3 iron and 1/3 vinyl;
- 1/3 vinyl and 2/3 cellulose;
- 2/3 vinyl and 1/3 cellulose;

5.2.1 Matrices with iron and vinyl

Using the obtained résults, we are able to draw reponse curves with a slide known as a function of vinyl proportion. Those lines will be directly used to identify undefined matrices.

5.2.2 Matrices with cellulose and vinyl

The discrimination vinyl - cellulose being not be made with counts from monitor m_3, we are only interested by signals from monitor m_2 for calculate mass proportion.

In a similar way as for the binary system iron/vinyl, a group of lines counting versus matrix mass is established for proportions like 1/3 vinyl and 2/3 cellulose or opposite.

So, by the same principle than before, a simple curve gives ability to achieve a practical determination.

5.3 Proposal of method for heterogeneous matrices identification

- Phase 1 : Weight measurement of drum for mass détermination.

- Phase 2 : 100 seconds measurement and recording of counts on monitors m_2 and m_3, for precise numerical knowledge of values.

Counts from m_2 allow to determine if there is a significant part of vinyl, while counts from m_3 give indications about graphite and metal.

In such a way, heterogeneous matrices we are dealing with can be identified and we are able to fit the mass calibration curves for waste drums including any of our seven usual homogeneous matrices. Software are now being updated for that.

CONCLUSION

The actual state of the study can be summarized by :

- measurement méthods and results analysis;
- fully automatized process of control-command and transfers;
- counting monitors m_1, m_2 and m_3 choosen;

That allow now to qualify the CELINE station for :

- nuclear mass determination between 0 and 300 mg of the ^{239}Pu, in pseudo-homogeneous contamination in seven genuine matrices;

- with gamma irradiations up to 2 R.h^{-1} close to the counters;

- with interfering neutron fluences up to 4.10^4 n.s^{-1};

- an heterogeneous matrices identification is at the actual state of software development on master microcomputer.

FIGURE 1

SYNOPTIC OF CELINE

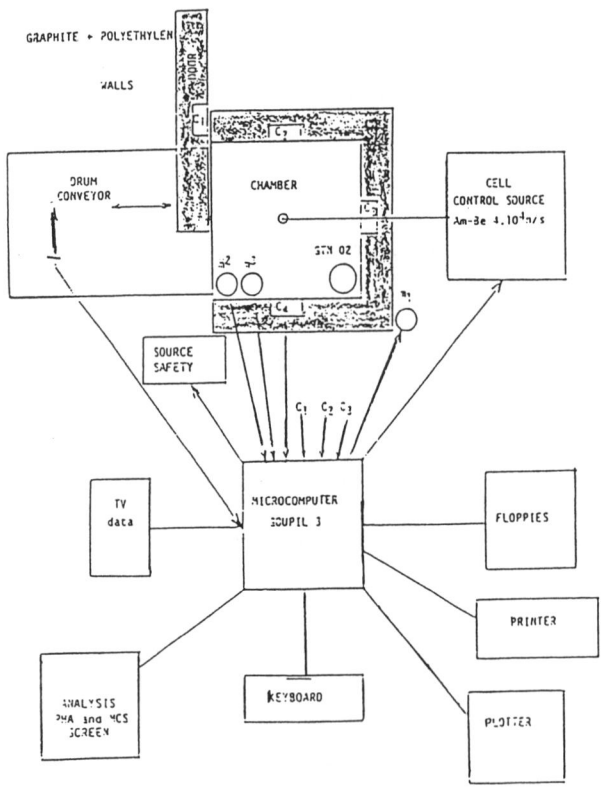

SOFT "MATRICE 1"   FIGURE 2

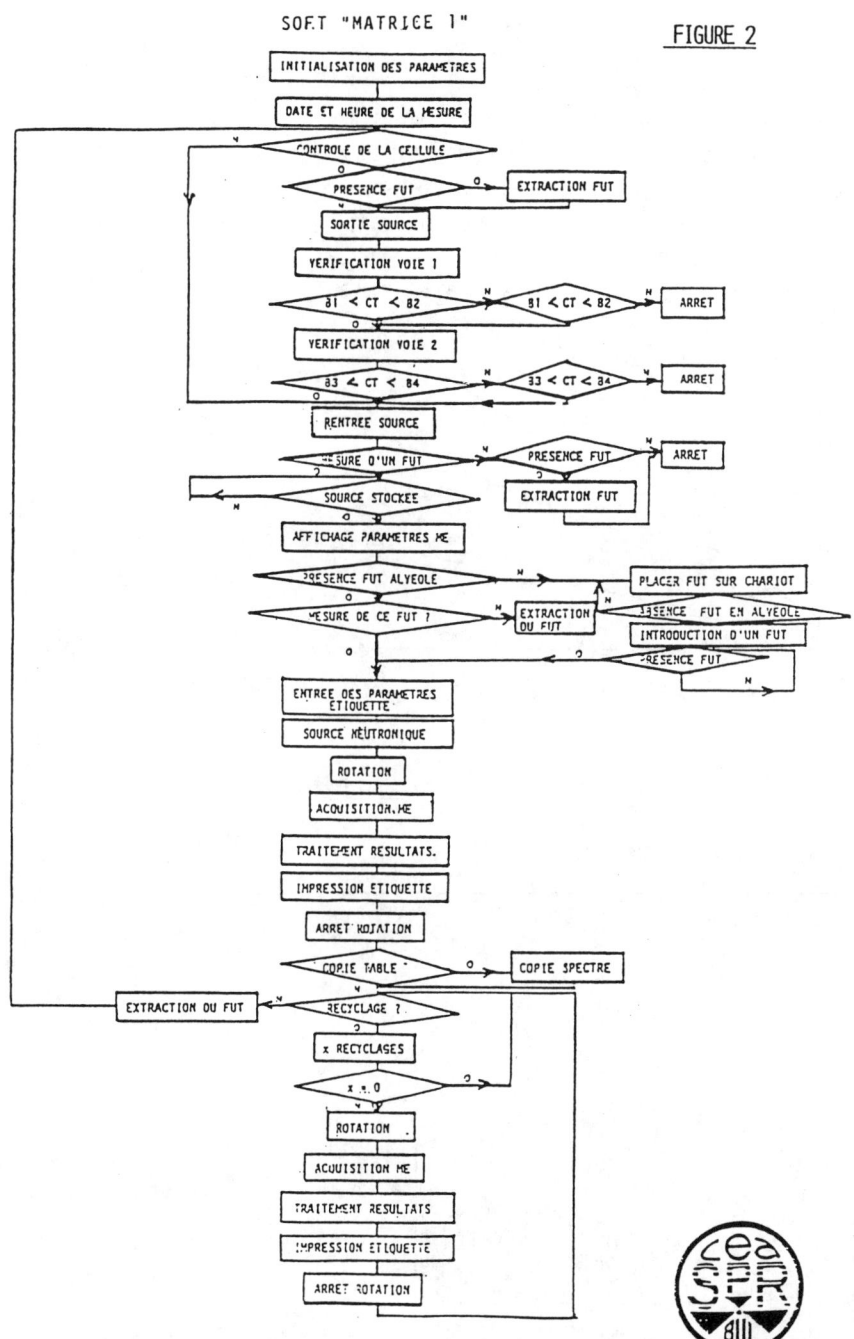

- 175 -

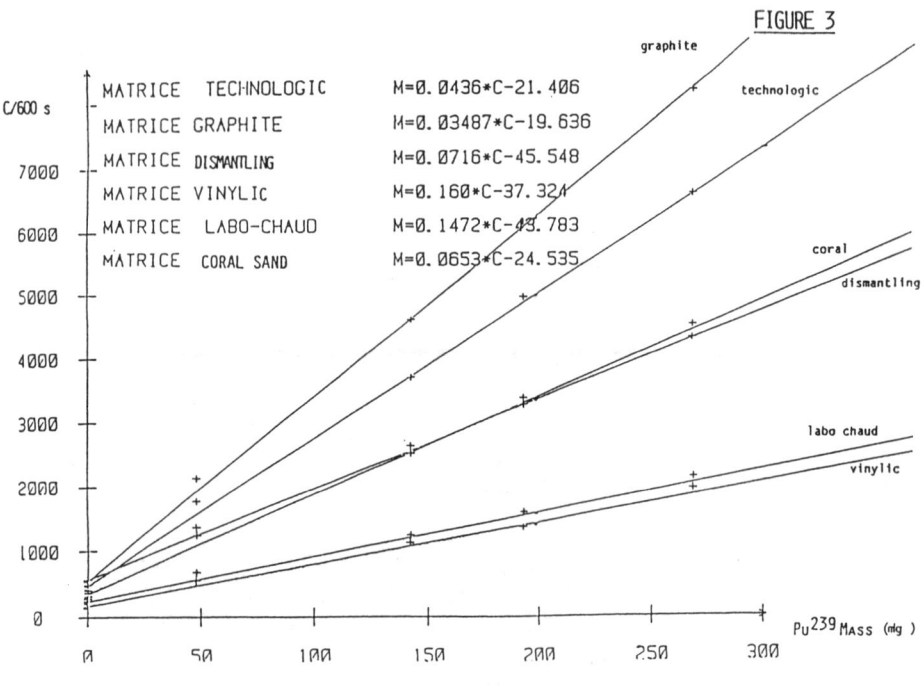

FIGURE 3

MATRICE	TECHNOLOGIC	M=0.0436*C-21.406
MATRICE	GRAPHITE	M=0.03487*C-19.636
MATRICE	DISMANTLING	M=0.0716*C-45.548
MATRICE	VINYLIC	M=0.160*C-37.324
MATRICE	LABO-CHAUD	M=0.1472*C-43.783
MATRICE	CORAL SAND	M=0.0653*C-24.535

C/600 s

graphite

technologic

coral

dismantling

labo chaud

vinylic

Pu239 MASS (mg)

FIGURE 4

CA# 0   CA# 255

HO FUT : TECHNO + 60Co (8 R/H)   DATE : 1989

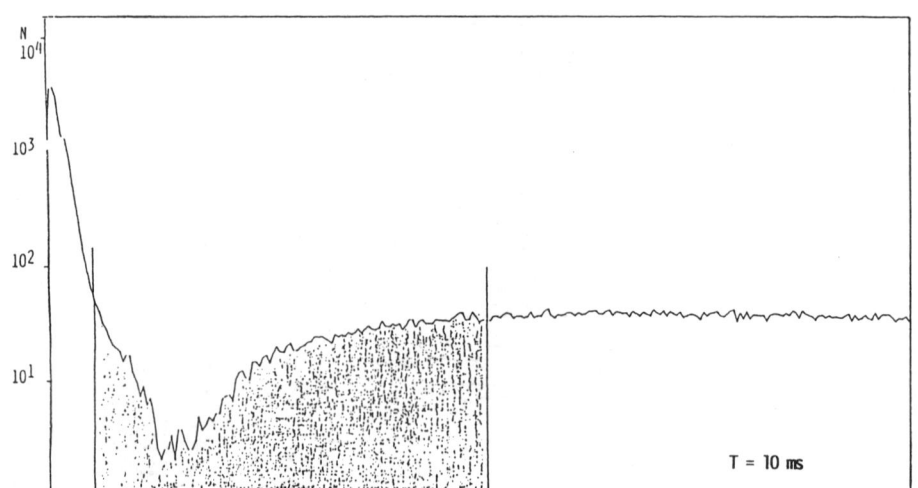

T = 10 ms

CURS=0   DU   CA# 14   AU CA# 120   PSEL(S) 600
COUPS= 192647   INT 9960   ACQ (S) 600

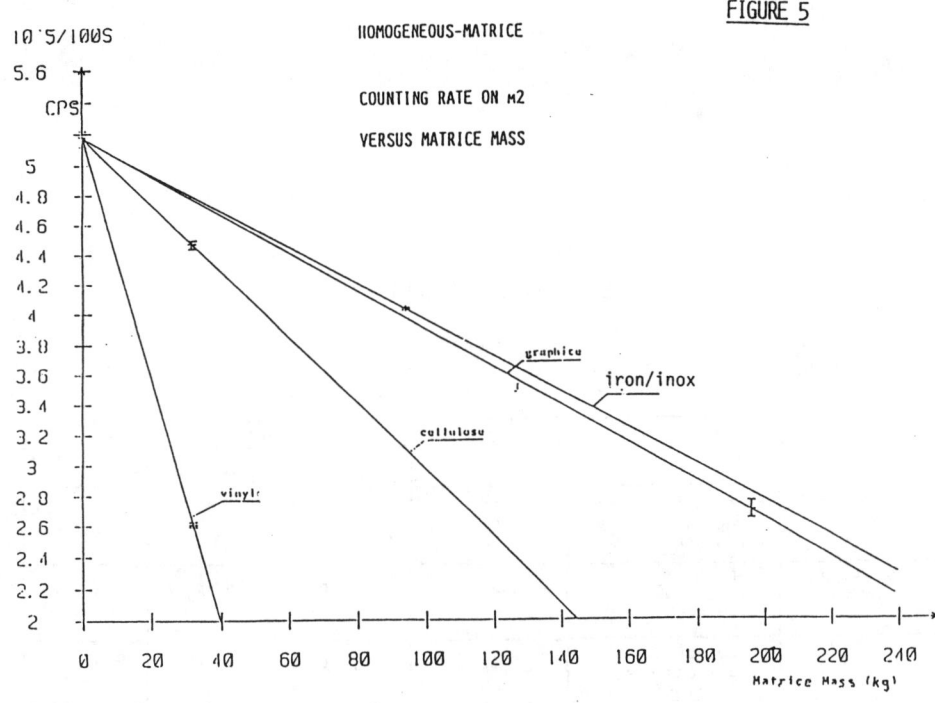

FIGURE 5

HOMOGENEOUS-MATRICE

COUNTING RATE ON M2
VERSUS MATRICE MASS

10^5/100S

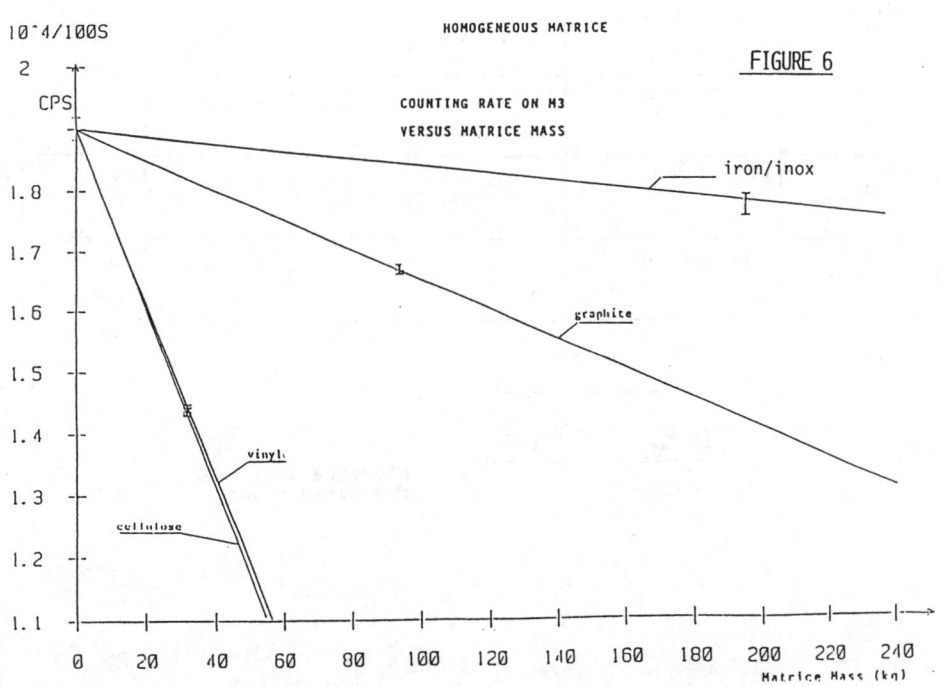

HOMOGENEOUS MATRICE

FIGURE 6

COUNTING RATE ON M3
VERSUS MATRICE MASS

10^4/100S

TABLE 1

PERFORMANCE OF CELINE

Matrice	Composition	Density	Dispersion		Détection
			Unit mass	contamination homogeneous	* limit 9 Pu mg
Technologic 49 kg	- coton - papers - few vinyl	0,16	± 41 %	± 7 %	2,0
Labo-chaud 67 kg	from BAG : - coton - red sack	0,25	± 65 %	± 11 %	6,0
Vinylic 49 kg	pièces of - soil carpets - altuglass...	0,16	± 67 %	± 12 %	7,5
Dismantling 182 kg	pieces and parts - iron - steel	0,03	± 49 %	± 8 %..	4,5-
Graphite 137 kg	pieces and parts of graphite	·0,60	± 35 %	± 6 %	1,7
Coral 1 137 kg	Coral sand	1,29	-	± 8 %	2,0
Coral 2 138 kg	Coral agregates	1,10	-	± 8 %	2,0

* Using

$$L.(m_{mg}) = 3 \sqrt{D} \frac{m}{C^x - D}$$

m : reference mass : 32 mg
C^x : associated count

EUROPEAN INTERLABORATORY TEST MEASUREMENTS ON ALPHA-CONTAMINATED WASTE

R. Dierckx, JRC Ispra

and

L. Bondar	JRC Ispra
R. Bosser	CEA Cadarache
W.B. Bremner	AEA Technology Dounreay
P. Cresti	ENEA Casaccia
P. Filss	KFA Jülich
J.W. Leake	AEA Technology Harwell
H. Ottmar	KfK Karlsruhe
C. Vicini	ENEA Casaccia
H. Würz	KfK Karlsruhe

The authors dedicate this report to the memory of Jean Ley.

SUMMARY

The Commission of the European Communities has organized interlaboratory test measurements on alpha-contaminated waste. The intend was to compare methods and tools, used in different laboratories, to determine the plutonium content of real waste drums. Seven laboratories from the Community participated.
The exercise consisted in measuring drums in which sources of known Pu-content and composition were placed in 'defined positions. Furthermore, a total of nine sealed drums with unknown Pu-content and composition had to be measured and the Pu-content determined.
The measurements show that more work has to be done. The stated uncertainties are not always realistic. A good calibration is necessary, taking into account matrix and source position effects. The correction for matrix and source position effects proves to be very important and has to be investigated further.

1. INTRODUCTION

In 1981 the Commission of the European Communities decided to organize interlaboratory test measurements on alpha contaminated waste. The goal was to compare methods and tools used in different laboratories to determine the plutonium content of real waste drums. Seven laboratories (Table I) from the Community participated, measuring with different techniques.
The first part of the exercise consisted in measuring non-sealed drums of synthetic waste (polyethylene foam) and a non-sealed concrete drum into all of which sources of known Pu-content and composition were placed in defined positions. In these non-sealed drums the Pu sources could be placed heterogeneously or such that a pseudo-homogeneous distribution was obtained.
In the second part of the exercise a total of nine sealed drums with

unknown Pu-content and composition had to be measured and the Pu-content determined. The measurements lasted from 1982 till 1986. A great part of this time was necessary because of transport difficulties between the different countries involved.

This paper summarizes the main results of this interlaboratory exercise. A full report, containing a detailed description of the exercise and of the analysis of the results will appear as an external Euratom report /1/.

TABLE I: Participants at the interlaboratory test. The numbers refer to the type of experiment for each laboratory, as used in the different tables.

Laboratory	Number	Experiment description
CEA Cadarache	1	Pu isotopic composition
	12	Pu-mass by neutron counting (VDC)
	13	Pu-mass by gamma counting
ENEA Casaccia	11	Pu-mass by gamma counting
UKAEA Douneay	6	Pu-mass by neutron counting (VDC)
	14	Pu-mass by gamma counting
Harwell Laboratory	4	Pu-mass by neutron counting (VDC)
	5	Pu-mass by neutron counting (SR)
	10	Pu-mass by gamma counting
JRC-Ispra	7	Pu-mass by neutron counting (SR)
	15	Pu-mass by neutron counting (TCA)
KFA-Jülich	8	Pu-mass by neutron counting (Sb-Be)
KfK Karlsruhe	2	Pu isotopic composition
	3	Pu-mass by neutron counting (SR)
	9	Pu-mass by neutron counting (DDT)

2. THE Pu ISOTOPIC COMPOSITION

For the non-sealed drums three types of Pu-sources are used: type C or PuO_2, type D or $(Pu+U)O_2$, and type E or $(Pu+Am)O_2$. The sealed drum 9R contained also natural uranium while 10R contained enriched uranium. Drums 12R, 13R and 14R contained only PuO_2. The reference composition of the Pu is given by the producers /1/.

Two laboratories did gamma-spectroscopy on the nine sealed drums and on three of the reference sources used to assemble the waste drums with known Pu-content. In general the measured isotopic weight ratios agree within their uncertainty with the reference composition as declared by the Pu producer. Table II gives the ^{240}Pu eq and ^{239}Pu eq weight percents calculated from the isotopic weight ratios.

We can conclude that the measured ^{240}Pu eq and ^{239}Pu eq weight percents contribute with an uncertainty between 1% and 5% to the total Pu-mass determination. It has to be noted that the isotopic measurements as done here, based on the analysis of the less abundant gamma rays above 100 keV, required long counting times (20 to 60 hours) to obtain a reasonable precision. Analyzing the more abundant, but also more complex gamma rays in the 100 keV range could reduce the counting times by a factor of at least 10.

TABLE II: Equivalent Pu weight percent as measured by laboratory 2 at 3/86

	$^{240}m_{eq}/^{240}m$	Weight % of $^{240}Pu\ eq$	$^{239}m_{eq}/^{239}m$	Weight % of $^{239}Pu\ eq$
Ref. PuO_2	1.470	35.53	1.194	73.03
C45	1.47±0.01	36.2±0.9	1.194±0.001	75.5±0.5
5S	1.51±0.07	35.8±2.6	1.193±0.002	73±2
6S	1.48±0.04	35.0±1.6	1.193±0.002	73.5±1.2
Ref. $(Pu+Am)O_2$	1.225	63.07	1.271	44.17
E43–E44	1.24±0.02	3±2	1.268±0.002	44±2
7S* {	1.23±0.02	62±2	1.269±0.002	45±2
	1.22±0.02	64±3	1.270±0.02	43±2
Ref. $(Pu+U)O_2$	1.082	25.24	1.0582	76.79
D44–D46	1.077±0.001	26.0±0.3	1.0568±0.0002	76.3±0.5
8S	1.078±0.002	25.4±0.6	1.0568±0.002	76.7±0.5
Ref.	1.480	37.52	1.234	71.77
9R	1.51±0.02	37.7±1.3	1.189±0.001	71.2±0.8
Ref.	1.050	19.85	1.0372	81.37
10R	1.049±0.001	20.0±0.5	1.0372±0.0002	81.2±0.3
Ref.	1.329	26.77	1.0972	78.34
12R	1.29±0.02	31±1	1.121±0.002	75±1
Ref.	1.392	35.35	1.150	71.84
13R	1.37±0.01	36±1	1.151±0.001	71.1±0.6
Ref.	1.225	18.29	1.0570	84.57
14R	1.216±0.007	18.3±0.7	1.0575±0.0003	84.1±0.4

*repeated measurements

3. ANALYSIS OF THE NEUTRON MEASUREMENTS WITH KNOWN Pu-CONTENT

3.1 Drums with synthetic waste matrix
    In drums 1 to 4, Pu-sources of known content and composition (type C (PuO_2), D $((Pu+U)O_2)$ and E $((Pu+Am)O_2)$) could be placed in different locations from a central to border position. Also with more sources a pseudo-homogeneous Pu-distribution could be realized.
    In a first analysis we started from the basic coincidence count rates, disregarding the own calibration of some laboratories.

3.1.1 Measurements with drum 1
    Extended measurements were done with drum 1 (220 litre, matrix density 0.26 g/cm^3 of polyethylene). In Table III the characteristics of the different neutron-interrogating instruments used in the different laboratories are summarized. The different instruments and analysis algorithms are described in Ref./1/. SR means shift register, VDC variable dead time counter, and TCA time correlation analyzer. Two different active

TABLE III: Characteristics of the neutron interrogating instruments used in different laboratories for drum 1 (220 litres, $\delta = 0.262$).

Lab.	Background (c/s)	Background (mg ^{240}Pu eq)	Mean squared efficiency (mg.s/c) Centered source	Homo-geneous	Min.	Max.	Minimum** detectable (mg ^{240}Pu eq)	Counting time Nx(s/N)	Method
3	0.010±0.005	4±2	570±50	380±30	184	770	3±3	20x1800	SR
4	0.18±0.05	50±20	360±40	270±30	160	5000	50±50	10x300	VDC
5	0.18±0.05	60±20	690±50	330±30	160	690	90±90	10x600	SR
6	0.005±0.002	10±4	4450±450	2200±200	1200	5000	60±60	1x1800	VDC
7	0.25±0.05	35±7	390±40	141±4	66	390	10±10	8x511	SR
12	0.025±0.005	16±3	960±60	640±30	340	1900	7±7	16x3120	VDC
		(mg ^{239}Pu eq)					(mg ^{239}Pu eq)		
8	0.18±0.02	300±40	∿2800*	1850±200	∿3000*	∿6000*	50^{+100}_{-50}	1x1000	Sb-Be
9	90±2	80±2	2.4±0.2	0.89±0.01	0.96	3.6	30±30	10x180	DDT

* At 560 mg ^{239}Pu (not corrected for self-shielding).
** The minimum detectable quantity is a function of the efficiency, background count-rate, and to a lesser extent the counting time. Definition of the minimum detectable quantity has to be based on the statistical probability of the detection of the stated quantity, together with the probability of false readings due to background alone. Values declared above are determined in a consistent manner.

neutron interrogation methods are used, a static interrogation with a Sb-Be source (Sb-Be) and a pulsed interrogation with a 14 MeV source applying the differential die-away technique (DDT).

Both active interrogation methods show the effect of sample self-shielding. The sample self-shielding becomes obvious for ^{239}Pu eq masses larger than 100 mg. In order to detect Pu masses of a few mg, long counting times are necessary. Typical detection limits for Pu in the 220 litre waste drum 1, as achieved with the different equipments for the operation conditions in the exercise, are deduced from the counting of small quantities of Pu, taking into account the background and the uncertainties of these measurements.

3.1.2 Measurement with drums 2, 3 and 4 (polyethylene matrix)

Some characteristics of these drums are reported relative to drum 1. The drums are characterized by the product of half the mean chord length (R_{av} = 2V/S with V the volume and S the surface of the drum) and the density of the matrix δ. The homogeneous squared efficiency (Fig. 1) shows an exponential behaviour as a function of $R_{av}.\delta$. Up to about $R_{av}.\delta = 5$ the centre squared efficiency equals the homogeneous squared efficiency for all coincidence counting instruments within the uncertainty limits. As a consequence the minimum detection limits decrease reaching values for long counting times of the order of 1 mg for the 25 litre drum 4.

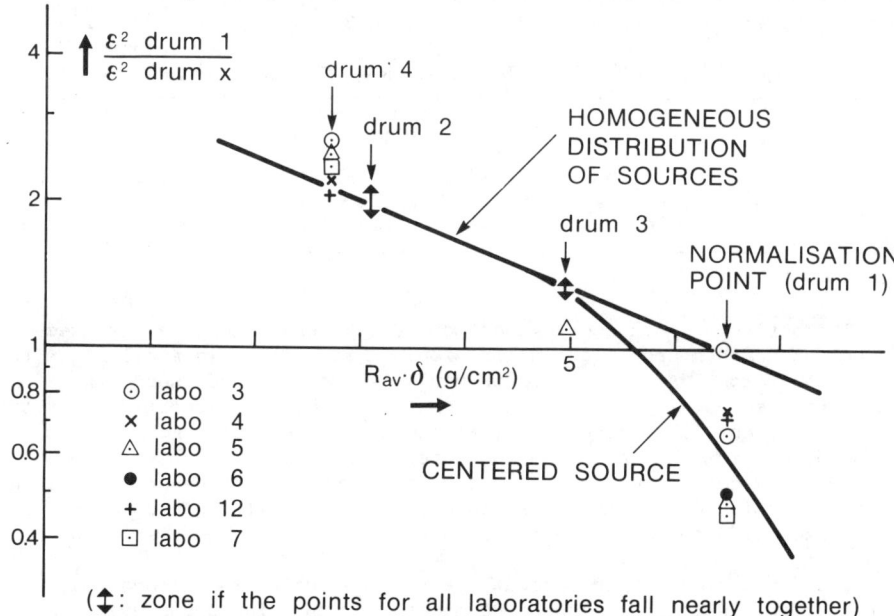

(⇕: zone if the points for all laboratories fall nearly together)

Fig.1: Normalised mean squared efficiencies (ϵ^2 drum 1/ϵ^2 drum x) as functions of $R_{av}.\delta$.

Qualitatively, the same correlations hold for the active DDT and Sb-Be methods. For the Sb-Be method the self-shielding effect tends to decrease for smaller $R_{av}.\delta$ values.

3.1.3 Use of measurements with known Pu-content
    Only a few laboratories used their own calibration and compared their determined Pu-content with the real one.
    Laboratories 4 and 5 show clearly a negative bias and measured too small Pu masses. For the 25 litre drum 4 the measurements were within 10% of the real values. The calibration for these laboratories was based on a bare plutonium source with no correction for matrix effects. Underestimation is, therefore, to be expected.
    Laboratory 7 used the measurement with a homogeneous distribution of Pu as a calibration of the instrument. The ^{240}Pu eq mass of the drums with centered or extreme source positions is then determined as if its distribution were homogeneous (Table IV). For the 25 litre drum 4 the measurements are within 10% of the reference values, as already observed for laboratories 4 and 5. On the same data, with an additional information of triplets (later referred to as measurement 15), laboratory 7 is able to determine independently the ^{240}Pu eq content using the TCA technique. The so determined ^{240}Pu eq masses are all within 10% of the reference values.

TABLE IV: Measured ^{240}Pu eq masses for heterogeneous source distributions by laboratory 7 (SR) and 15 (TCA) (source: 0.270 g; counting time: 8x511 s).

Source	Position	Drum 1 (220 l)		drum 3 (100 l)		drum 11 concrete	
r	h	SR	TCA	SR	TCA	SR	TCA
0	0	0.098	{<0.360 >0.190	0.178	0.244	0.050	{<0.360 >0.190
R/2	0	0.153	0.274	0.212	0.243	0.129	0.212
R	0	0.372	0.277	0.328	0.278	0.433	0.265
R	-H/2	0.574	0.284	0.447	0.303	0.642	0.341

    The analysis method of laboratory 6 is very extensive. It used its own Pu standard source for calibration. A correction is applied for the source position. Also the instrument is constructed such that the detectors are put in a position such that a minimum sensitivity of the source position is obtained for their normal waste drums. For drums 1 and 2, which they measured although the detector position was not optimum, and applying the correction for source positioning, their measured Pu masses are all within 10% of the real masses. Of course, if the source position is not known, with a non-optimized detector position the results will be worse.
    The results of these laboratories, using their own calibration, will be reflected in their measured Pu-masses of the nine drums with unknown Pu-masses.

3.2 Drum with concrete matrix
    In this case of a high density matrix, mainly the neutrons emitted near the surface are counted. A source positioned at the surface gives a count-rate about 7 times higher than in the centre for the passive methods, and even 20 times higher for the active DDT method. Using the TCA technique (laboratory 15, utilizing the additional information of triplets), the so independently determined ^{240}Pu eq masses in a concrete

drum lie all within 25% of the real values (Table IV). The drawback of the TCA method is, however, that it cannot be used to measure masses smaller than 100 mg in a 220 litre drum with concrete matrix.

4.   ANALYSIS OF THE GAMMA MEASUREMENTS WITH KNOWN Pu-CONTENT

Four laboratories determined the Pu-mass by gamma measurements, and compared their results with the reference Pu-content. Laboratories 10 and 14 used the segmented gamma scanner and measured the ^{239}Pu-content. Laboratory 13 measured the ^{239}Pu- and ^{241}Pu-content using a continuous segmentation. Laboratory 11 determined the ^{241}Pu-content only on the 25 litre drum 4.

The best results were obtained for the 25 litre drum 4 which was measured by all four laboratories. Laboratory 13 measured all the drums. In general the agreement with the reference values is within 20%. The results of laboratory 10 for the 220 litre drums are not very satisfactory, and only quantities above a few hundred of mg are measurable. Gamma measurements on drum 11 with a concrete matrix were without success.

5.   ANALYSIS OF THE DRUMS WITH UNKNOWN Pu-CONTENT

A total of nine drums (5S, 6S, 7S, 8S, 9R, 10R, 12R, 13R and 14R) with unknown Pu-content and composition had to be measured and the Pu-content determined. The Pu-content is measured with different techniques: passive neutron measurements resulting in the ^{240}Pu eq mass, active neutron interrogation giving the ^{239}Pu eq mass and gamma measurements in which the ^{239}Pu-mass is determined (Table I). Using the isotopic composition as measured by laboratory 2 (Table II), the total Pu-mass can be calculated. Results are listed in Table V.

Some remarks on each measurement concerning calibration and uncertainties.
1) The calibration of laboratories 3, 8 and 9 is done with the drums with known Pu composition and content and a quasi-homogeneous distribution of Pu. An a-posteriori analysis, taking into account the differences in matrix density, brought their measured data in agreement with the reference data /1/.
2) Laboratories 4 and 5 used their own calibration and did not apply a matrix or sample distribution correction. In general their measured Pu eq masses are underestimated.
3) Laboratory 6 used its own calibration and its own sophisticated analysis method.
4) Laboratory 12 made calibrations relative to the drums with known Pu-content, distributed homogeneously in a synthetic waste matrix.
5) Laboratory 7 measurements were calibrated relative to the drums with known Pu-content homogeneously distributed. Measurement 15 uses the same data, with in addition the triplets, analyzing them with the TCA technique. It is an absolute measurement.
6) Drum 8S is systematically underestimated by all laboratories, and drum 9R is systematically overestimated. Drum 9R was opened in 1988 to control the Pu content and its composition. It was found that the ^{241}Am content was higher and the ^{241}Pu content lower than originally stated, thus confirming the Pu-composition measurements of laboratory 2. The total Pu-content as stated was confirmed.

TABLE V: Total Pu weight measured by different laboratories and different methods

| | | Measured Pu weight/Reference Pu weight | | | | | | | | | | |
| | | Passive neutron measurements | | | | | | | Active neutron interrogation | | Gamma measurements | | Mean per sample |
	(g)	(3)	(4)	(5)	(6)	(7)	(12)	(15)	(8)	(9)	(10)	(13)	
5S	0.955	1.4±0.3	0.62±0.03	0.41±0.02	0.96±0.09	0.93±0.13	1.06±0.08	0.89±0.13	1.26	1.4±0.3	1.2±0.3	1.0±0.3	1.0±0.3
6S	3.487	1.4±0.4	0.71±0.02	0.44±0.02	0.92±0.06	1.4±0.2	0.80±0.06	0.86±0.11	0.40	0.55±0.12	0.43±0.14	1.0±0.3	0.8±0.4
7S	4.132	1.0±0.2	0.73±0.01	0.49±0.01	-	1.06±0.14	0.73±0.05	0.77±0.12	0.44	0.51±0.12	1.1±0.3	0.8±0.2	0.8±0.2
8S	2.402	0.9±0.2	0.57±0.02	0.38±0.03	-	0.92±0.12	0.75±0.08	0.62±0.08	0.50	0.50±0.12	0.63±0.08	1.1±0.3	0.7±0.2
9R	4.800	1.6±0.5	1.81±0.04	1.00±0.06	1.83±0.14	1.4±0.1	1.4±0.1	1.4±0.2	1.7	1.9±0.7	1.2±0.1	0.7±0.2	1.5±0.4
10R	5.211	1.7±0.5	1.11±0.04	0.86±0.02	-	1.1±0.2	1.24±0.09	0.94±0.13	1.0	1.7±0.6	0.81±0.04	1.2±0.5	1.2±0.3
12R	10.403	1.3±0.4	0.62±0.01	0.52±0.02	-	1.2±0.2	0.81±0.09	0.82±0.12	0.89	1.3±0.4	0.8±0.2	0.4±0.1	0.9±0.3
13R	7.597	1.4±0.4	0.58±0.03	0.54±0.03	-	1.2±0.2	1.3±0.2	1.01±0.14	0.53	0.54±0.17	0.87±0.08	1.0±0.3	0.9±0.3
14R	3.83	1.4±0.3	1.52±0.03	0.78±0.03	-	1.3±0.2	1.2±0.2	1.3±0.2	0.52	1.2±0.4	0.84±0.15	0.84±0.2	1.1±0.3
Mean labortory set value		1.4±0.3	0.9±0.4	0.6±0.2	-	1.2±0.2	1.0±0.2	1.0±0.2	0.8±0.4	1.0±0.5	0.8±0.3	0.9±0.3	

6.   CONCLUSIONS

A Round-Robin exercise of this type needs to be viewed as an exercise which can produce only a limited test of NDA systems. This arises from the limitations imposed during the preparation of the measurement items. The most useful information probably arose from the forty-one measurements carried out on drums of known source distribution, plutonium mass and matrix composition.

6.1 Pu isotopic composition

The measurements of the isotopic composition of the Pu-content by gamma-counting in a large waste drum (more than 100 litre) with low density matrix is possible but impracticable for small Pu amounts.

Reasonable results for the isotopic ratios can be obtained (normally within 5% of the destructive measured reference values).

Furthermore, the isotopic ratio for ^{242}Pu cannot be measured and has to be taken from destructive measurements, or deduced from isotopic correlations. The contribution of the uncertainty on the isotopic ratios in the calculation of the total Pu-mass lies between 1 and 5%.

In pratice, isotopic measurements in most cases will not be necessary since the operator usually has a good knowledge of the isotopic composition of the plutonium present in his waste items. This will enable him to establish isotope correlations for a reliable estimate of the total Pu from the measured isotope fractions (^{240}Pu eq, Pu$_{fiss}$).

6.2 Pu-mass determination

6.2.1 Gamma measurements

For low density waste without fission products, gamma counting represents a practical alternative to the neutron counting with comparable reliability and accuracy. Reliable gamma assays require:
- a segmented gamma scan;
- a matrix correction using an external gamma transmission measurement;
- a correction for self-attenuation based on the measured differential absorption of plutonium gamma rays at different energies.

In general the gamma measurements are less accurate than neutron measurements and only the 25 litre drum is measured within 10%. Pu amounts of less than a few hundreds of mg are best measured by neutron techniques.

6.2.2 Passive neutron measurements

The neutronic measurements on waste drums with known Pu-content, Pu-composition, matrix and positioning of the Pu-contaminant in the matrix lead to the following conclusions:
1) A good calibration is necessary, taking into account matrix and source position effects.
2) A correction for the matrix composition and density is necessary when different from the calibration item.
3) The source distribution can affect significantly the measured Pu-content, but should be corrected for when possible.
4) In order to be able to make the corrections for matrix and source distribution effects, additional measurements and/or a-priori knowledge are necessary.
5) Without corrections only small drums with a $R_{av}.\delta < 3$ g/cm^2 can be measured correctly (within 10%) for hydrogeneous waste matrices.

6) The minimum detectable limit for NDA techniques is a function of background, detector geometry and counting times. Under fully optimized count-rate conditions, mg levels of plutonium can be detected in low density waste streams.
7) The TCA method has an intrinsic correction for matrix and source position effects, due to the use of the measured triplets, leading for 220 litre and hydrogeneous matrix density until 0.1 g/cm^3 to reasonable results. The TCA technique has proven to be more powerful for measuring concrete drums.

6.2.3 Active neutron measurements

The advantage of active neutron interrogation is the direct determination of all fissile nuclides. The method also works in case of MAW waste containing curium as a main spontaneous neutron emitter. The disadvantage of the active systems is the necessary transport of interrogating neutrons into the drum thus increasing the influence of the matrix composition.

From the exercise the following conclusions can be drawn:
1) A good calibration is necessary taking into account matrix, source position and sample self-shielding effects.
2) A correction for the matrix composition and density is necessary when different from the calibration item.
3) A correction for the source distribution is necessary when different from the calibration item.
4) Sample self-shielding has to be taken into account for sample sizes above 100 mg Pu_{tot}. It leads to a systematic underestimation. However, the effect remains unknown and cannot be evaluated. Sample self-shielding is occurring for the 3 waste drums 6S-8S. The effect decreases the signal count-rate for 6S by a factor of 1.9 and for 7S and 8S by a factor of 1.5.
5) Inhomogeneous neutron poisons also lead to a systematic underestimation. Again its influence cannot be evaluated.
6) The presence of uranium cannot be discriminated. ^{235}U contributes to ^{239}Pu eq with 0.68 mU[5]. This leads to possible sources of error in the Pu-mass determination.
7) In order to be able to make the corrections for matrix and source distribution effects, additional measurements and/or a-priori knowledge are necessary. For the DDT measurement such additional measurements were performed.
8) The DDT method should not be applied for measuring plutonium in a concrete waste matrix because the sensitivity is rather low and the inner drum region (diameter $<$ 20 cm) is almost completely shielded.
9) The minimum detectable limit for the active interrogation methods is a function of background, count-rate, detector geometry and counting times. Under fully optimized conditions, mg levels of plutonium can be detected in low density waste systems by the DDT and some 10 mg by the Sb-Be method.

7. GENERAL RECOMMENDATIONS

1) The measurements of the waste drums with unknown Pu-content show that more work has to be done. The stated uncertainties are not always realistic. In addition to quality assurance procedures, a fully documented and complete uncertainty analysis should be done.
2) The overall conclusion is that large (> 100 litre) drums can be measured with confidence only under favourable conditions. If the

primary measurement on waste can be carried out on 25 litre containers then the possibility of large errors is very much reduced.

3) Future Waste Management should be documented such that sufficient information about Pu and matrix composition is available. Any new intercomparison exercise should be based on this situation, and include more recently developed techniques.

REFERENCES

1. DIERCKX, R. et al., European Interlaboratory Test Measurements on Alpha Contaminated Waste, EUR report (to be published, 1989).

MEASUREMENTS OF PLUTONIUM RESIDUES
FROM RECOVERY PROCESSES

S. -T. Hsue, D. G. Langner, V. L. Longmire,
H. O. Menlove, P. A. Russo, and J. K. Sprinkle, Jr.
Los Alamos National Laboratory, USA

Summary

Conventional methods of nondestructive assay (NDA) have accurately assayed the plutonium content of many forms of relatively pure and homogeneous bulk items. However, physical and chemical heterogeneities and the high and variable impurity levels of many categories of processing scrap bias the conventional NDA results. The materials also present a significant challenge to the assignment of reference values to process materials for purposes of evaluating the NDA methods.

A recent study using impure, heterogeneous, pyrochemical residues from americium molten salt extraction (MSE) has been aimed at evaluating NDA assay methods based on conventional gamma-ray and neutron measurement techniques and enhanced with analyses designed to address the problems of heterogeneities and impurities. The study included a significant effort to obtain reference values for the MSE spent salts used in the study. Two of the improved NDA techniques, suitable for in-line assay of plutonium in bulk, show promise for timely in-process assays for one of the most difficult pyrochemical residues generated as well as for other impure heterogeneous scrap categories.

1. INTRODUCTION

Plutonium in the bulk forms generated by scrap recovery operations is often chemically impure and physically and chemically heterogeneous. The pyrochemical residues, in particular, are lean and highly impure chloride-salt-based materials in which the plutonium can coexist in both metallic and salt forms, the americium content is typically high (from a few to tens of weight percents relative to plutonium), and the typical bulk residue consists of heterogeneous and nonrepresentative chunks of various sizes. This describes the americium molten salt extraction (MSE) spent salts, in particular. Minimizing the handling of such highly radioactive materials requires assigning accountability values to these residues without removing them from the process line; they can then be immediately routed either to the next stage of processing or to waste disposal.

For the MSE spent salts (and for other pyrochemical residues) plutonium accountability values assigned by difference (residue value equals feed value minus product value) are grossly in error because they are the small difference of two large numbers, each with its own uncertainty. Routine crushing, blending, and sampling of combined MSE spent salt items for accountability is prohibited by the extreme inhomogeneities and high contact radiation dose rates. If calorimetric measurements, interpreted by gamma-ray isotopics for the bulk items, are used to obtain plutonium assays, pre-equilibration procedures must be incorporated as a result of the extremely long (12- to 20-h) thermal equilibration times required for the chunky residues. Moreover, when the americium content exceeds several percent (by weight relative to plutonium), even the state of the art in gamma-ray isotopics for heterogeneous bulk materials may not provide the required accuracy for the specific power (watts per gram of plutonium) because most of the heat is provided by the americium. Furthermore, the practicality of in-line calorimetry has not been demonstrated to date.

Segmented gamma-ray scanning (SGS) and passive neutron coincidence counting (PNCC) are two methods of measuring bulk plutonium that have been used in glove-box lines. Recent advances in plutonium nondestructive assays (NDAs) based on SGS[1,2] and

PNCC[3,4] show promise for the bulk assay of plutonium in MSE spent salts and other pyro-chemical residues. A recent Los Alamos study[5] using MSE spent salts has evaluated these two new NDA approaches, in addition to the NDA of plutonium by calorimetry/gamma-ray isotopics. This study has two distinct phases: evaluating the NDA methods, and assigning reference values (for total plutonium and isotopic composition) to the 14 process residues (MSE spent salts) used in the study.

The first efforts to obtain reference values for the MSE spent salts involved completely dissolving the individual bulk items and sampling the dissolver for destructive analysis (DA).[5] This method was extremely difficult to control and was abandoned after it was applied to four of the residues used in the study. Subsequently, each remaining bulk residue was pulverized, sieved, blended, and sampled for DA. The metallic portion (remaining in the sieve) of each residue item was stoichiometrically oxidized and reintroduced into the pulverized residue salt phase for blending. Because this second approach to obtaining reference values sacrifices only a few grams (out of 1 to 3 kg) of the bulk mass for DA, the method leaves behind ho-mogenized reference materials with the bulk chemical composition of the process residues. It also provides an additional phase for the NDA evaluation in that each NDA method can be used again on the homogenized residues to independently evaluate the sensitivity of the method to the heterogeneities of the process materials.

A separate report on the details and results of the physical and chemical procedures used to characterize the residues is being prepared.[6] This report also includes the results of the evaluation of the NDA of the MSE spent salts by calorimetry/gamma-ray isotopics, and con-clusions on its use as a reference technique for evaluating or verifying other NDA methods applied to these residues. Table I lists the bulk masses and the reference values for total plu-tonium mass and isotopic composition for each of the 14 MSE spent salts used in the study. The following sections describe the evaluations of the new SGS and PNCC methods for the NDA of spent MSE salts.

2. SGS ASSAYS WITH LUMP CORRECTIONS

2.1 Description of Method

The traditional SGS procedure was developed in the early 1970s.[7] In this procedure, transmission-corrected gamma-ray assays are performed on a rotating bulk item, segment by segment, to minimize some of the effects of heterogeneity. For each segment, the measured transmission is used to derive a self-attenuation correction. The assumption is that the bulk attenuation can be characterized by a single linear attenuation coefficient, which, in general, is

Table I. Reference Values for Total Plutonium and Isotopic Composition of the Original (Uncrushed) MSE Spent Salts

Sample ID	*	Pu Weight Fractions from Chemistry					^{241}Am (ppm)	Total Pu (g)
		^{238}Pu	^{239}Pu	^{240}Pu	^{241}Pu	^{242}Pu		
XBLP120	D	0.0151	93.732	5.9297	0.2754	0.048	26 081	112
XBLP267	D	0.013	93.872	5.9097	0.1842	0.021	35 142	126
XBLPs300	LD	0.013	93.642	6.0497	0.2473	0.048	3 036	199
XBLP270	D	0.009	93.992	5.8297	0.1506	0.019	33 737	99
XBLP121		0.0103	93.987	5.7256	0.2403	0.037	36 168	155.4
XBLP278		0.0077	94.599	5.2541	0.115	0.024	35 413	90.4
XBLPs301	L	0.0129	93.647	6.0431	0.247	0.05	2 519	247
RFMSE1		0.0067	93.973	5.853	0.147	0.02	53 633	243.8
RFMSE2		0.0101	93.805	5.9536	0.2032	0.028	47 482	372.7
RFMSE3	L	0.0097	93.826	5.86	0.2633	0.04	7 103	55.6
RFMSE4		0.0093	93.889	5.8719	0.2052	0.025	45 486	408.7
RFMSE5	L	0.0094	93.832	5.8576	0.2698	0.032	3 351	141.2
ARF595		0.0097	93.688	5.9354	0.3359	0.031	43 918	263.6
ARF642		0.0075	94.438	5.418	0.1213	0.015	45 777	219.5

*L = Lower Am content; D = Characterized following complete dissolution of bulk item.

dominated by the matrix. This procedure can provide unbiased assays if the item contains SNM in powders or fine particle sizes.

Unfortunately, much of the scrap and waste generated in the facilities contains SNM in the form of lumps, and if the SGS procedure is applied to these items, the assay results will be biased, usually in the negative direction. To minimize the bias caused by the lumps, lump-corrected segmented gamma-ray scanning (LCSGS) was developed. Details of the technique can be found in Ref. 1.

2.2 Comparison

The MSE spent salts are ideal for studying the differences between traditional SGS and LCSGS analyses because these salts are known to contain metallic lumps of plutonium. Fourteen MSE spent salt items were measured with the SGS, and the spectra for each item were analyzed by both methods. The traditional SGS assay results are summarized in Table II.

Table II. Summary of Traditional SGS Measurements of MSE Salt Samples

Sample ID	SGS (129) (g)	Sig (129) (g)	SGS (203) (g)	Sig (203) (g)	SGS (345) (g)	Sig (345) (g)	SGS (414) (g)	Sig (414) (g)
XBLP120	78.7	0.9	90.8	0.9	91.8	0.5	90.7	0.2
XBLP267	76.7	0.7	104.5	1.0	116.3	0.5	116.7	0.2
XBLPS300	108.2	0.8	147.6	0.9	164.0	0.4	163.9	0.2
XBLP270	60.5	0.6	81.5	0.8	89.8	0.4	89.7	0.2
XBLP121	83.9	0.6	116.7	1.2	127.0	0.8	128.6	0.5
XBLP278	52.8	0.5	68.3	0.7	73.3	0.4	74.0	0.2
XBLPS301	92.7	0.9	134.2	1.2	179.2	0.6	186.1	0.3
RFMSE1	172.2	4.2	212.0	2.7	220.2	0.7	218.5	0.4
RFMSE2	149.4	14.6	241.5	0.6	322.6	0.8	326.3	0.7
RFMSE3	35.0	0.2	43.6	0.4	47.5	0.3	48.2	0.1
RFMSE4	156.6	7.7	262.1	5.0	357.2	0.9	358.6	0.7
RFMSE5	55.2	1.2	82.1	0.9	109.1	0.4	112.1	0.2
ARF595	176.7	14.3	227.8	7.5	248.1	0.6	244.2	0.3
ARF642	182.4	10.2	202.6	5.7	198.7	0.5	197.5	0.3

Notice that for some items there is a substantial increase in the SGS assay with increasing assay energy; this indicates that the items do contain SNM in lumps. The lump-corrected SGS assay results are summarized in Table III. The lump correction can be applied to the 414-keV assay, based on the difference between 129- and 414-keV assay results; these are listed under LCSGS(129) in the table. Similarly, the lump correction can be applied to the 414-keV assay based on the difference between 203- and 414-keV assays [listed under LCSGS(203)], as well as between 345- and 414-keV assays [listed under LCSGS(345)].

The lump correction can be applied in two ways: to the data, segment by segment, or to the entire item (the sum of all the segments). We found that the magnitude of the lump corrections for these MSE spent salts varies from segment to segment. The correction in this table has been made to the data from each segment individually. The comparison to the reference values is shown in Table IV and also in Fig. 1.

From this study, the following conclusion can be made. The traditional SGS assay is biased an average of 8.7% for these MSE spent salts. The LCSGS assay significantly reduces the bias from 8.7% to 4%. The remaining bias could arise from three sources. Some of these samples contain relatively large amounts of ^{237}Np, which emits a 415.76-keV

Table III. Lump-Corrected SGS Measurements of MSE Salt Samples

Sample ID	LCSGS (129) (g)	LCSGS (203) (g)	LCSGS (345) (g)	LCSGS Average (g)
XBLP120	93.4	94.0	91.5	92.97
XBLP267	124.5	123.5	119.4	122.47
XBLPS300	173.5	171.2	165.9	170.20
XBLP270	94.9	92.2	91.1	92.73
XBLP121	135.8	132.5	133.0	133.77
XBLP278	77.8	76.2	74.6	76.20
XBLPS301	206.6	214.0	209.2	209.93
RFMSE1	224.8	224.1	225.0	224.63
RFMSE2	353.5	358.3	342.8	351.53
RFMSE3	50.7	49.8	49.1	49.87
RFMSE4	394.4	399.2	373.5	389.03
RFMSE5	119.9	123.9	120.2	121.33
ARF595	252.5	250.5	251.5	251.50
ARF642	206.9	209.6	202.0	206.17

Table IV. Comparison of SGS Measurements with the Reference Values

Sample ID	Ref ^{239}Pu (g)	SGS (g)	LCSGS (g)	Lump Corr (%)	Ratio (SGS/Ref)	Ratio (LCSGS/Ref)
XBLP120	104.98	90.7	92.97	2.44	0.864	0.886
XBLP267	118.28	116.7	122.47	4.71	0.987	1.035
XBLPS300	186.35	163.9	170.20	3.70	0.880	0.913
XBLP270	93.05	89.7	92.20	2.71	0.964	0.991
XBLP121	146.03	128.6	133.77	3.86	0.881	0.916
XBLP278	85.51	74.0	76.20	2.89	0.865	0.891
XBLPS301	231.26	186.1	209.93	11.35	0.805	0.908
RFMSE1	229.13	218.5	224.63	2.73	0.954	0.980
RFMSE2	349.65	326.3	351.53	7.18	0.933	1.005
RFMSE3	52.13	48.2	49.87	3.35	0.925	0.957
RFMSE4	383.69	358.6	389.03	7.82	0.935	1.014
RFMSE5	132.50	112.1	121.33	7.61	0.846	0.916
ARF595	246.93	244.2	251.50	2.90	0.989	1.019
ARF642	207.29	197.5	206.17	4.21	0.953	0.995
				Average	0.913	0.959
				1σ	0.056	0.052

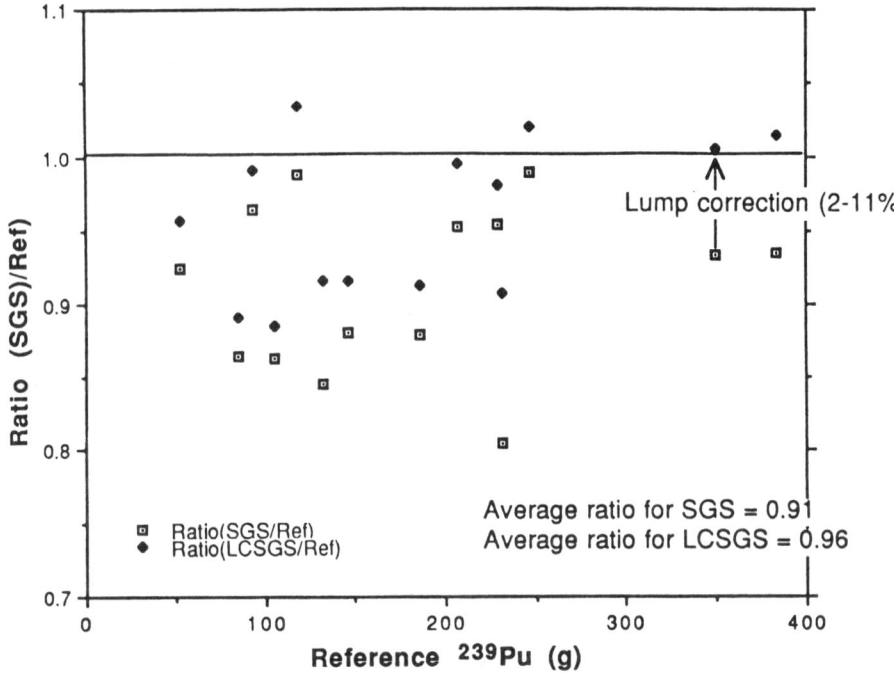

Fig. 1. SGS and LCSGS assays of the 14 MSE samples. The reference values are obtained by destructive analyses.

gamma ray through its ^{233}Pa daughter. This peak interferes with the 414-keV peak determination, and in the region-of-interest method of obtaining the peak area, the peak area is underestimated. Therefore the 414-keV assay could be biased low. A second possibility is a similar interference in the 414-keV background region caused by pulse pileup of the intense 208-keV gamma-ray photopeak. The third possible source of bias is that the LCSGS assumes that all the SNM lumps are the same size. If the SNM particle sizes are distributed, the correction may be biased, and the bias will depend on the particle size distribution.

3. PNCC MULTIPLICATION-EQUIVALENT-FISSILE (M-E-F) ASSAYS

3.1 Description of Method

The three unknown quantities in plutonium assays based on PNCC measurements are the mass of the plutonium, the neutron multiplication (M) of the unknown item, and the ratio (α) of α,n to spontaneous fission (SF) neutrons. For relatively pure materials, the value of α can be assigned based on the chemical composition so that the two measured quantities, the real coincidence (R), and total neutron (T) count rates can be used to solve both for M and for the effective ^{240}Pu mass, ^{240}Pu$_{eff}$, the spontaneously fissioning component of the plutonium.[8] The known isotopic composition of the plutonium-bearing item is used to obtain the total plutonium mass from the assay result. Because impurities in the residue categories are large and variable, values for α cannot be assigned to residue materials. An alternate approach must be taken to perform the assay when α is unknown. The assays described use PNCC data obtained with the HLNCC-II.[9]

In the point model of passive neutron coincidence counting as described by Stewart,[10] T and R are expressed as α- and M-dependent quantities. By substituting T into the equation for R to give

$$R = k_0(M^2)\,{}^{240}Pu_{eff} + k_1\,M(M - 1)\,T \ , \tag{1}$$

the quantity α becomes included in the measured T. The constants in Eq. (1) can be defined, using Stewart's notation,[10] as

$$k_0 = \frac{\varepsilon^2 f}{2}\,\overline{\nu(\nu - 1)}^{SF}\,n_{SF} \tag{2}$$

and

$$k_1 = \frac{\varepsilon f}{2}\,\frac{\overline{\nu(\nu - 1)}^{(I)}}{\overline{\nu}^{(I)} - 1} \ , \tag{3}$$

where ε is the neutron detection efficiency, f is the fraction of neutrons counted in the coincidence gate, n_{SF} is the spontaneous-fission decay rate per unit mass of ${}^{240}Pu_{eff}$, and the quantities $\overline{\nu}$ and $\overline{\nu(\nu - 1)}$ are the first and second moments (respectively) of the SF and induced (I) fission neutron multiplicity distributions. Note that for the HLNCC-II, k_0 is a well-established quantity ($18.14 \cdot s^{-1} \cdot g\ {}^{240}Pu_{eff}$) for assays of moisture-free bulk items.

The two unknowns in Eq. (1) can be reduced to one by describing the (unknown) M in terms of a multiplication-equivalent-(effective) fissile mass or M-E-F mass, ${}^{239}Pu_{eff}$, and a sample geometry factor, G:

$$M = \frac{k_2\,{}^{239}Pu_{eff}}{G} + 1 \ , \tag{4}$$

where k_2 is a constant and G is defined by the bulk shapes and dimensions of the unknown items. [For cylindrical packages of radius r and fill height h, G is equivalent to $r(r + h)$].[4] Similar to ${}^{240}Pu_{eff}$, the ${}^{239}Pu_{eff}$ is defined by the total plutonium mass and the isotopic composition of the plutonium-bearing item.[11] The ratio of the effective masses is γ such that

$${}^{240}Pu_{eff} = \gamma \cdot {}^{239}Pu_{eff} \ , \tag{5}$$

where γ is determined by isotopic composition alone, independent of mass. Substitution of Eqs. (4) and (5) into Eq. (1) gives an equation in the measured quantities, R and T, with only one unknown, ${}^{239}Pu_{eff}$. In the limit, as (M - 1) approaches zero, appropriate for residues, the equation is simplified for two general cases: (i) very large α and (ii) undefined α.

3.2 Assay of M-E-F Mass When α is Very Large (Simple Self-Interrogation Approximation)
Substituting Eqs. (4) and (5) into Eq. (1) and rearranging gives

$$\frac{R_{IF}}{T} = A\left[\frac{{}^{239}Pu_{eff}}{G}\right]^2 + B\left[\frac{{}^{239}Pu_{eff}}{G}\right] + C \ , \tag{6}$$

where

$$R_{IF} = R - \gamma\,k_0\,{}^{240}Pu_{eff} \ , \tag{7}$$

$$A = k_1 k_2^2 , \tag{8}$$

and

$$B = k_1 k_2 . \tag{9}$$

For residues such as the first extraction MSE spent salts, (M - 1) approaches zero, and the α values are very large because of the high (>1%) americium content. In this limit, the simple self-interrogation (SI) approximation is

$$C = \gamma k_0 (M^2 - 1)\, {}^{239}Pu_{eff} , \tag{10}$$

where C is shown[12] to be sufficiently small to be approximated by a constant in Eq. (6). The residues that qualify will have characteristically small values of R/T.

Calibration of the SI assay is performed with R and T data obtained from the high-americium crushed MSE spent salts by fitting a quadratic function to Eq. (6) to solve for A, B, and C. The results of this fit are shown in Fig. 2.

The assays of the original (uncrushed) MSE spent salts are based on a rearrangement of Eq. (6):

$$0 = a({}^{239}Pu_{eff})^2 + b({}^{239}Pu_{eff}) + c \tag{11}$$

where the coefficients,

$$a = \frac{AT}{G} , \tag{12}$$

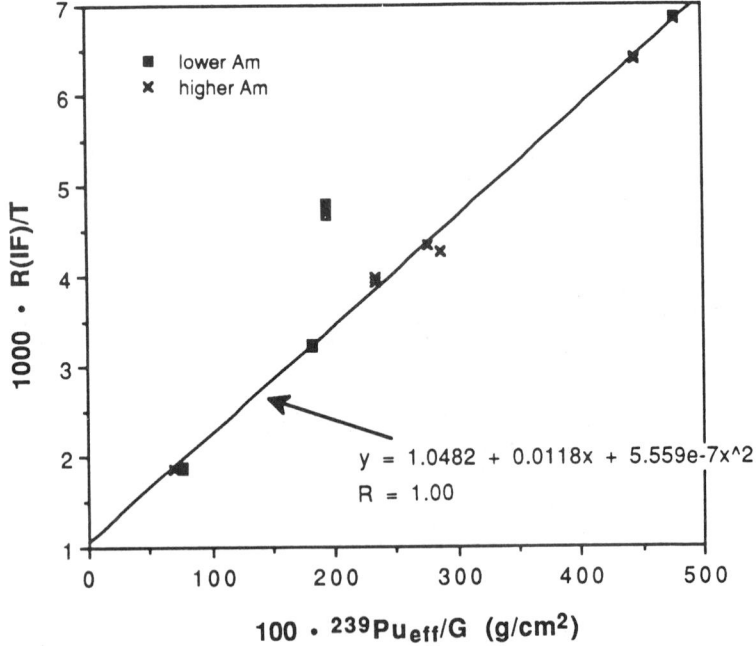

Fig. 2. *Quadratic fit to crushed salt measurement results for calibration of SI assay.*

$$b = \frac{BT}{G} + k_0\,\gamma \; , \tag{13}$$

and

$$c = CT - R \; , \tag{14}$$

are unique to each unknown item. The SI assay result is

$$^{239}\text{Pu}_{\text{eff}} = \frac{-b + \sqrt{b^2 - 4\,ac}}{2a} \; . \tag{15}$$

The values of R, T, G, and γ are given elsewhere[4,12] for each MSE spent salt measurement (including multiple measurements of given salts in different containers corresponding to different G values). The ratio of the SI assay result to the reference value for each measurement is given in Table V. The average ratio is biased high if the (four) lower-americium items are included, consistent with expectations for these items. The ratios are also plotted vs $^{239}\text{Pu}_{\text{eff}}$ and vs percent americium in Fig. 3.

3.3 More General Assay of M-E-F Mass, Including Lower-α Cases

Some MSE spent salts resulting from a second extraction of americium from impure metal, and other plutonium residues with large but less than 1% americium content do not satisfy the criteria for SI. In this case, the measured ratio, R/T, is larger for a given residue category. One result of a decrease in α is that ε in k_1 [Eq. (3)] is no longer constant because the (generally higher energy) fission neutrons become a significant fraction of the neutrons produced. This limitation can be more significant for the chloride-based salts for which average α,n neutron energies are ≤ 1 MeV compared to 2 MeV for fission neutrons. However, the americium content of the second extraction MSE spent salts is still relatively high ($\geq 0.3\%$) so that typical α values are ≥ 5 for these materials. In these cases, the neutron energy spectrum is still dominated by α,n neutrons. Therefore, Eq. (1) is again used, assuming k_1 to be constant, with an approximation based on the limit as M - 1 approaches zero. The approximation includes

$$M^2 \cong 2M - 1 \tag{16}$$

and

$$M(M - 1) \cong M - 1 \; . \tag{17}$$

Substituting Eqs. (4), (5), (16), and (17) into Eq. (1) gives the M-E-F mass relationship to R and T that includes lower-α cases:

$$\frac{R}{T} = A'\left(\frac{^{239}\text{Pu}_{\text{eff}}}{G}\right)\left(\frac{^{240}\text{Pu}_{\text{eff}}}{T}\right) + B'\left(\frac{^{240}\text{Pu}_{\text{eff}}}{T}\right) + C'\left(\frac{^{239}\text{Pu}_{\text{eff}}}{G}\right) \; , \tag{18}$$

where

$$A' = 2k_0\,k_2 \; , \tag{19}$$

$$B' = k_0 \tag{20}$$

and

$$C' = k_1\,k_2 \; . \tag{21}$$

Table V. Comparison of M-E-F Assay Results with Reference Values

Sample ID	Reference ^{239}Pu$_{eff}$ (g)	Can Type	^{239}Pu$_{eff}$/Reference from R(IF)/T vs ^{239}Pu/G (SI)	Can Average	^{239}Pu$_{eff}$/Reference from Small M Approximation (More General)	Can Average
XBLP120	110.5	#20	0.969	0.969	0.998	0.998
XBLP267	124.9	Tall	1.025		1.069	
		Tall	1.013	1.026	1.059	1.070
		Tall	1.040		1.083	
		#20	1.041	1.041	1.079	1.079
XBLPs300	193.6	Tall	1.047	1.047	0.977	0.977
		#20	1.083	1.083	1.021	1.021
XBLP270	98.1	Tall	1.068	1.068	1.080	1.080
		#20	1.116	1.116	1.126	1.126
XBLP121	153.9	Tall	1.010	1.002	1.040	1.032
		Tall	0.994		1.025	
		#20	1.068	1.077	1.088	1.097
		#20	1.087		1.106	
XBLP278	89.8	Tall	0.890	0.890	0.971	0.971
		#20	0.936		1.009	
		#20	0.959	0.949	1.030	1.021
		#20	0.951		1.023	
XBLPs301	240.3	Tall	1.095		1.030	
		Tall	1.099	1.108	1.034	1.043
		Tall	1.130		1.065	
		#20	1.147	1.153	1.108	1.115
		#20	1.159		1.121	
RFMSE1	244.3	P	0.929	0.933	0.936	0.940
		P	0.937		0.944	
RFMSE2	372.1	P	1.040	1.063	1.048	1.073
		P	1.086		1.098	
RFMSE3	54.2	P	1.003		0.937	
		P	1.021	1.018	0.953	0.950
		P	1.031		0.961	
RFMSE4	407.8	P	1.022	1.032	1.024	1.034
		P	1.042		1.045	
RFMSE5	137.4	P	1.084	1.019	1.047	1.022
		P	0.954		0.998	
ARF595	263.0	P	0.969	0.979	0.973	0.982
		P	0.988		0.991	
ARF642	218.8	P	0.926	0.935	0.937	0.946
		P	0.944		0.955	

14 higher Am items average (1σ) = 1.006 (0.065)   1.032 (0.058)

All 20 items average (1σ) = 1.025 (0.068)   1.029 (0.057)

Calibration for this more general approach that includes lower α values uses R and T data from all of the crushed MSE spent salts in a least squares fit to Eq. (18) to obtain A′, B′, and C′.

For assay of the original (uncrushed) MSE spent salts, Eq. (18) is rearranged to give

$$0 = a'\left(^{239}Pu_{eff}\right)^2 + b'\left(^{239}Pu_{eff}\right) + c' ,\qquad (22)$$

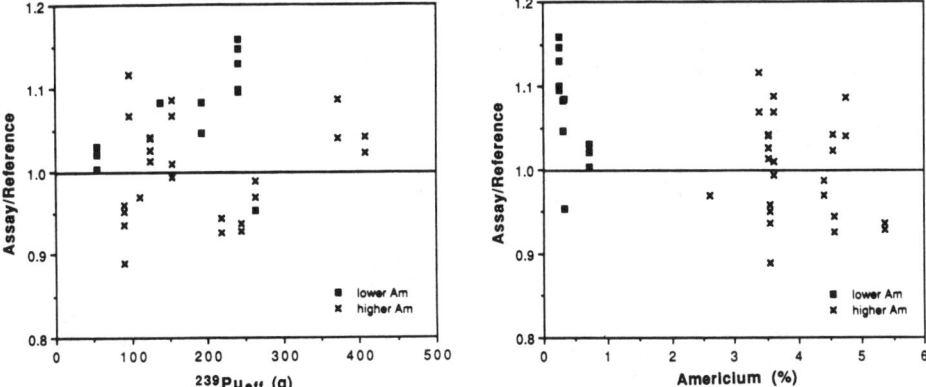

Fig. 3. The SI M-E-F assay results divided by the reference value vs $^{239}Pu_{eff}$ (left) and vs americium weight percent (right) for the original MSE spent salts.

where the coefficients,

$$a' = \frac{A' \gamma}{G} ,$$ (23)

$$b' = B' \gamma + \frac{C' T}{G} ,$$ (24)

and

$$c' = R$$ (25)

are unique to each unknown item. The assay result for this more general approach is

$$^{239}Pu_{eff} = \frac{-b' + \sqrt{b'^2 - 4 a' c'}}{2a'} .$$ (26)

The ratio of the assay result to the reference value for each assay is given in Table V. A 3% bias appears in the average result, but the relative bias for the lower-americium MSE spent salts has vanished. The ratios are also plotted vs $^{239}Pu_{eff}$ and vs percent americium in Fig. 4.

4. CONCLUSIONS

Both the LCSGS and the M-E-F PNCC assays show promise for timely in-process assays of pyrochemical residues such as the MSE spent salts. The study demonstrates that 5% (1σ) assay results are achieved with the LCSGS. Further investigations will determine the cause of a 4% negative bias in the assay. Evaluation of the LCSGS results for the crushed MSE spent salts should help distinguish between ^{237}Np or 208-keV pileup interference and the variability of lump sizes as postulated causes of the bias. Elimination of the bias may also improve the assay uncertainty.

The M-E-F PNCC results obtained in the simple SI approximation are 7% (1σ) assay results, which are unbiased for those residues for which the americium content exceeds 1 wt% (relative to plutonium). A more general approach to the M-E-F PNCC assays eliminates the

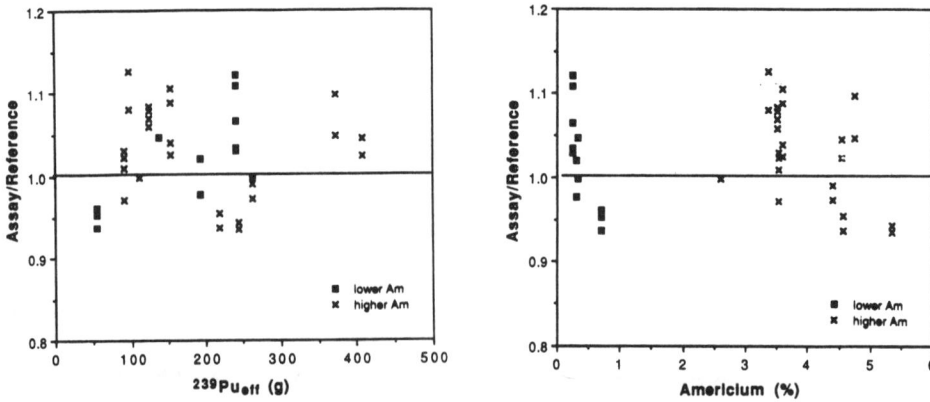

Fig. 4. The more general M-E-F assay results divided by the reference value vs $^{239}Pu_{eff}$ (left) and vs americium weight percent (right) for the original MSE spent salts.

positive bias that occurs in the SI results for the lower americium residues. The more general approach gives 6% (1σ) assays, but in this case the assays have a 3% positive bias. This bias may be caused by the energy-dependent neutron detection efficiency coupled with shifts in the neutron energy spectrum that result from variable α values and chemical differences in the matrix composition. Further investigations will include measurements designed to provide energy information that might be used to identify and correct these sources of bias. A longer term improvement is the use of a coincidence counter designed for minimum neutron energy dependence of the detection efficiency.

REFERENCES

1. J. K. Sprinkle, Jr., and S. -T. Hsue, "Recent Advances in Segmented Gamma Scanner Analysis," in *Proc. 3rd International Conference on Facility Operations--Safeguards Interface*, San Diego, California, November 29-December 4, 1987, ANS (1988), pp. 188-193.

2. J. K. Sprinkle, Jr., G. E. Bosler, S. -T. Hsue, M. P. Kellogg, M. C. Miller, S. M. Simmonds, and A. R. Smith, "Nondestructive Assay of Plutonium-Bearing Scrap and Waste," Los Alamos National Laboratory document LA-UR-89-2373 (1989).

3. H. O. Menlove, "Neutron Counting for Confirmation Measurements," in *Proc. 26th Annual Meeting, Institute of Nuclear Materials Management,* Albuquerque, New Mexico, July 21-25, 1985 (J. Inst. Nucl. Mater. Manage. **XIV**(3), Proc. Issue), pp. 589-594.

4. P. A. Russo, H. O. Menlove, K. W. Fife, M. H. West, and B . L. Miller, "Evaluation of the Neutron Self-Interrogation Approach in High-α,n Materials," in *Proc. 3rd International Conference on Facility Operations--Safeguards Interface*, San Diego, California, November 29-December 4, 1987, ANS (1988), pp. 176-187.

5. V. L. Longmire, W. E. Sedlacek, J. R. Hurd, and A. M. Scarborough, "A Comparison of Conventional and Prototype Nondestructive Measurements on Molten Salt Extraction Residues," in *Proc. 3rd International Conference on Facility Operations--Safeguards Interface*, San Diego, California, November 29-December 4, 1987, ANS (1988), pp. 313-315.

6. V. L. Longmire, et al., private communication (to be published).

7. E. R. Martin, D. F. Jones, and J. L. Parker, "Gamma-Ray Measurements with the Segmented Gamma Scan," Los Alamos Scientific Laboratory report LA-7059-M (December 1977).

8. N. Ensslin, "A Simple Self-Multiplication Correction for In-Plant Use," in *Proc. Seventh Annual ESARDA Symp. Safeguards Nucl. Mater. Manage.*, Liege, Belgium, May 21-23, 1985 (ESARDA 1985), pp. 233-238.

9. H. O. Menlove and J. Swansen, "A High-Performance Neutron Time-Correlation Counter," *Nucl. Technol.* **71**, p. 470 (November 1985).

10. J. E. Stewart, "A Hybrid Monte Carlo/Analytical Model of Neutron Coincidence Counting," Los Alamos National Laboratory document LA-UR-86-2290 (1989), presented at ANS 1986 Winter Meeting, Washington, DC (November 1989).

11. J. E. Stewart, H. O. Menlove, and N. Ensslin, "Definition of an Effective Plutonium Mass for Induced Fission," Los Alamos National Laboratory draft report (September 1987).

12. D. G. Langner, et al., private communication (to be published).

MEASUREMENTS OF URANIUM WASTE
USING A CALIFORNIUM SHUFFLER

J. K. Sprinkle, Jr., H. O. Menlove,
N. Ensslin, and T. W. Crane
Los Alamos National Laboratory, USA

Summary

We describe a passive/active neutron counter (PAN) based on a ^{252}Cf shuffler for 208-L drums. It is a flexible instrument that can be used to measure the nuclear material content of large containers. This instrument is installed in the Portsmouth Gaseous Diffusion Plant in Piketon, Ohio. This paper describes the results of a calibration for an iron matrix. For 0 to 100 g ^{235}U, the PAN meets our accuracy goal of 10% and our precision goal of 1% for 100 g ^{235}U. With its passive and active capability, this shuffler addresses future needs for materials control and accountability, and health, safety, and environment. The hardware portion of the counter is a good candidate for transfer to the commercial sector. We plan to focus our future waste assay efforts on developing more sophisticated analysis techniques for this generic hardware, rather than developing customized hardware for each application.

1. INTRODUCTION

Uranium contaminated waste is often packaged in large containers, such as 208-L drums, because process operators find drums more economical to handle than small containers. These drums often contain large quantities of matrix materials and small quantities of uranium. We measure the uranium content of these drums to satisfy nuclear material safeguards and accounting requirements and provide improved criticality safety. Waste does not require measurements of the same accuracy as those performed on the refined product, because waste contains nuclear material in a less desirable form. However, measurements of waste with uncertainties of 10-20% are more useful than measurements of waste with uncertainties of 50-100%, especially for accounting purposes.

Because the drums often contain 1 to 100 g of ^{235}U embedded in 10 to 100 kg of waste, the assay method must be extremely sensitive to the uranium but not sensitive to the matrix components. The passive radiation from the ^{235}U is not adequate for an assay because the neutron signal is weak and the gamma rays do not penetrate well. Therefore, we must use an active technique; preferably one that is less sensitive to the matrix. We give up some sensitivity to the ^{235}U to have a method that is less sensitive to the matrix effects.

Deciphering the signal from mixtures of special nuclear material (SNM) has been a problem, in part, because some facilities have traditionally mixed their waste streams without properly segregating the nuclear material and matrices. Gamma-ray-based measurements can easily distinguish between isotopes, but do not provide accurate measurements of scrap and waste in large containers, unless the matrix and SNM are uniformly distributed, the matrix consists of low-density and low-atomic-number constituents, and the SNM loading is small compared with the matrix. In general, if these conditions are not met, the gamma-ray assay underestimates the uranium content. A low bias in the uranium result is often not acceptable for diversion or criticality safety decisions. If the waste is of high density or contains any metal, active-neutron nondestructive assay techniques must be used. Epithermal or fast neutrons penetrate the contents of the drums better than low-energy ^{235}U passive gamma rays. However, neutron measurements are not without difficulties. Neutron poisons, (α,n) contaminants, multiplication, the inability to distinguish between isotopes, and moderation can significantly bias the measurements. Neutron measurements usually overestimate the amount of uranium, unless one corrects for these effects. Segregating the waste according to its neutron transport properties allows the operator to apply unique calibration standards to each waste category. When these standards have been applied, the ^{252}Cf shuffler has performed well and has the best potential to assay uranium in large, dense containers.[1] This potential has been confirmed by Monte Carlo calculations.[2]

2. DESCRIPTION OF THE PASSIVE/ACTIVE NEUTRON COUNTER

This paper describes a passive/active neutron (PAN) counter based on a californium shuffler that has been delivered to the Portsmouth Gaseous Diffusion Plant in Piketon, Ohio. A californium shuffler has four major components: the storage module for the californium source, the sample chamber and detectors, the control electronics (including the drive system for the californium source), and the software. Figure 1 is a conceptual drawing of the instrument. The shuffler is based on the principle that about 1% of the neutrons released following fission of ^{235}U come from the fission products with half-lives ranging from a fraction of a second to a few minutes (delayed neutrons). We interrogate the sample for several cycles; in each, the californium is moved close to the sample for 10 s, it is withdrawn, and the delayed neutrons are counted for 7 s. The source storage module shields both the instrument operator and the sample chamber from the californium source neutrons.

We designed this shuffler to have a detection efficiency for fission neutrons of about 20%, large enough to make the counter suitable for passive coincidence counting. We also included a shift-register circuit for passive neutron coincidence counting. To better correct for matrix effects, we use neutron-flux monitors. To correct for isotopic variations, we can use two modes of interrogation. Either thermal or epicadmium neutrons can be used in the active interrogation. The epicadmium mode is achieved with a steel reflector behind the ^{252}Cf source and a cadmium liner completely covering the inner surface of the counting chamber. In the thermal mode we add a 2.5-cm-thick polyethylene liner between the drum and the cadmium liner. Dual-energy interrogation allows us to correct for the ^{238}U effects in samples of unknown uranium isotopic distribution.

The Portsmouth shuffler is improved over previous 208-L drum shufflers in several ways. One of our objectives has been to make the new shuffler so versatile that private companies would find it an attractive instrument to market. We added a dual interrogation mode to determine the ratio of ^{235}U to ^{238}U when the uranium enrichment is unknown. The inside of the sample cavity has been completely lined with cadmium to enhance the differences between the two interrogation modes. A better reflector design and higher detection efficiency each contributed 20 to 30% to reducing the source size without decreasing the sensitivity of the measurement. The instrument has more ^3He detector tubes than earlier designs and faster AMPTEK-based electronics. The polyethylene shielding has been simplified. The source shield is now an integral unit that can be lifted into place with a forklift. The innovative use of borated materials in the source shield eliminated the need for a 2.5-cm-thick lead outer cover on the shield. Figure 2 is a cross section of the (4 ft on a side) source storage cube. The tungsten core absorbs 10% of the neutron flux. More neutrons are captured near the center; consequently this design has more boron there to reduce the intensity of the hydrogen capture gamma rays.

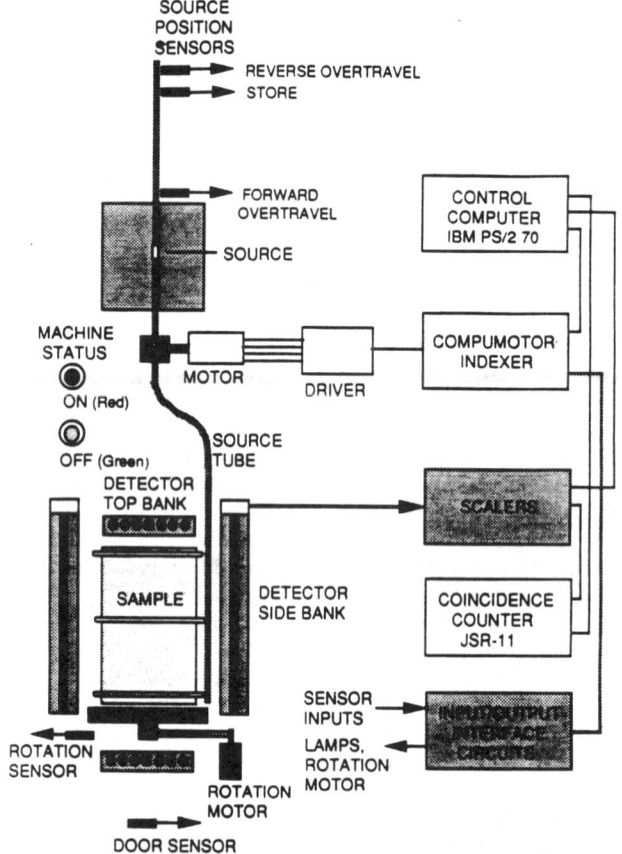

Fig. 1. The PAN controls diagram. Three major components of the PAN are shown; the sample chamber, the californium source storage module, and the electronics/controls
(including those for the californium source transfer). The
fourth component is the software.

Improvements in the electronics and software make the user interface more pleasant for
the operator. The modular software package uses the more versatile and portable C language.
This software package will have built-in hardware diagnostic capabilities and automatically
archived parameters, results, and data; and will perform more sophisticated data analyses.
The computer now performs more of the functions previously left for the operator.

3. RESULTS AND PERFORMANCE

First we measured the passive spatial response of the PAN to a fission source, californium, and a lower-energy neutron source, AmLi. Then we mapped the active response of
the shuffler as we moved a small quantity of uranium in the sample chamber. Because of the
large sample size, we found it necessary to move the californium source alongside the drum
during the 10-s irradiation to get a similar response from a small amount of ^{235}U placed at the

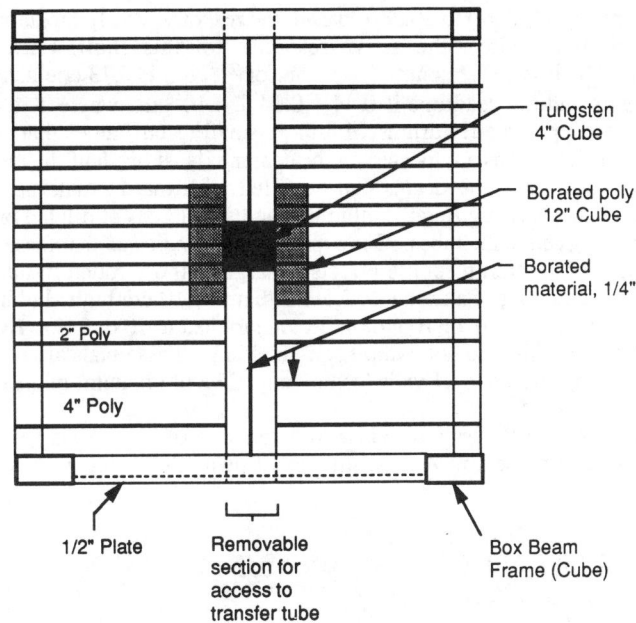

Fig. 2. The ²⁵²Cf source storage module. This integral unit is
assembled once and moved as a unit if necessary. It is a 4-ft
cube (1.22 m on a side) that reduces the radiation from a 300-
µg californium source to less than 5 mR/h when in contact with
the shield's outer surface.

top, center, or bottom of the drum. The scan is 33 in. long, with a pause at the top and bottom. In 10 s, three up-and-down scans are performed with a total pause of 4 s. The response to all samples was normalized to 1.000 at the center. Table I lists the detection efficiencies and summarizes the response for the two passive surveys and the active survey. The standard deviation of the response indicates that the chamber's efficiency is relatively flat over the sample volume. The minimum response is typically at the top of the drum.

TABLE I. Spatial Response Summary for Passive/Active Maps of the Portsmouth Shuffler

Neutron Source	Detection Efficiency (%)	Flatness Response				
		Center	Average	Std. Dev.	Maximum	Minimum
Californium	17.4	1.00	0.98	0.06	1.08	0.90
AmLi	19.5	1.00	0.92	0.07	1.04	0.81
Uranium	------	1.00	1.01	0.04	1.06	0.95

For an iron matrix (130 kg) in a 208-L barrel, the response was flat to about 20%. Using a 300-μg californium source, the active response for this shuffler for ^{235}U is 1.46 counts/(s·g), for ^{238}U it is 0.062 counts/(s·g) and for ^{239}Pu it is 0.73 counts/(s·g). The die-away time is 70 μs and the deadtime is $0.44 + 0.14$ T x 10^{-6} μs, where T is the total count rate. We have developed matrix drums with three small (3-cm) tubes that run the vertical length of the drum. The tubes are located on the center axis, at one half the radius, and at the outer radius of the drum. The drums can be filled with matrix materials such as iron, graphite, polyethylene, Raschig rings, or alumina, and small vials of diluted or pure uranium compounds can be placed in the tubes. Our experience has indicated that these matrix materials exhibit a range of neutron response effects as large as those found in nuclear facilities. The small vials contain 40 g of alumina in 2.5-cm-diam by 10-cm-high aluminum cylinders. Each vial contains 0.1, 1.0, 5.0, or 10 g of uranium enriched to 10 or 94%. Even 5 g of 94% uranium is sufficiently dilute to not exhibit self-shielding. These vials are placed at various positions in the drum, with up to 21 vials (containing 105 g of uranium) in the chamber at one time.

Figure 3 shows the shuffler's response as a function of ^{235}U mass for a matrix drum containing 130 kg of iron scrap. A least squares fit using the Deming approach[3] gives the following fit:

$$\text{Response} = -2.2567 \times 10^{-4}(m_{235})^2 + 1.4586\, m_{235} + 6.2472 \times 10^{-2}\, m_{238} . \qquad (1)$$

Note that the nonlinear term in Eq. (1) is 4 orders of magnitude smaller than the linear term. This is demonstrated more clearly in Fig. 4, where the response per gram is plotted as a function of ^{235}U mass. The scatter in the data for Fig. 4 is consistent with the counting statistics.

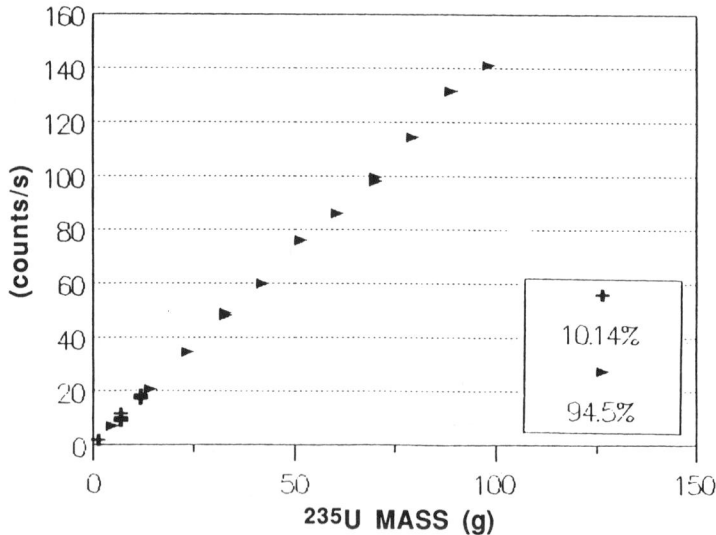

Fig. 3. *The shuffler response as a function of ^{235}U mass. Small vials of 10- and 94%-enriched uranium were placed in a 208-L drum filled with 130 kg of iron scrap. The multiple data points at 6, 12, 21, and 65 g ^{235}U illustrate the effects of measuring the same quantity of uranium at different positions in the drum.*

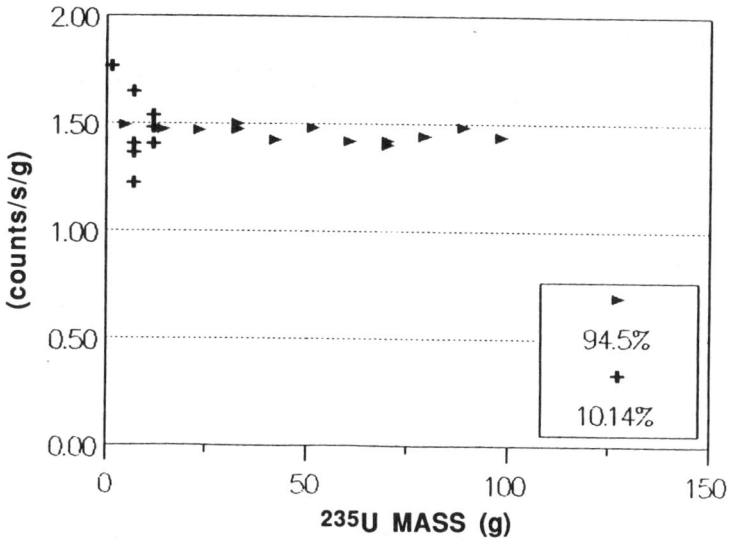

Fig. 4. The shuffler response per gram as a function of ^{235}U mass. Small vials of 10- and 94%-enriched uranium were placed in a 208-L drum filled with 130 kg of iron scrap. The multiple data points at 6, 12, 32, and 65 g ^{235}U illustrate the effects of measuring the same quantity of uranium at different positions in the drum. The scatter in the points is consistent with counting statistics.

Figure 5 compares the shuffler response as a function of mass for three sample types. The small sample response curve is independent of enrichment (0.2-93%) and the matrix (graphite, low-enriched oxide, alumina). The small sample response is similar to the lead matrix response discussed above. The alumina matrix has a much higher response per gram of ^{235}U. We plan to use the flux monitor response to correct for this effect. This barrel contains 400 lb of Al_2O_3 with a nominal 3 wt% water. The scatter in the alumina data is caused by spatial response variations. The response at the bottom and edge of the drum is much lower than the response over the interior of the drum.

Table II summarizes some preliminary results that show the effects of various matrices on the two flux monitors. A 5-g 94% vial was placed in different locations in the matrix drum to measure the active response. Bare Flux Monitor and Cd Flux Monitor indicate the responses from the bare and cadmium-covered flux monitors, respectively. The number of points indicates how many different positions in the drum were sampled. At a minimum, the center and the upper and lower outside edges were always sampled. The average response column lists the average and standard deviation of the responses from different positions. The variation in the average response standard deviation indicates that it is not easy to achieve uniform detection capability over the entire drum volume. Altering the californium source scan parameters had no substantive effect on the response characteristics for the alumina matrix. It was encouraging to see the flux monitor signals track the average response.

4. CONCLUSIONS

Our initial investigations clearly showed the major source of error when measuring large waste containers is the influence of the samples rather than the sample chamber on the neutron

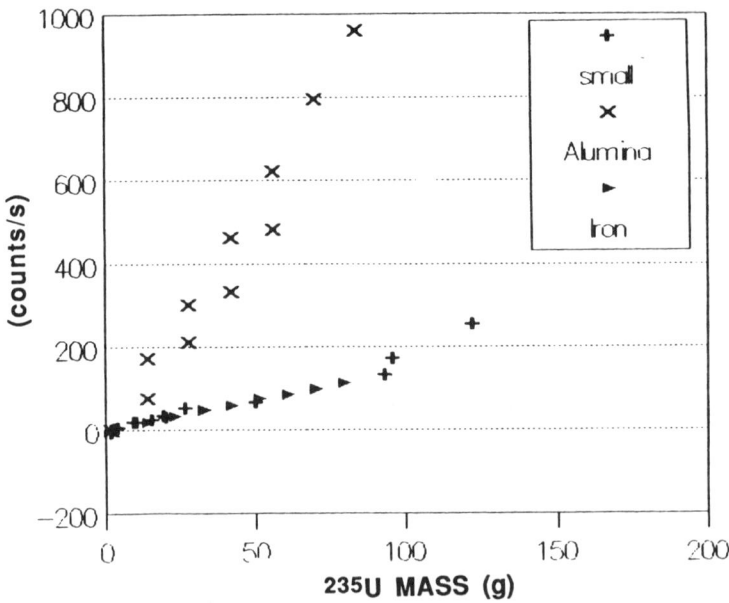

Fig. 5. The shuffler response as a function of ^{235}U mass for three sample types. The response from assorted small samples of many enrichments and matrices is consistent with the response from the iron matrix also shown in Figs. 3 and 4. The alumina (Al_2O_3) matrix demonstrated a higher response per unit mass and much greater spatial sensitivity.

TABLE II. Matrix Effects vs Flux Monitor

Matrix Material	Average Response	Number of Points	Bare Flux Monitor	Cd Flux Monitor
None	1.01 ± 0.04	6	1.00	1.00
Iron	1.12 ± 0.17	3	1.16	1.12
Alumina	6.70 ± 2.5	21	1.67	1.41
Graphite	3.68 ± 1.44	6	1.44	1.40
12-lb poly	2.26 ± 0.65	6	1.24	1.14
Poly chunks	4.06 ± 1.59	6	2.02	1.08

transport characteristics. We are transferring the shuffler technology to the commercial sector. We have developed active analysis procedures for two sample types that yield measurement accuracies of 10%. We plan to focus our future waste assay efforts (on large containers) on developing more sophisticated analysis techniques based on this generic hardware, instead of developing customized hardware for each application. With its passive or active capability, this instrument can measure SNM in a variety of matrices and containers.

ACKNOWLEDGMENTS

The authors are indebted, for the use of the californium source, to the US DOE Californium Industrial Loan Program administered by the Office of Nuclear Materials Production through the facilities of Oak Ridge National Laboratory.

The authors are also indebted to Paul Henriksen for editing this manuscript, and to the Safeguards Assay Group personnel at Los Alamos National Laboratory for fabricating, assembling, and testing the PAN at Los Alamos and for installing it at the Portsmouth Gaseous Diffusion Plant in November 1989.

REFERENCES

1. D. B. Smith, comp., "Safeguards and Security Progress Report, January-December 1985," Los Alamos National Laboratory report LA-10787-PR (March 1987).

2. J. F. Briesmeister, ed., "MCNP--A General Monte Carlo Code for Neutron and Photon Transport, Version 3A," Los Alamos National Laboratory report LA-7396-M, Rev. 2 (September 1986).

3. P. M. Rinard and A. Goldman, "A Curve-Fitting Package for Personal Computers," Los Alamos National Laboratory report LA-11082-MS (November 1987).

ELABORATION AND CHARACTERISATION OF PLUTONIUM WASTE REFERENCE MATERIALS

Commission d'Etablissement des Méthodes d'Analyses - CETAMA
COMMISSARIAT A L'ENERGIE ATOMIQUE - FRANCE
Centre d'Etudes Nucléaires de FONTENAY AUX ROSES
BP N° 6
92265 - FONTENAY AUX ROSES CEDEX

Presented by J.P. PEROLAT

Summary.

The Analysis Methods Establishment Commission (CETAMA) has set up a program for the elaboration and characterisation of plutonium waste reference materials.

The object of this program is to give laboratories the possibility to test and calibrate apparatus used in non-destructive methods for the analysis of plutonium waste.

The different parameters of this program are presented :

- characterisation of plutonium,
- type and number of containers,
- plutonium distribution inside the different containers,
- description of the matrix.

1.   INTRODUCTION.

The fuel cycle generates waste which has to be measured, not only upstream for nuclear material balance purposes but also downstream to ensure that the nuclear waste packages delivered to the permanent disposal and interim storage facilities contain quantities of radioactive materials which do not exceed the standards defined by currently enforced regulations (0.1 Ci/metric ton for alpha-emitter permanent storage).

These quantities have to be controlled by the waste producers, who are responsible for setting up quality assurance systems, liable to auditing by the national competent authority for waste management.

For the products discussed in this paper, i.e. low activity waste containing plutonium, control of the finished product (generally 100 or 200 l drums) is mandatory. With this type of package, significant errors may occur owing to effects related to the matrix, blend non-homogeneity and geometrical features.

Let us call to mind certain of the risks run by the waste producer:

- delivery of packages where the regularory Pu contents were exceeded,

- delivery of too large a number of packages for the above risk to be prevented. This could have been avoided by judicious management,

- fabrication of drums difficult to monitor or not in compliance with specified requirements, resulting in "in situ" interim storage pending the requisite unpackaging and repackaging, which is both very time consuming and penalizing in terms of productivity.

In the circumstances, on both deontological and good management grounds, the waste producer is well advised to have assays and measurements performed upstream, on smaller volumes of waste in 15 ℓ bags, where measurements can be more accurate than on the drums [1]. The bags are then packaged in the drums prior to disposal.

Reducing measurement uncertainty obviously implies research and development on measurement methods and systems, notably as regards more accurate determination of the effects of the waste matrix.

The CETAMA contibution in this field would be focused on optimization of measurement quality assurance on the waste producer side, on a national or even international scale.

For this purpose, we recommend :

- that the producers be provided with referenes for measurement system calibration. These references should be accepted as national standards,

- that measurement quality control be entrusted to systems of the "EQRAIN type", as is the case in other fields.

EQRAIN is the acronym for Evaluation de la Qualité du Résultat d'Analyse dans l'Industrie Nucléaire.

In a preliminary fact-finding stage, the CETAMA organized two intercomparisons :

- the first, in 1985, concerned four 15 ℓ bags of waste, each containing 700 mg of plutonium,

- the second, in 1986/87, concerned six 100 ℓ drums, containing 10 to 70 mg of plutonium.

The conclusions of the two intercomparisons may be summarized as follows :

- variable but satisfactory results for the best equipped laboratories,

- for the bags, the best results differed by less than 5% from the reference value,

- drum results more divergent, owing to the volumes and the smaller quantities of plutonium,

- systematic errors difficult to assess, owing to lack of knowledge regarding the distribution in the matrix of the waste containing plutonium.

In the light of the results obtained and the requirements voiced by the CETAMA "waste measurement" working group, it was decided to elaborate plutonium containing waste reference materials.

2. DEFINITION OF THE REFERENCE MATERIALS.

Elaboration of reference materials for the NDA is a very time consuming and expensive process. For this reason, we restricted our activity, to begin with, to 100 liter drums, which are easier to transport than 200 ℓ drums, and 15 ℓ bags, representative of the products with which the drums are filled.

The following characteristics were selected as regards isotopic composition and mass :

100 ℓ drums :

- 4 drums containing PuO_2 (PWR-type), where the approximate Pu contents were respectively 20, 50, 500 and 1000 mg.

- 5 drums containing PuO_2 (7% Pu 240 content), where the approximate Pu contents were 50, 100, 200, 300, 400, 1000 mg.

15 ℓ bags :

- 3 bags containing PuO_2 (PWR-type), where the approximate Pu contents were 10, 20 and 50 mg.

- 7 bags containing PuO_2 (7% of Pu 240), where the approximate Pu contents were 10, 50, 100, 200, 300, 400 and 1000 mg.

Each container is filled with a light waste matrix, which will be describeb later.

The material described here is obviously not a primary reference material. The methodology adopted for its elaboration and the expected applications fall between those corresponding to secondary reference materials and working reference materials.

3. ELABORATION PROCESS.

Elaboration of the reference material, entrusted to the MARCOULE/ COGEMA laboratory department, comprised several stages : characterisation of the plutonium, preparation of plutonion sources, source positioning and packaging, construction of the waste matrix.

3.1 Plutonium characteristics.

In order to elaborate references where traceability would be irrefutable and to facilitate preparation from the material standpoint, it was decided to use PuO_2.

Three batches of this oxide were used, two of which were PWR derived, with very similar isotopic compositions.

For all the batches, the oxide was previously calcined at 950° C for 4 hours.

Plutonium content :

After sampling and dissolution, the plutonium content was determined by titrigravimetric analysis [2], a well proven method, where result accuracy is 0.1%, or even better, on a 95% confidence level.

Isotopic composition :

The isotopic composition is determined by thermo-ionization mass spectrometry. Relative uncertainty obtained on the Pu 240 is 0.1% for PWR type plutonium.

The americium content is also determined by mass spectrometry.
Results of characterisation of three batches of PuO_2 are given below :

Pu	% in mass on 08.07.87	Pu	% in mass on 08.07.87	Pu	% in mass on 08.07.87
238	0.24 ± 0.01	238	0.22 ± 0.01	238	< 0.03
239	67.71 ± 0.04	239	67.62 ± 0.01	239	92.60 ± 0.01
240	25.74 ± 0.02	240	25.92 ± 0.02	240	6.59 ± 0.02
241	5.12 ± 0.01	241	5.04 ± 0.01	241	0.70 ± 0.01
242	1.19 ± 0.01	242	1.16 ∓ 0.01	242	0.11 ± 0.01

Pu content :
(88.01 ± 0.09)%
Am 241 content :
(250 ± 12.5) ppm/Pu
on 08.07.87

Pu content :
(87.96 ∓ 0.09)%
Am 241 content :
(270 ± 13) ppm/Pu
on 24.06.88

Pu content :
88%
Am 241 content :
(541 ± 27) ppm/Pu
on 19.05.87

3.2 Source preparation and packaging.

The oxide is placed in a small 2 cm³ plastic bottle, with a stopper.

The oxide mass is determined by comparing the difference in the weight of the bottle before and after insertion of the PuO_2. A SARTORIUS 1712 MP6 balance is used for the weighing. The PuO_2 mass is determined with an accuracy of 0.1 mg.

The bottle is then filled with epoxy resin. After complete solidification of the resin, this first bottle is placed inside a larger bottle (h : 37 mm, i.d. : 20 mm, vol : 10 cm³), fitted at the bottom with a 3 mm diameter PVC spacer to facilitate complete encapsulation of the first bottle by the same epoxy resin (cf figure 1).

3.3 Source positioning.

Each drum comprises five holders for five sources. These holders consist of vertical PVC tubes, 27 mm in i.d., numbered S1 to S5, closed at the top by screw-on PVC stoppers.

The sources are positioned inside the tubes by means of PVC spacers, 24 mm in diameter .

The first tube is centered on the drum axis and the other four are positioned radially 14.5 cm from the center tube.

The source layout is as follows :
1   Central source,   317 mm from the bottom.
2   Lower sources,   177 mm from the bottom.
3   Upper sources,   395 mm from the bottom.

The drum dimensions are dia : 460 mm
                              h : 670 mm.

The bags are designed on similar principles, but only contain 3 sources.

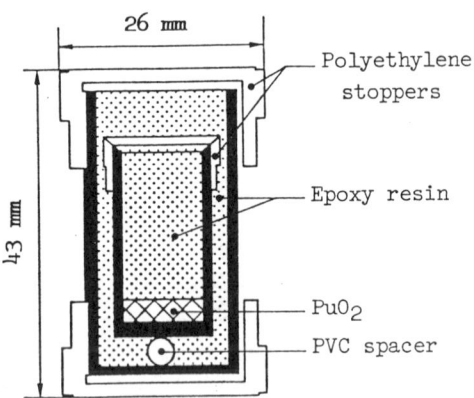

Figure 1 : Source packaging

3.4 Composition of the waste matrix.

The empty drums weigh 10.9 kg. When filled, their total weight is 23.9 kg.

The waste matrix comprises the following materials :
- paper : 10 %
- rubber gloves : 20%
- plexiglass, polyethylene : 5%
- PVC (glove box filling sleeves) : 60%
- cotton : 5%.

The products are shredded so that they can be tightly packed in the drum with a view to obtaining a homogeneous, stable mixture.

4. CERTIFICATION.

Certification of each reference material will cover the following aspects :

* the mass of each source and   !
  whole mass of plutonium   !
* the isotopic composition   !values to be certified
* the Am 241 content   !

The matrix composition and the position of each source are indicated.

5. CONCLUSIONS.

Laboratories concerned will be able to request that the reference materials produced be made available to them, on a temporary basis, for calibration of their installations.

After circulation of the materials to the various laboratories, results will be compared as a basis for decisions as to appropriate follow-up. Plans have already been made to organize an "EQRAIN" [3] set (CETAMA acronym for systems for analytical result quality evaluation). Arrangements are also made whereby certified references of this type of similar configuration would be made available to users requesting them.

Finally, the elaboration of equivalent reference materials with "heavy" matrices is envisaged at a later stage.

REFERENCES.

[1] R. DIERCKX and al, 11th ESARDA Symposium on Safeguards and Nuclear Material Management. LUXEMBOURG, 30 May/01 June 1989, 195-199.

[2] J.P. MULLER, J. BLACHERE, J. RIVIERE, J.P. JOSIEN
     Analusis 1980, V.8, N°. 8, 362 - 366.

[3] C. HOUIN, 9th ESARDA Symposium on Safeguards and Nuclear Material Management. London U.K., May 1987, 381 - 384.

S E S S I O N   B 2   –   C O R R E L A T I O N

Chairman : R. Estep – Los Alamos National Laboratory

The principle behind correlation measurements is that the activities of fissile, alpha-emitting, or other controlled radionuclides in waste can be measured indirectly by measuring the activities of "key" radionuclides that occur in combination with the controlled nuclides. This allows, for example, the determination of Pu isotopes using ^{137}Cs gamma rays in circumstances where the Pu gamma rays themselves are masked or are too weak to be seen. Thus, the approach is well-suited for characterizing low-level waste, provided that appropriate key radionuclides can be identified and the scaling factors are known.

The three papers presented here on correlation range in subject from the computer modelling study of M. Adler and C. Degueldre (Switzerland) is a feasibility study to dertermine the utility of using correlation x-ray spectroscopy to characterize a variety of low-level waste forms. This was accomplished using computer simulations of expected gamma spectra to estimate the observation limits for a suite of key radionuclides.

The paper presented by C.W. Fern et al. (UK, presented by C.R. Harvey) details the gamma ray spectroscopy system now being used by the Central Electricity Generating Board to characterize their low level waste. This system, which is currently being expanded, relies on gamma-ray "fingerprints" to first identify the contamination type (and thus the correct scaling factors to use) and then to estimate the amount of contamination.

The paper by J.F. Montigon (France, presented by A. Saas) summarizes their experiences in dealing with a variety of waste forms using correlation techniques.

Basicly the correlation of measurable nuclides, preferably by non-destructive methods, allows to estimate the content of critical nuclides for storage or disposal. Emphasis should be put on measurement of primary wastes originated from reproducible nuclear waste streams to get better and exacter correlations between the nuclides.

CONTROL OF RADIOACTIVE WASTE PACKAGES

FOR DISPOSAL USING ACTIVITY SCALING FACTORS

J. C. ALDER

National Cooperative for the Storage of Radioactive
Waste (NAGRA). Baden, Switzerland

and

C. DEGUELDRE

Paul Scherrer Institute (PSI). Würenlingen, Switzerland

Summary

The limiting concentrations of key radionuclides (KN), named Co-60, Cs-137 and Ce-144, which can be detected by measurements with a Ge-Li detector on a 200 l-drum homogeneously conditioned Swiss NPP wastes (resins and concentrates) have been calculated with a gamma transport code. They have been multiplied by the highest value of published scaling factors to obtain conservative detection limits of other radionuclides which are relevant for safe disposal of these waste packages (safety relevant nuclides - SRN). The purpose of the work was to assess whether inventories of SRN can be controled using the correlation technique for conditioned and packaged wastes assuming that the assumed correlations can be later validated. One conclusion is that Ce-144 would be difficult to use as a KN. Control of nuclides based on Co-60 and Cs-137 measurement would be relatively straightforward for the existing disposal limits. Cs-137 scaling factors for actinide isotopes should be developed to replace the more numerous Ce-144 ones for such a control.

1. INTRODUCTION

Before disposal of NPP operational waste, assessment or control of radionuclide contents may be required. Nagra studies provide tentative maximum allowable specific activities of nuclides in the waste packages which are derived from safety analyses of projected repositories. However, many of these nuclides are impossible to measure by direct counting techniques. Their accurate determination would require destructive measurements of the waste packages and/or complex, expensive analyses. Correlation between measureable gamma-emitting nuclides, called "key nuclides" (KN), and these disposal "safety-relevant radionuclides" (SRN) have been derived in various published studies and could be used under certain conditions for assessment or control.

A preliminary study was performed to assess the applicability of this method to the bulk of Swiss operational wastes after they have been conditioned and packaged. The question adressed was the following. Assuming the validity of the scaling factors for these wastes can be proven in the future, considering the uncertainties in published scaling factors, taking into account the waste package configuration (i.e. geometry and materials of drum, binding agent, raw waste and internal shielding if any) and its activity content, is it possible by measuring the key nuclides to assess or control the safety-relevant nuclides with respect to the maximum allowable specific nuclide activities for disposal? In this study wastes homogeneously distributed in the matrix were investigated. Only after a positive answer to this question, it is justifiable to proceed to the determination and validation of the scaling factors and, further, to experimental trials if the need for such detailed controls is accepted.

2. STUDY METHOD

The first step is to determine the detection limits DL(KN) of the key nuclides KN (KN = Co-60, Cs-137, Ce-144) in the waste packages when using a typical GeLi gamma detector. These detection limits are expressed as specific activities. In the second step the deduced detection limit DL(SRN) of the safety-relevant nuclides SRN is obtained by multiplying the detection limit of the key nuclides DL(KN) by the scaling factor SF(SRN, KN), where the scaling factor is defined as being the ratio of the specific activity SA of the safety-relevant nuclide by that one of the key nuclide. In this feasibility study the highest published values of the scaling factors for the considered wastes are used to obtain conservative detection limits of the safety-relevant nuclides.

In the last step a comparison is made between magnitudes of the derived detection limits of the safety-relevant nuclides in the different waste packages and the requirements which must be met to ensure that disposal limits are satisfied.

3. WASTE PACKAGES CONSIDERED

Typical 200 l and 100 l-drum packages for various NPP operational wastes with or without internal shielding and with cement, bitumen or plastic binding agents were selected. The different package geometries are given in Fig. 1 and the wastes and binding agents considered are given in Table I.

Type a   Type b   Type c

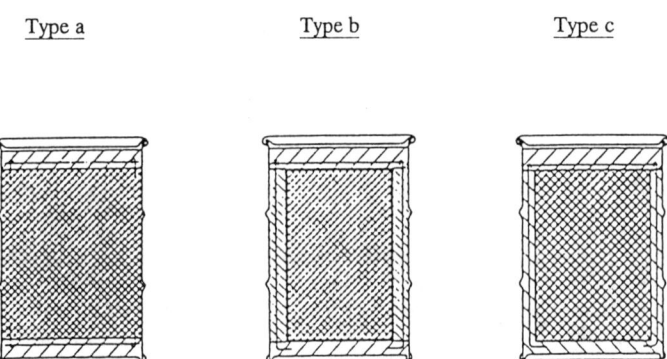

FIGURE 1. Waste package geometries
               Type a: 200 l-drum
               Type b: 100 l-drum with internal cement shielding and internal 55 l-drum
               Type c: 200 l-drum with internal cement shielding

TABLE I: Waste types considered.

Waste type*	Waste	Binding agent	Package geometry	Activity [Ci] and [Bq]
BA-KKB-1B NPP Beznau (PWR)	Beaded-resin from RCC	Plastic (Polystyrol-DVB)	Type b	10 $3.7 \ 10^{11}$
BA-KKM-1A NPP Mühleberg (BWR)	Powdered-resin from CPS	Cement + additives	Type c	0.079 $2.9 \ 10^{9}$
BA-KKL-1C NPP Leibstadt (BWR)	Beaded resin from filters + concentrates	Cement + additives	Type a	0.10 $3.7 \ 10^{9}$
BA-KKG-2 NPP Gösgen (PWR)	Concentrates	Bitumen (Ebano B15)	Type a	0.44 $1.6 \ 10^{10}$

* A full description of the waste package can be found in [9]

4.   DERIVATION OF THE KEY NUCLIDE DETECTION LIMITS

4.1 Detection limit definitions

The gamma measurement is assumed to be carried out with a generic GeLi detector located in the mid-plane of the drum (no scanning).

The measurement detection limit of a key nuclide gamma ray is defined by the number of gamma counted in the corresponding peak of the spectrum being 3 times the square root of the total background events counted under this peak. The derived detection limit DL(KN) of the specific activity SA(KN) in the waste package can therefore be obtained from the expected gamma spectrum as a function of the experimental configuration and the measurement time.

4.2 Base case spectrum calculation

The gamma spectrum in the waste-matrix is evaluated using waste composition data derived in earlier NAGRA work on characterization of operational wastes [9]. The "measured" gamma spectrum of a waste package is calculated using the gamma transport code MELANIE [10]. The base case distance drum-detector is 5 cm. The characteristics of the GeLi detector are as follows: surface: 1 cm², efficiency: parametrised as a function of the gamma energy (1 % at 1 MeV). The energy resolution of the detector is not included in the spectrum calculation but used to define the energy window (6 KeV for the 662 KeV of Cs-137) for counting the background. A natural background spectrum is finally added to the waste package gamma spectrum. The base case acquisition time is 10'000s (2,8 hours).

To illustrate this procedure, the emitted gamma spectrum of the waste package BA-KKM-1A (Table I) is given in Fig. 2. The corresponding calculated spectrum of the detected gammas is shown in Fig. 3.

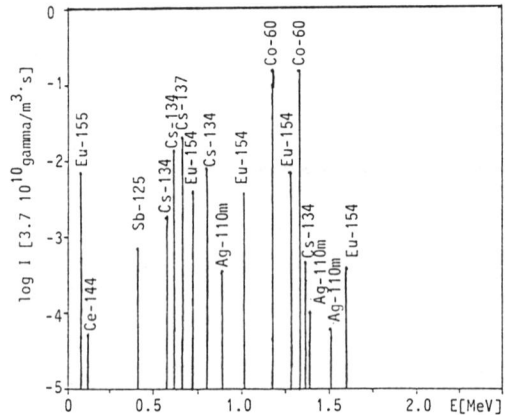

FIGURE 2. Spectrum of the gamma emitted by the waste matrix of package BA-KKM-1A
(reference time: 2 years after discharge)

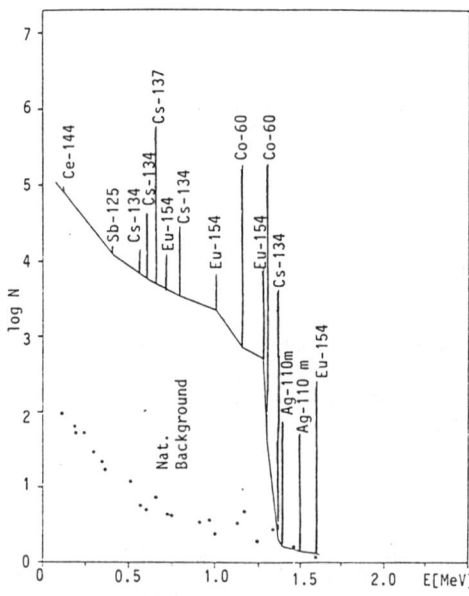

FIGURE 3. Calculated gamma spectrum for measurement of BA-KKM-1A waste package (for conditions see text)

4.3 Effect of acquisition time and drum-detector distance

The detection limit is proportional to the inverse of the square root of the acquisition time. A time reduction of a factor 100 as compared to the base case (leading to an aquisition time of 100s) will raise the detection limit by a factor 10. Calculations carried out for an increase of the base case drum-detector distance of 5 cm show only a very slight increase of the detection limit. This parameter is to be fixed from other experimental conditions.

5. KEY NUCLIDE DETECTION LIMITS

The key nuclide detection limits DL(KN) determined by the procedure given in paragraph 4 are summarized in Table II.

TABLE II: Key nuclide detection limits DL(KN) and
waste-matrix specific activities (SA(KN) in Bq/cm³.

Waste type (see Table I)	BA-KKB-1B	BA-KKM-1A	BA-KKL-1C	BA-KKG-2
DL (Co-60) SA (Co-60)	2.3 $5.2\ 10^6$	$1.1\ 10^{-1}$ $6.2\ 10^3$	$6.1\ 10^{-2}$ $6.6\ 10^3$	$1.2\ 10^{-1}$ $3.6\ 10^4$
DL (Cs-137) SA (Cs-137)	$1.1\ 10^2$ $6.8\ 10^5$	2.5 $9.2\ 10^2$	2.0 $1.1\ 10^3$	4.4 $3.6\ 10^3$
DL (Ce-144) SA (Ce-144)	$1.4\ 10^3$ $1.7\ 10^3$	$3.3\ 10^1$ $1.8\ 10^1$	$2.2\ 10^1$ 3.5	$5.4\ 10^1$ $7.2\ 10^1$

Measurement conditions: 10'000 s with GeLi detector of 1 cm² surface and
1 % efficiency at 1 MeV, placed 5 cm away from drum.

The key nuclide detection limit increases with the total activity of the waste package due to gamma of higher energy and to scattering materials whereas for a single gamma source it would decrease with its activity. The key nuclide detection limit increases with decreasing gamma energy (Co-60: 1.33 MeV, Cs-137: 0.662 MeV, Ce-144: 0.134 MeV).

The Ce-144 detection limit is about equal to or lower than its specific activity in the waste packages. For BA-KKM-1A and BA-KKL-1C Ce-144 is not measurable (see Fig. 3). For BA-KKL-1C, a 40 times longer acquisition time would bring the detection limit down to the matrix specific activity. The main conclusion is that the use of this key nuclide is not appropriate for these conditioned and packaged NPP reactors wastes. The fast decay of Ce-144 ($t_{1/2}$ = 285 days) is a further drawback.

6. SCALING FACTORS

To derive the detection limit of the safety-relevant nuclides, available or published scaling factors are considered. They are summarized in Table III.

TABLE III: Scaling factors.

Key nuclide KN	Safety relevant nuclide SRN	PWR Resins	PWR Dry waste and trash	BWR Resins	BWR Dry waste and trash
Co-60	C-14	$1\ 10^{-1}$ [3] $3\ 10^{-3}$ [5] $1\ 10^{-4}$ [4]	$1\ 10^{-1}$ [3] $1\ 10^{-3}$ [4]	$3\ 10^{-2}$ [3]	$3\ 10^{-2}$ [3] $4\ 10^{-3}$ [4]
	Ni-59	$2\ 10^{-2}$ [3] $6\ 10^{-4}$ [4]	$2\ 10^{-2}$ [3,8] $6\ 10^{-4}$ [4]	$3\ 10^{-3}$ [3] $6\ 10^{-4}$ [4] $3\ 10^{-4}$ [8]	$3\ 10^{-3}$ [3] $3\ 10^{-4}$ [8]
Cs-137	C-14		$4\ 10^{-3}$ [4]	$6\ 10^{-4}$ [4]	$6\ 10^{-4}$ [4]
	Cl-36	$6\ 10^{-6}$ [7]		$6\ 10^{-6}$ [7]	
	Sr-90	$5\ 10^{-2}$ [8] $3\ 10^{-2}$ [3] $8\ 10^{-3}$ [4] $3\ 10^{-3}$ [2]	$5\ 10^{-2}$ [8] $3\ 10^{-2}$ [3] $3\ 10^{-3}$ [2]	$7\ 10^{-2}$ [3] $5\ 10^{-2}$ [8] $2\ 10^{-2}$ [2]	$5\ 10^{-2}$ [8] $2\ 10^{-2}$ [2] $1\ 10^{-2}$ [3]
	Tc-99	$1\ 10^{-4}$ [2,6] $1\ 10^{-4}$ [3] $4\ 10^{-5}$ [4]	$1\ 10^{-4}$ [2,6] $4\ 10^{-5}$ [4]	$2\ 10^{-3}$ [4] $1\ 10^{-4}$ [2] $1\ 10^{-4}$ [6] $7\ 10^{-5}$ [3]	$3\ 10^{-4}$ [3] $1\ 10^{-4}$ [2] $1\ 10^{-4}$ [6]
	I-129	$3\ 10^{-5}$ [5] $8\ 10^{-6}$ [3] $1\ 10^{-6}$ [2]	$2\ 10^{-3}$ [3] $1\ 10^{-4}$ [4] $1\ 10^{-6}$ [2]	$7\ 10^{-4}$ [3] $2\ 10^{-4}$ [8] $5\ 10^{-5}$ [2] $4\ 10^{-5}$ [4]	$5\ 10^{-4}$ [3] $1\ 10^{-4}$ [4] $5\ 10^{-5}$ [2]
	Cs-135	$6\ 10^{-6}$ [3] $3\ 10^{-6}$ [7]	$4\ 10^{-5}$ [4] $6\ 10^{-6}$ [3]	$4\ 10^{-5}$ [4] $3\ 10^{-6}$ [7]	$4\ 10^{-5}$ [4]
	Pu-241	$5\ 10^{-3}$ [2]	$5\ 10^{-3}$ [2]	$8\ 10^{-3}$ [1]	
	Cm-242	$2\ 10^{-4}$ [2]	$2\ 10^{-4}$ [2]	$3\ 10^{-4}$ [1]	$2\ 10^{-4}$ [1]
Ce-144	Pu-238	$1\ 10^{-2}$ [3]	$1\ 10^{-2}$ [3]	$3\ 10^{-2}$ [3]	$3\ 10^{-2}$ [3]
	Pu-239	$1\ 10^{-2}$ [3,8]	$1\ 10^{-2}$ [3,8]	$2\ 10^{-2}$ [3]	$2\ 10^{-2}$ [3]
	Pu-241	$4\ 10^{-1}$ [3]	$4\ 10^{-1}$ [3]	$7\ 10^{-1}$ [3]	$7\ 10^{-1}$ [3]
	Am-241	$7\ 10^{-3}$ [3]	$7\ 10^{-3}$ [3]	$2\ 10^{-2}$ [3]	$2\ 10^{-2}$ [3]
	Cm-242	$2\ 10^{-3}$ [3]	$2\ 10^{-3}$ [3]	$4\ 10^{-2}$ [3]	$4\ 10^{-2}$ [3]

The range of the scaling factors quoted for the same pair of radionuclides extend up to a factor 100, with a single exception of 2000 for I-129 in PWR dry active waste and trash, and shows the different NPP operational conditions.

7. SAFETY-RELEVANT NUCLIDE DETECTION LIMITS

7.1 Results

These limits are obtained by multiplying the key nuclide detector limits for the waste packages given in Table II by the scaling factors of Table III. For a conservative assessment of the safety-relevant nuclide detection limits, the highest value of the scaling factors quoted is considered. The results are presented in Table IV.

TABLE IV: Safety-relevant nuclide detection limits DL(SRN) for the waste type package considered in Bq/cm^3.

KN	SRN	BA-KKB-1B	BA-KKM-1A	BA-KKL-1C	BA-KKG-2
Co-60	C-14	$2.3 \cdot 10^{-1}$	$3.3 \cdot 10^{-3}$	$1.8 \cdot 10^{-3}$	$1.2 \cdot 10^{-2}$
	Ni-59	$4.6 \cdot 10^{-2}$	$3.3 \cdot 10^{-4}$	$1.8 \cdot 10^{-4}$	$2.4 \cdot 10^{-3}$
Cs-137	C-14	--	$1.5 \cdot 10^{-3}$	$1.2 \cdot 10^{-3}$	$1.8 \cdot 10^{-2}$
	Cl-36	$6.6 \cdot 10^{-4}$	$1.5 \cdot 10^{-5}$	$1.2 \cdot 10^{-5}$	--
	Sr-90	5.5	$1.8 \cdot 10^{-1}$	$1.4 \cdot 10^{-1}$	$2.2 \cdot 10^{-1}$
	Tc-99	$1.1 \cdot 10^{-2}$	$5.0 \cdot 10^{-3}$	$4.0 \cdot 10^{-3}$	$4.4 \cdot 10^{-4}$
	I-129	$3.3 \cdot 10^{-3}$	$1.75 \cdot 10^{-3}$	$1.5 \cdot 10^{-3}$	$8.8 \cdot 10^{-3}$
	Cs-135	$6.6 \cdot 10^{-4}$	$1.0 \cdot 10^{-4}$	$8.0 \cdot 10^{-5}$	$1.8 \cdot 10^{-4}$
	Pu-241	$5.5 \cdot 10^{-1}$	$2.0 \cdot 10^{-2}$	$1.0 \cdot 10^{-2}$	$2.2 \cdot 10^{-2}$
	Cm-242	$2.2 \cdot 10^{-2}$	$7.5 \cdot 10^{-4}$	$6.0 \cdot 10^{-4}$	$8.8 \cdot 10^{-4}$
Ce-144	Pu-238	$1.4 \cdot 10^{1}$	Ce-144 activity	Ce-144 activity	$5.4 \cdot 10^{-1}$
	Pu-239	$1.4 \cdot 10^{1}$	is not measurable	is not measurable	$5.4 \cdot 10^{1}$
	Pu-241	$5.6 \cdot 10^{2}$			$2.2 \cdot 10^{1}$
	Am-241	9.8			$3.8 \cdot 10^{-1}$
	Cm-242	2.8			$1.1 \cdot 10^{-1}$

Waste types: see Table I
Measurement conditions: 10'000s with GeLi detector of 1 cm^2 surface and 1 % efficiency
at 1 MeV, placed 5 cm away from drum

7.2   Consequences for declassification of wastes

The use of Ce-144 as a key nuclide for conditioned and packaged NPP operational waste is, as already mentioned, not appropriate. If measurable, Ce-144 would lead to detection limits of safety-relevant nuclides which would be more than 1000 times higher than those based on Cs-137 (see results for Pu-241 and Cm-242). Ce-144 should be used for assessing raw wastes as the correlation of actinides with Ce-144 is considered more accurate than with Cs-137. This is reflected in more actinide scaling factors published for Ce-144 than for Cs-137.

The safety-relevant nuclide detection limits in Table IV are below the values considered today for declassifying radioactive waste, the lowest of which is around 1 Bq/cm³ for actinides and high energy gamma emitters. (Therefore the procedure proposed here are in principle sufficiently accurate and sensitive to discrimate between conditioned wastes being classified in this way for these nuclides.) Exceptions are some detection limits based on Ce-144.

The list of the disposal safety-relevant radionuclides in Tables III and IV is not complete. Other nuclides to be considered would be H-3, Se-79, Zr-93, Pd-107, Sn-126, Ra-226, U-235, U-238, Np-237, Pu-240. For the first 5 of them, further work using scaling factors related to Cs-137 and Co-60 (Zr-93) indicates that the detection limits also lies below the values for declassifying radioactive waste. For the last 5 of these nuclides, and also for the other actinide isotopes, there are scaling factors related to Cm-244 which cannot be measured in packaged wastes. The detection limit of the actinide isotopes in such wastes would also be under the declassifying values when using scaling factors based on Cs-137.

7.3   Consequences for allocation of wastes in disposal facilities

Preliminary limits for waste disposal in a planned horizontally accessed underground repository in Switzerland [11] are several orders of magnitude higher than the detection limits of Table IV and the detection limits of those further safety-relevant radionuclides and actinide isotopes mentioned above. The lowest disposal limits are for I-129: 4 Bq/cm³ and Cl-36: 16 Bq/cm³, which are still more than 400 time higher than the detection limits for the waste packages considered. The effective acquisition time for a waste package control can be reduced well below 1 hour.

For a near surface repository with administrative control which is not considered in Switzerland, the disposal limits for I-129 and Cl-36 are of the same order as in the above case, for the alpha emitting actinides of the order of 1000 Bq/cm³, and for the other radionuclides yet higher. Control of the waste packages considered would also last less than 1 hour.

8.   CONCLUSIONS

1. This work has shown that the correlation technique can be used in principle for conditionned and packaged Swiss NPP homogeneous operational waste (resins, concentrates and burned waste) to assess and control the disposal safety-relevant radionuclides (listed in Table IV and point 3 of paragraph 7.2). The specific activity disposal limits are higher than the safety-relevant nuclide detection limits.

2. One problem identified is the difficulty to measure the key nuclide Ce-144 in such conditioned and packaged wastes. For actinide isotopes, scaling factors related to Cs-137 should be used instead. Ce-144 could be measured on raw waste if absolute necessary but this was not the scope of this work.

3. The question of validating specific scaling factors for the different NPP waste sorts (validation of the correlations and accuracy of the scaling factors) has to be investigated further, especially for
the Cs-137 actinide isotope correlation, if the method is to be implemented.

4. Co-60 and Cs-137 measurement could be carried out in less than 1 hour for estimating all
safety-relevant nuclides.

5. The results of this study are sufficiently encouraging to justify study of the feasiblity of
applying the correlation technique to heterogenous waste (trash) where scanning of the waste package
is to be used.

REFERENCES

1. LIEBERMAN, J.A.; MC ILAINE, J.B; MILLER, A.D. and RODGER, W.A. (1983).
    Methodology for the classification of low-level radioactive wastes from nuclear power plants.
    AIF/NESP-027.

2. STANFORD, R.E.L. (1984). Classification of low-level radioactive wastes from nuclear power
    plants. Proceedings of the sixth Annual Participants' Information Meeting of the DOE Low-
    level Waste Management Program. Denver, Colorado, Sept. 11-13. CONF-8409115. p. 389-398.

3. CLINE, J.E.; NOYCE, J.R.; COE, L.J. and WRIGHT, K.W. (1985). Assay of long-lived
    radionuclides in low-level wastes from power reactors. NUREG/CR-4101.

4. WILD, R.E. et al. (1981). Data base for radioactive waste management. Waste source options
    report. NUREG/CR-1759. 2.

5. Characterisation of selected low-level radioactive waste generated by four commercial light-
    water reactors (1977). ORP/TAD-77-3.

6. MILLER, A.D. and LEVENTHAL, L. (1984). Data base for 10CFR61. Trans. Am. Nucl.
    Soc. 46.

7. Values available to NAGRA.

8. CLINE, J.E. and COE, L.J. (1983). Long-lived radionuclides in low-level waste. SAI-83/1075.

9. NAGRA (1984). Inventar und Charakterisierung der radioaktiven Abfälle in der Schweiz.
    Nagra NTB 84-47. Nagra, Baden, Switzerland.

10. LAFRANCHI, M. (1987). Code MELANIE. EIR Report 632. Paul Scherrer Institut (PSI),
    Würenlingen, Switzerland.

11. NAGRA (1985). Projekt Gewähr 1985. Radioaktive Abfälle: Eigenschaften und Zuteilung auf
    die Endlager-Typen. NAGRA NGB 85-02. p. 79. Nagra, Baden, Switzerland.

THE CENTRAL ELECTRICITY GENERATING BOARD APPROACH TO
LOW LEVEL WASTE ACTIVITY ASSESSMENT

C.W. Fern, C.E. Fincher, C.R. Harvey and P.B. Woollam
Technology Division, CEGB, National Power Division, UK

Summary

Assessment of the activity of low level waste (LLW) within Central
Electricity Generating Board (CEGB) has historically been achieved by
a combination of dose rate measurements and empirical isotopic
composition data. Because of changes in regulatory requirements and
the increased costs of disposal, it became clear in 1986/87 that
fundamental changes in the method of assay of LLW were needed. This
paper describes the approach adopted by CEGB to introduce these
changes. Details of the philosophy of 'area fingerprints' and the
programme of work for their determination are outlined together with a
description of an on-site monitoring system which represents a
practical implementation of the approach.

1. INTRODUCTION

    Previous practice within Central Electricity Generating Board (CEGB)
for the assay of low level waste (LLW) was based on package dose rate
measurements together with a knowledge of whether the contamination was
fission or activation product dominated. Application of empirical
relationships then enabled an assessment of package activity.
    However, because of recent regulatory changes following a
re-assessment of the radiological impact of the principal disposal site
and increased costs associated with the disposal of LLW in the UK, CEGB
instituted a critical review of LLW assay procedures. The major changes
identified as necessary related to the need to provide more detailed data
concerning the radionuclide content of LLW. These were both because of
the additional regulatory requirements and a disposal cost component
associated with individual nuclides.
    While the incentive for change was there, the means of compliance were
somewhat more difficult to define, bearing in mind that the problem has
several intractable components:
        1) LLW composition is very variable
        2) the activity distribution is inhomogeneous
        3) remote analysis of low energy X-ray, pure beta and alpha
            emitting nuclides is not possible
    Consequently, the only reliable way of complying with the requirements
would be either to analyse each individual item of LLW, or to completely
homogenise the waste and provide a detailed radiochemical analysis for each
disposal consignment. Neither of these options represents a practical
proposition, hence an approach was sought that met the requirements by
establishing a LLW database that allowed accurate assessment of disposals
when averaged over a period of time.

The basis of the CEGB approach is the concept and use of 'area fingerprints'. The approach assumes that the radionuclide inventory of low level waste can be assessed from knowledge of the isotopic ratios of contamination in individual plant areas where the waste arises. This relies on the assumption that the waste results from contact with the contaminated area or contaminated items within it and that, to a first approximation, the transfer coefficient for all radionuclides is similar. Consequently, while only a fraction of the activity may be transferred to the waste during the contamination process, the ratio of nuclides will remain constant. The ratio for a given area is called the 'area fingerprint' and as most areas on a power station are used routinely for only one type of work a single fingerprint will characterise the waste from that area. This can be used in conjunction with gamma spectrometry, as outlined later in this paper, to provide an assay system.

2. OVERALL APPROACH

2.1 Outline

In 1986/7 a programme of work was planned with the eventual aim of providing a radionuclide fingerprint for every area generating LLW on each nuclear power station. This was to be achieved by first identifying the areas concerned then by extensive sampling of each area and comprehensive radiochemical analysis of the samples to arrive at the characteristic ratios or 'fingerprint' of nuclides for that area.

A large number of separate areas are involved at each site and if waste were to be segregated on an individual area basis, with a fingerprint for each, the resulting waste management scheme would be difficult and costly to administer. However it was considered likely that there would be sufficient similarity between waste from groups of areas to enable combination into single waste streams. Such combination would be mainly on the basis of similar fingerprints, however knowledge of the volumes and activities of area arisings might allow combination of wastes with different fingerprints provided that the effect is quantifiable. A fingerprint for the combined waste stream would then be derived from the contributing area fingerprints.

Waste from each stream would be packaged in separate drums and labelled with the stream identifer. The major gamma emitting species in each drum would be determined utilising a drum monitoring system. It was envisaged that the system would verify that the waste has arisen from a particular stream by checking for readily identifiable gamma nuclide ratios. The system would then apply the full waste stream fingerprint in order to provide a complete radionuclide inventory for each drum of waste. Drum data would be retained by the system and aggregated to provide details for a full consignment of waste.

2.2 Sampling Programme

The accuracy of an area fingerprint depends upon the quality and consistency of the sampling of that area. It was decided that these could be best achieved using a designated team for all sampling. Samples are taken mainly in the form of swabs as it is believed that this method provides the most representative samples. It is assumed that contaminating nuclides are transferred to the swab in much the same ratio as to the waste itself.

The overall programme involves 12 stations with the number of plant areas averaging 15 per station. Sampling under both outage and normal operating conditions is considered necessary to ensure characterisation of the waste. This programme was bound to generate a considerable number of samples requiring comprehensive radiochemical analysis which is both time consuming and labour intensive. Therefore, a staged programme was devised which represents a phased but continual improvement in the supply of information to the disposal site management.

The first stage of the sampling programme is complete and involved the collection of samples from three typical stations both during normal operation and a reactor outage. All samples have been sent to the CEGB Central Radiochemical Laboratory (CRL) for radiochemical analysis. These samples are being used to provide site specific waste stream data for those stations sampled and 'global' waste stream data for the remainder. The data will be used in conjunction with the existing drum activity assessment technique to provide the LLW consignment radionuclide inventories for all sites in the short term.

The second stage will extend the sampling programme to the remaining stations, thus replacing 'global' data with site specific information as it becomes available. Gamma spectrometry based drum monitors will be gradually introduced and used with the fingerprint data at each site, phasing out the existing drum activity assessment technique.

Finally a quality assurance programme of sampling and analysis will be established at each site. This is expected to involve checks on an annual or bi-ennial basis to ensure the continued validity of waste stream fingerprints derived from the initial survey.

2.3 Analytical Programme

The determination of alpha, beta and X-ay emitting species is, by far, the most labour intensive part of the overall programme and governs the timescales involved in production of the area and waste stream fingerprints. The radionuclides appearing in UK gas-cooled reactor waste streams and requiring determination for compliance with the conditions of acceptance for LLW disposal and UK transport regulations are shown below in Table I. The Table also includes nuclides with half-life less than 3 months, not specifically required by the disposal authorisation, but of some significance for waste studies.

Table I: Radionuclides requiring determination

Gamma Nuclides	Beta Nuclides	X-ray Nuclides	Alpha Nuclides
Sc-46 Ag-110m	H-3	Fe-55	Ra-226
Cr-51 Sb-124	C-14	I-129	Th-232
Mn-54 Sb-125	S-35		U-234
Co-58 Sn-126	Cl-36		U-235+236
Fe-59 Cs-134	Ca-45		U-238
Co-60 Cs-137	Ni-63		Np-237
Zn-65 Ce-144	Sr-90		Pu-238
Se-75 Eu-152	Pm-147		Pu-239+240
Nb-94 Eu-154	Pu-241		Am-241
Nb-95 Eu-155			Cm-242
Zr-95 Ta-182			Cm-243+244
Ru-103 Hg-203			
Ru-106			
Ag-108m			

The analysis includes gamma emitting nuclides as these must be present in the fingerprint data to allow the assay of non-externally detectable nuclides in waste packages by any monitoring system.

In most cases considerably more than one sample has been received from each area; the number being dependent on the size and complexity of the area. It is not practical to perform radiochemical analysis on every sample due to the manpower and timescales concerned, hence area bulks have been prepared. However, every sample has been subjected to gamma spectrometry prior to bulking in order to obtain an indication of the degree of variation in nuclide ratios across the area.

Solutions are required for the radiochemical analysis, hence a total dissolution technique has been developed to solubilise all activity whilst ensuring the retention of potentially volatile species such as H-3, C-14, S-35 and Cl-36. The C-14 is released as CO_2 during the dissolution process and is trapped using sodium hydroxide which is treated separately.

The radiochemical analysis of area bulk samples from the first stage of the sampling programme is in progress but full area fingerprint data is not yet available for any of the three stations sampled.

3. LOW LEVEL WASTE ASSAY SYSTEM

The CEGB Berkeley Nuclear Laboratories (BNL) have been involved in designing a system for non-destructive assay of LLW in standard containers to allow full implementation of the disposal strategy. The system operates automatically, minimising operator time and the potential for human error. The system described below has been purchased for use by BNL and as a prototype for future use at all CEGB sites.

3.1 Waste and Waste Packages

The main function of the system is to assay radioactive waste in standard 200 litre steel drums, handling a maximum drum weight of 350kg. It will assay a wide range of media including hydrocarbons, concrete and metals and can cope with non-uniform waste density. The maximum acceptable surface dose equivalent rate at the package surface is 2 mSv hr^{-1}.

Purpose built jigs allow packages smaller than 200 litre drums to be handled by the system automatically. Larger packages are assayed individually using a separate detector system measuring items rotating on a turntable but data are still fed to the control computer and handled in the an indentical fashion to that from the automatic system.

3.2 Detector and Nucleonics

The primary detector is hyperpure germanium of 30% relative counting efficiency equiped with a high count rate preamplifier. The latter, together with a gated integrator type amplifier, allows an input count rate of around 300 KHZ to be maintained without degrading the performance of the detector. A 10 cm cylindrical lead shield around the detector incorporates X-ray shielding, minimises background effects and defines the assay collimation.

A digital gain stabilizer is incorporated in the system to prevent energy drift. Pulses from the detectors are handled by high speed ADCs into a standard multichannel analyser controlled by an IBM PS/2 model 60 computer. This computer also deals with the logic of the drum handling system, gamma spectrum analysis and all data handling.

3.3 Drum Handling

Each 200 litre drum is assayed in eight horizontal slices to reduce errors due to density or radioactivity distribution within the package and the drum is also rotated to minimise any further distribution effects. The geometry of the slices is defined by the 10 cm lead shield. Up to ten packages may be loaded onto a roller driven feed conveyor for sequential assay and are automatically fed onto the turntable. The system measures the weight of every package, together with the average dose rate over each of the eight slices.

3.4 Software

Specialist software has been written to:
- analyse the gamma spectra into isotope dependent drum activity content
- allow data input by unskilled operators
- determine the drum inventory of non gamma emitting nuclides
- input all necessary set up data
- calculate all of the information required for the consignment documentation
- undertake regular quality control checks

In particular the system has been designed to determine accurate transmission corrections to compensate for the self-shielding effects of waste in the drums. This is achieved primarily by measuring the differential attenuation of gamma ray photopeaks from the same isotope, however if no multi-peaked isotopes are present the drum weight and volume are used to determine the mean density and hence attenuation.

The system measures accurately at gamma ray energies in excess of about 150 KeV. The transmission correction algorithm becomes increasingly less accurate when the atomic number of the constituents of the waste averages above about 40. Hence, the system is not suitable for waste containing predominantly heavy metals such as cadmium, tin, tungsten or lead.

3.5 Consignment Information

The system is told which package are to be loaded into a transport container for dispatch. A calculation is then made to determine the total mass of waste and the mass and volume specific radioactivities of each isotope present within the container. Disc files are kept of all packages currently on site and of the contents of each dispatched container. The access to these data and operating systems is password controlled.

3.6 Calibration and Validation

The detector has been carefully calibrated to define the efficiency, energy, peak shape and count rate/random summing losses. Extensive checks have shown that the transmission correction algorithms are accurate. A quality control check is undertaken before the assay of each batch of drums which ensures that all important functions are operating correctly. A record is automatically kept of all the results in each check.

The assay system has undergone initial validation using a sample drum of incinerator ash from a CEGB power station. Ash was chosen for this purpose as it was considered to be the most suitable, reasonably homogeneous material available. A 200 litre drum of the ash was measured using the system and the CRL analysed five samples taken from the incinerator ash as the drum was filled. The CRL analysis included X-ray, pure beta and alpha emitting species together with gamma measurements and the results were used to provide a full fingerprint for the ash with each nuclide activity as a ratio to that of Co-60. The fingerprint was than used in the system assay of the waste ash; all non-gamma emitting nuclide activities being scaled from the system Co-60 result. Gamma results from the CRL analysis were used to cross-check the assay system results for the same nuclides.

A comparison of the CRL analysis results and those determined by the waste assay system are shown below in Table II. All results are decay corrected to the date that the ash was removed from the incinerator and the uncertainties shown with the CRL data are population standard deviations over the five samples analysed. Gamma results form the waste assay system are quoted with one sigma counting errors based on a single measurement.

The CRL gamma measurements identified a number of nuclides not detected by the waste assay system but levels were below the minimum required for disposal accountability and for simplicity these are not given in the Table. CRL performed radiochemical analyses for the majority of nuclides listed in Table I and those below the limit of detection are similarly not quoted below.

Table II: Measurements on incinerator ash

| Isotope | Drum content (MBq) determined by | |
	CRL sample analysis	Waste Assay System
Mn-54	1.2 +/- 0.85	1.5 +/- 0.1
Co-60	28.0 +/- 11	37 +/- 1.9
Zn-65	1.2 +/- 0.5	1.2 +/- 0.1
Ag-110m	0.25 +/- 0.13	0.62 +/- 0.03
Cs-137	1.2 +/- 0.7	0.32 +/- 0.02
C-14	0.020 +/- 0.038	0.027
S-35	8.7 +/- 5.8	11
Ca-45	0.47 +/- 0.39	0.62
Fe-55	450 +/- 360	590
Ni-63	0.63 +/- 0.19	0.82
Sr-90	0.14 +/- 0.08	0.18
Pu-238	0.0039 +/- 0.0029	0.0052
Pu-239+240	0.011 +/- 0.008	0.014
Pu-241	0.43 +/- 0.30	0.57
AM-241	0.012 +/- 0.012	0.016

4. DISCUSSION

The first stage of the sampling and analysis schedule was expected to extend over a two year period. Although sample collection from the three stations was completed within about a year the radiochemical analysis programme suffered delays. As a consequence the full area fingerprint data for three typical stations are not yet available but are expected, on a site by site basis, during the next twelve months.

It is anticipated that the second stage of the programme will be complete by the middle of 1993, providing station specific waste stream data and automatic drum gamma monitors.

The initial validation exercise for the waste assay system using incinerator ash shows reasonable agreement between the system and CRL gamma spectrometry data. Only for Ag-110m and Cs-137 is the system result outside one standard deviation of the CRL values. Measurements performed by BNL on 20 samples from the drum suggest that this is likely to be due to non-uniformity in the source material. The waste assay system averages the total contents of the drum far more effectively than a small number of discrete samples.

Since the data from the waste assay system for X-ray, pure beta and alpha emitting species require the CRL results the comparison shown in Table II demonstrates the correct function of the fingerprint software used by the assay system.

The overall approach to determination of LLW activity inventories for CEGB sites cannot give accurate data for non gamma emitting nuclides on an individual drum basis because of the short term variation in radionclide composition even within a given waste stream. However, it is anticipated that reasonable estimates will be achieved for each consignment of waste drums as required by UK legislation. It is certainly expected that the approach will provide a good assessment for LLW disposal from each station averaged over a year.

5. CONCLUSIONS

The use of area waste stream fingerprints is a practical solution to the problem of assessment of the X-ray, pure beta and alpha nuclides in LLW arising from nuclear sites.

The prototype waste assay system used in conjunction with station specific waste stream fingerprints will provide a reasonable assessment of LLW consignment activity inventories from CEGB sites in compliance with current UK legislation.

6. ACKNOWLEDGEMENTS

This paper is published with the permission of National Power which is currently a division of the CEGB.

NONDESTRUCTIVE MEASUREMENTS OF NUCLEAR WASTES :
VALIDATION AND INDUSTRIAL OPERATING EXPERIENCE

J.F. MONTIGON - V. GUERIN - R. LALANDE - A. SAAS

CEA : Institut de Recherche et de Développement Industriel (IRDI)
Division d'Etudes de Retraitement et des Déchets et de Chimie
Appliquée (DERDCA)
Département de Recherche et Développement Déchets (DRDD)
Service de Caractérisation, d'Evaluation des Confinements
et d'Analyse (SCECA)
CEN Cadarache 13108 - St Paul-Lez-Durance - Cédex

ABSTRACT

After a short survey of the means employed for the nondestructive
measurement of specific activities (γ and X-ray) in waste packages
and raw waste, the performances of the device and the ANDRA
requirements are presented.
The validation of the γ and X-ray measurements on packages is
obtained through determining, by destructive means, the same
activity on coring samples. The same procedure is used for
validating the homogeneity measurements on packages (either
homogeneous or heterogeneous).
Different operating experiences are then exposed for several kinds
of packages and waste.
Up to now, about twenty different types of packages have been
examined and more than 200 packages have allowed the calibration,
validation and control.

1. INTRODUCTION

Characterization and control assays performed on request of the
waste producer, waste manager or safety authorities, include the
determination of mass activity either in raw waste or in conditioned
or immobilized waste packages.
For the purpose, the Service de Caractérisation, d'Evaluation des
Confinements et d'Analyses (SCECA), through the Laboratoire
d'Expertise et de Caractérisation des Confinements (LECC) and the
Section d'Analyses des Effluents et Déchets (SAED), has developed,
for its own use and for producers, some mobile devices suitable for
all waste production, regardless of the type of waste, matrix and
package.

Operating experience is available on several dozen packages :

- CEN-Cadarache packages (homogeneous and heterogeneous cemented waste),

- STE$_3$ bitumen packages,

- STEL bitumen packages,

- SSM packages (ANDRA).

This paper briefly discusses the means employed, the French requirements for radioactive homogeneity and mass activity measurement ; then describes the validation tests required for all types of waste and finally presents some examples of industrial measurements.

2. EXAMINATION MEANS

The means used comprise mainly :

- a "high purity Germanium" gamma detector (range : 10 keV to 10 MeV ; resolution : 1 keV at 122 keV and 2 keV at 1.33 MeV),

- a "low-energy germanium" gamma detector (range : 3 keV to 1 MeV ; resolution : 230 eV at 5.9 keV and 540 eV at 122 keV),

- a 4096-channel analyzer,

- a microcomputer,

- specialized spectrum analysis software ,

- an optical detector alignment system,

- accessories for liquid nitrogen,

- lead shielding for collimation,

- a radioactive sealed source for calibration,

- a turnable for waste packages (permissible load = 10 tons ; rotation speed = 1 revolution per minute).

All these examination means are mobile and can thus be used to perform measurements directly in the producer's facilities, reducing measurement costs by eliminating the need to ship radwaste packages to the laboratory.
The turntable is specifically used for heterogeneous packages and to test the containment properties of overpacks and concrete shells.
Table I shows the application scope and the main device performance specifications.

TABLE I

APPLICATION FIELDS AND PERFORMANCE

1. APPLICATION FIELDS

   COMPLETE PACKAGES :

   - Drums,

   - Shells,

   - Homogeneous or heterogeneous wastes in different matrices.

   SAMPLES :

   - Core-samples, filters, control rods, unconditionned dismantling
     wastes, graphite, raw wastes, wastes with radium, etc.

2. $\beta\gamma$ - X-RAY ACTIVITY - PERFORMANCE

   $\beta\gamma$: up to 1000 Ci/package or raw waste,

   X-ray : up to 50 Ci/package or raw waste.

3. DETECTION LIMITS :

   $\beta\gamma$ activity : 10^{-3} to 10^{-4} Ci/t per emitter,

   X-ray   : (^{241}Am) : 10^{-2} to 10^{-3} Ci/t.

3. FRENCH HOMOGENEITY AND MASS ACTIVITY DETERMINATION REQUIREMENTS FOR RADWASTE TO BE DELIVERED TO ANDRA

Current ANDRA regulations define homogeneity requirements, embedding thresholds, and the maximum permissible activities for the main isotopes.

Waste package homogeneity

The homogeneity of a homogeneous radwaste package is defined by a maximum mass concentration difference of $\pm$ 25 % from the mean activity.
To be acceptable, the mass activity in any 100-litre volume inside a heterogeneous waste package must not exceed one-fifth of the total activity.

Embedding threshold

This is the mass activity above which waste must be conditioned ; two types of thresholds are specified :

- a total emission threshold :
  e.g., $\alpha = 5 \times 10^{-3}$ Ci/t
  $\beta\gamma = 1$ Ci/t

- individual thresholds for each isotope with a half-life exceeding 180 days.

Maximum permissible activity

This activity is defined for each radionuclide ;

e.g. 1300 Ci/t of ^{60}Co
130 Ci/t of ^{137}Cs
total $\alpha = 0.1$ Ci/t.

The current requirements on measurements levels for raw waste, packages and leachates are shown in table II for homogeneous waste.

4. VALIDATION TESTS ON MEASUREMENTS

These validation tests were carried out concurrently on raw waste and conditioned waste ; the validation concerns both $\beta\gamma$ and α (^{241}Am) emitters and emission ratios for pure β and α emitters, e.g. ^{60}Co/^{63}Ni ; ^{137}Cs/^{90}Sr ; ^{241}Am/total α.

4.1. Validation on raw waste

Two objectives were assigned :

- validation of sampling and of the destructive techniques necessarily used for pure β emitters in order to define the ratios ;

TABLE II

SENSITIVITY AND DETECTION LIMIT REQUIRED FOR ACTIVITY MEASUREMENTS FOR RAW AND CONDITIONNED WASTES

SPECIFICATION	ACTIVITY LEVEL	EMITTER	DETECTION LIMIT	UNCERTAINTY OF RESULTS
EMBEDDING LIMIT	10^{-3} Ci/t	α	5×10^{-9} Ci/m^3	50 %
ACCEPTABILITY LIMIT	0.1 Ci/t	α	5×10^{-7} Ci/m^3	10-20 %
MAXIMUM LIMIT	1 Ci/t	α	5×10^{-6} Ci/m^3	5-10 %
EMBEDDING LIMIT	0.1 Ci/t	$\beta\gamma$	5×10^{-7} Ci/m^3	50 %
EMBEDDING LIMIT	1 Ci/t	$\beta\gamma$	5×10^{-6} Ci/m^3	10-20 %
ACCEPTABILITY LIMIT (MEDIUM)	10 to 20 Ci/t	$\beta\gamma$	10^{-4} Ci/m^3	5-10 %
MAXIMUM LIMIT	1000 Ci/t	$\beta\gamma$	10^{-2} Ci/m^3	5 %
ACCEPTABILITY LIMIT	2 Ci/t	H_3	5×10^{-7} Ci/m^3	20 %
EMBEDDING LIMIT	10^{-4} Ci/t	γ LONG HALF-LIFE	$(10^{-10} - 10^{-11})$ Ci/m^3	50 %
ACCEPTABILITY LIMIT	10^{-1} Ci/t	γ LONG HALF-LIFE	$(10^{-7} - 10^{-8})$ Ci/m^3	10-20 %
EMBEDDING LIMIT	10^{-3} Ci/t	β LONG HALF-LIFE	10^{-9} Ci/m^3	50 %
ACCEPTABILITY LIMIT	10^{-1} Ci/t	β LONG HALF-LIFE	$(10^{-7} - 10^{-8})$ Ci/m^3	10-20 %
MAXIMUM LIMIT	10 Ci/t	β LONG HALF-LIFE	$(10^{-5}$ à $10^{-6})$ Ci/m^3	5-10 %

- sample optimization for measurements (mass, measurement representativity, uncertainties).

These aims are illustrated by the results shown in tables III and IV. In table III, graphite is used as an example ; nondestructive measurements on about 30 samples were used to validate :

- the minimum number of samples to examine (3),

- the dissolution technique,

- the definition of the main ratios.

In table IV, the example chosen illustrates the optimization of sample mass for ion exchange resins and the processing of primary samples.
The following remarks are applicable to these examples :

- for each type of waste, the preparation and dissolution techniques of raw samples must be validated,

- the optimum mass for raw solid waste is about 10 grams.

4.2. Validation of conditioned wasté packages

Several steps were necessary to validate the nondestructive measurements on packages :

- comparison of measurements on the package and its core samples using the same technique,

- comparison of measurements on the package and the samples (taken from the coring samples) by nondestructive means,

- comparison of nondestructive and destructive results,

- comparison of different measurement procedures (immobilized package, rotating package, different detectors).

Comparison between the package and its core sample

By coring without water, the laboratory can take several samples from the package, to control the activity level and distribution, and to perform different tests (compression, leaching, etc).
An example of activity comparison on a cemented concentrates package is given in table V. These measurements were been carried out by nondestructive means. The good correlation of measurements can be seen.

TABLE III

GRAPHITE SAMPLE MEASUREMENTS

NONDESTRUCTIVE $\beta\gamma$ ACTIVITY MEASUREMENTS

^{60}Co :1 to 12   kBq/g (0.3 - 3.2 x 10^{-1} Ci/t)

^{133}Ba :0.05 to 0.1 kBq/g (0.15 - 0.3 x 10^{-2} Ci/t)

^{134}Cs :0.01   kBq/g (0.3 x 10^{-3} Ci/t)

^{137}Cs :0.01 to 0.15 kBq/g ((0.03 - 4.5) x 10^{-3} Ci/t)

^{154}Eu :0.4 - 0.8   kBq/g ((1.2 - 2.4) x 10^{-2} Ci/t)

DESTRUCTIVE MEASUREMENTS

^{3}H   : 340 - 400   kBq/g (9.2 - 10.8 Ci/t)

^{14}C   : 6 - 25   kBq/g (0.2 - 0.7 Ci/t)

^{63}Ni : 1 - 7.5   kBq/g (0.3 - 2.0 x 10^{-1} Ci/t)

^{93m}Nb : < 0.1   kBq/g (< 0.3 x 10^{-2} Ci/t)

^{36}Cl : 0.4 - 1.5   kBq/g ((1.2 - 4.5) x 10^{-2} Ci/t)

^{60}Co : 2 - 15   kBq/g ((0.6 - 4.0) x 10^{-1} Ci/t)

^{133}Ba : 0.05 - 0.2 kBq/g ((0.15 - 0.6) x 10^{-2} Ci/t)

^{134}Cs : 0.01 - 0.1 kBq/g ((0.3 - 3) x 10^{-3} Ci/t)

^{137}Cs : 0.01 - 0.1 kBq/g ((0.3 - 3) x 10^{-3} Ci/t)

^{154}Eu : 0.4 - 0.8   kBq/g ((1.2 - 2.4) x 10^{-2} Ci/t)

^{155}Eu : 0.15 - 0.4 kBq/g ((0.4 - 1.2) x 10^{-2} Ci/t)

RATIOS FOR DIFFERENT ISOTOPES

^{14}C/total $\beta\gamma$ activity : 1.7 $\pm$ 0.7

^{63}Ni/total $\beta\gamma$ activity : 0.5 $\pm$ 0.1

^{63}Ni/^{60}Co   : 0.58 $\pm$ 0.16

TABLE IV

VALIDATION AND OPTIMIZATION FOR SAMPLING

1. DESTRUCTIVE ET NONDESTRUCTIVE MEASUREMENTS FOR ION EXCHANGER RESINS (IER)

RADIONUCLIDE	TOTAL SAMPLE (NONDESTRUCTIVE)	SAMPLE FRACTIONS (DESTRUCTIVE) 10 g	SAMPLE FRACTIONS (DESTRUCTIVE) 13 g
^{54}Mn	25.5 $\pm$ 0.8	17.7 $\pm$ 1.3	37.6 $\pm$ 2.6
^{58}Co	28.0 $\pm$ 0.7	19.0 $\pm$ 6.8	40.0 $\pm$ 8.2
^{60}Co	54.0 $\pm$ 7.6	38.0 $\pm$ 2.3	79.3 $\pm$ 4.2
^{110m}Ag	52 $\pm$ 1.5	62.3 $\pm$ 4.8	63.0 $\pm$ 4.8

2. VALIDATION OF SAMPLING TO OPTIMIZE THE MEASUREMENTS (DESTRUCTIVE ANALYSIS)

DRY WEIGHT	^{54}Mn	^{60}Co	^{110m}Ag
6.4 g	36.6 $\pm$ 4.4	73.4 $\pm$ 6.2	71.0 $\pm$ 8.0
12.6 g	33.5 $\pm$ 3.3	76.4 $\pm$ 5.8	68.1 $\pm$ 6.8
14.6 g	34.5 $\pm$ 3.6	76.8 $\pm$ 5.5	64.8 $\pm$ 6.5
22.4 g	33.7 $\pm$ 3.0	72.2 $\pm$ 4.9	63.2 $\pm$ 4.9
31.9 g	32.2 $\pm$ 2.6	72.5 $\pm$ 4.7	64.2 $\pm$ 5.5
51.2 g	30.4 $\pm$ 2.6	66.2 $\pm$ 4.2	56.7 $\pm$ 4.6
MOYENNE	33.5 $\pm$ 3.0	72.9 $\pm$ 5.0	64.7 $\pm$ 6.0
DEVIATION FROM MEAN : MAXIMUM MINIMUM	1.09 0.91	1.05 0.91	1.10 0.88

Comparison between the package, the core samples and their slices

The purpose of these nondestructive validation tests is to confirm that the self-shielding coefficients, the geometry and the scale effect allow the technique to be used for any test of characterization, control and inspection.

An example of this type of validation is given in table VI. These results come from control measurements on cemented concentrates from a nuclear power plant. The deviations between the different measurements do not exceed 10 to 15 %.

Comparison between destructive and nondestructive measurements

The aim here is to allow a complete cross-checking of measurements on mass activities of all the isotopes concerned by the requirements on conditioned waste : $\beta\gamma$, pure β, and α emitters.

The example in table VII shows the results of control measurements on an old 700-litre package (cemented concentrates, weighing 1.4 ton). The measurements allow a good evaluation of the mass activity of this package.

Comparison between the different types and means of nondestructive measurements

Although the calibrations and validations above allow a good cross-checking of the measurements, the influence of the measurement mode on heterogeneous packages must be assessed.

Table VIII shows the results given by measurements on steady and rotating packages, both homogeneous and heterogeneous. It is clear that heterogeneous packages must be rotated to measure them. With overcoated packages, the containment must also be tested by measuring possible diffusion of activity in the overcoating.

The use of several types of detectors and the need for evaluating α activity through measuring ^{241}Am with low-energy detector also call for validation of this technique.

An example is given in table IX ; the cross-checking of ^{241}Am measurements by a γ-emitter such as ^{137}Cs validates this type of measurement.

These tests, calibrations, destructive and nondestructive measurements allowed validation of the device employed for determining the mass activity of packages and raw waste in compliance with ANDRA requirements.

The second step involves industrial operating experience.

5. INDUSTRIAL MEASUREMENT OPERATING EXPERIENCE

The measurements and operating experience presented below concern :

- homogeneity measurements on homogeneous and heterogeneous packages to be shipped to the surface storage site,

- determination of conversion tables for decommissioning waste,

TABLE V

COMPARISON BETWEEN FULL SCALE PACKAGE AND CORE SAMPLE
(ACTIVITY RELATIVE TO FULL SCALE PACKAGE IN mCi)

RADIONUCLIDE	CORE SAMPLE	FULL SCALE PACKAGE
^{60}Co	0.34 ± 0.051	0.30 ± 0.046
^{134}Cs	0.078 ± 0.012	0.080 ± 0.012
^{137}Cs	71 ± 11	80 ± 12

TABLE VI

NONDESTRUCTIVE MEASUREMENT VALIDATION FROM DIFFERENT SIZE OF SAMPLES
(CONCENTRATES IN CEMENT)

RADIONUCLIDE		FULL SCALE PACKAGE (220 1 - 370 kg)	CORE SAMPLE (800 x 80 mm - 15 kg)	LEACHING SAMPLING (80 x 80 mm - 0.6 kg)
^{54}Mn	Bq	$(4.2 \pm 0.6) \times 10^8$	$(3.1 - 3.2) \times 10^8$	$(4.5 - 4.8) \times 10^8$
	mCi	11.4 ± 1.6	$8.4 - 8.8$	$12.3 - 12.9$
^{60}Co	Bq	$(1.6 \pm 0.2) \times 10^9$	$(1.35 - 1.39) \times 10^9$	$(1.51 - 1.53) \times 10^9$
	mCi	43.2 ± 5.4	$36.5 - 37.4$	$40.8 - 41.4$
^{134}Cs	Bq	$(1.1 \pm 0.2) \times 10^8$	$(9.25 - 9.62) \times 10^7$	1.13×10^8
	mCi	3.0 ± 0.5	$2.5 - 2.6$	3.05
^{137}Cs	Bq	$(9.3 \pm 1.4) \times 10^8$	$(9.10 - 9.17) \times 10^8$	$(1.03 - 1.05) \times 10^9$
	mCi	25.1 ± 3.8	$24.6 - 24.8$	$27.8 - 28.4$
TOTAL	Bq	$(3.1 \pm 0.5) \times 10^9$	$(2.70 - 2.75) \times 10^9$	$(3.1 - 3.2) \times 10^9$
	mCi	84.0 ± 12	72.0 to 74 ± 7	84 ± 8 to 86 ± 8

TABLE VII

COMPARISON BETWEEN DESTRUCTIVE AND NONDESTRUCTIVE MEASUREMENTS

(ACTIVITY RELATIVE TO FULL SCALE PACKAGE IN mCi)

RADIONUCLIDE	FULL SCALE PACKAGE	CORE SAMPLE	SAMPLE 5 - 15 g	TEST SAMPLES $\simeq$ 10 g
^{60}Co	0.30 ± 0.046	0.34 ± 0.051	0.35 ± 0.05	0.30 ± 0.05
^{134}Cs	0.080 ± 0.012	0.078 ± 0.012	0.081 ± 0.012	0.075 ± 0.015
^{137}Cs	80 ± 12	71 ± 11	85 ± 12	78 ± 10
^{241}Am	1.4 ± 0.2	-	1.3 ± 0.1	1.6 ± 0.2
^{90}Sr	-	-	-	0.19 ± 0.03

TABLE VIII

TYPE OF MEASUREMENTS FOR HETEROGENEOUS AND HOMOGENEOUS PACKAGES

HOMOGENEOUS PACKAGE MEASUREMENT

RADIONUCLIDE	STATIC MEASUREMENT	MEASUREMENT WITH TURNTABLE
^{60}Co	0.30 ± 0.046	0.32 ± 0.048
^{134}Cs	0.08 ± 0.012	0.09 ± 0.016
^{137}Cs	80 ± 12	79.7 ± 12

HETEROGENEOUS PACKAGE MEASUREMENT

RADIONUCLIDE	STATIC MEASUREMENT	MEASUREMENT WITH TURNTABLE
^{60}Co	6.70 ± 0.6	7.20 ± 0.6
^{134}Cs	1.5 ± 0.1	1.9 ± 0.2
^{137}Cs	15.3 ± 1.5	37.9 ± 4

TABLE IX

EQUIPMENT AND SAMPLE COMPARISON

COMPARISON OF MEASUREMENTS BY DIFFERENT DEVICES : (^{241}Am)HOMOGENEOUS PACKAGE MEASUREMENT

SAMPLES (PACKAGES)	γ PROBE	X-ray PROBE
1	$1.4 \times 10^3 \pm 300$	$1.6 \times 10^3 \pm 350$
2	$2.7 \times 10^4 \pm 6000$	$2.5 \times 10^4 \pm 5500$

COMPARISON OF MEASUREMENTS FROM DIFFERENT TYPES OF SAMPLES (MBq/DRUM)

TYPE OF SAMPLES	^{241}Am	^{137}Cs
FULL SCALE PACKAGE (1.4 t)	1.4×10^3	2.9×10^3
CORE SAMPLE (16 kg)	NON MEASURED	2.66×10^3
SECTION OF CORE SAMPLE (1 kg)	0.9×10^3	2.84×10^3
DESTRUCTIVE MEASUREMENT OF SAMPLE (10 g)	1.6×10^3	2.80×10^3

- water filter measurements in routine operation before immobilization in concrete shells,

- special waste measurements : after incident, or waste waiting for transportation, old waste removed from storage,

- measurement for experimental operation when starting new facilities.

5.1. Homogeneity measurements on packages before shipping to surface storage sites

These measurements are mainly carried out at the Cadarache Nuclear Research Center but also for control. Two examples are given ; the first one concerns a package from Cadarache (homogeneous waste) ; table X illustrates the technique used and the results obtained for a $\beta\gamma$ emitter (^{137}Cs) and an α emitter (^{241}Am). For the latter radionuclide, figure 1 shows the ^{241}Am spectrum obtained by low-energy detector for one measurement point.

The second example concerns a heterogeneous package ; table XI shows an application of the regulation that limits the activity inside any 100-litre volume to one-fifth of total activity. The percentage of relative activity (100 %) corresponds to 5 % of total mass activity.

5.2. Conversion tables for decommissioning waste

In addition to actual dismantled waste, decommissioning generates technological waste (gloves, cotton, etc...). To ensure rational operation, routine controls are performed and it is very useful to define conversion tables relating routine dose rate measurements to total activity and typical spectrum.

An industrial example is given below, with all its steps :

- the first step consists in nondestructive measurements on several series of packages issued from the different zones of the facility being decommissioned, (table XII),

- the second step is to determine the typical spectra and the different waste characteristics (density, dose rate), (table XIII),

- the last step consists in computing the conversion coefficients applicable to each type of waste (table XIV).

These steps are carried out as the decommissioning work progresses.

5.3. Routine measurements on water filters before immobilization in concrete shells

Water filters from nuclear power plants are waste with activities varying over a wide range, from a few mCi to several tenths of Curies per filter.

TABLE X

RESULTS FOR HOMOGENEITY CONTROL (700 l PACKAGE N° 1901)

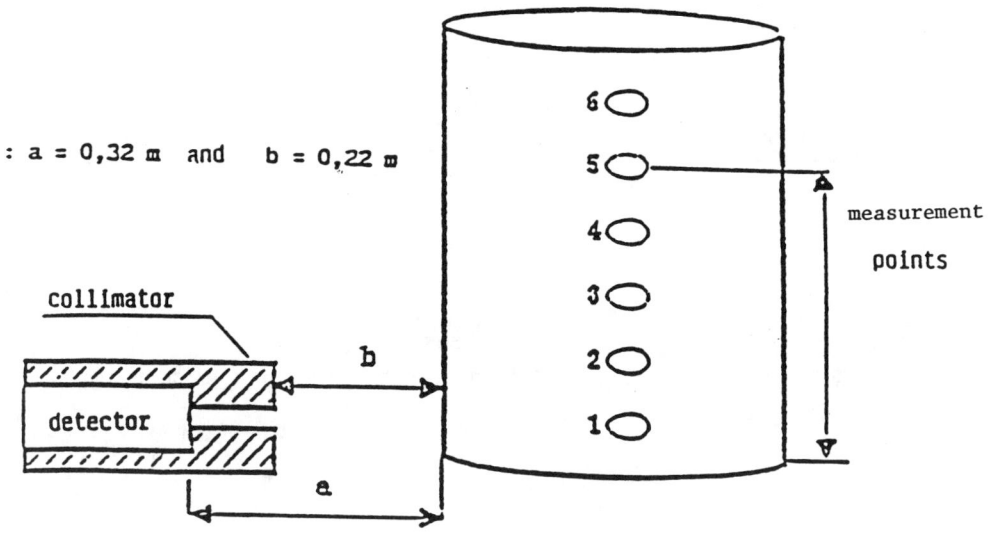

: a = 0,32 m and b = 0,22 m

MEASUREMENT POINTS	HEIGHT	RELATIVE ACTIVITY (%)	
	(m)	^{137}Cs	^{241}Am
1	0.15	82 ± 8	92 ± 9
2	0.30	76 ± 8	78 ± 8
3	0.45	87 ± 9	84 ± 8
4	0.60	100 ± 10	100 ± 10
5	0.75	91 ± 9	79 ± 8
6	0.90	97 ± 10	79 ± 8

TABLE XI

γ SCANNING OF HETEROGENEOUS PACKAGE
(% OF RELATIVE ACTIVITY)

MEASUREMENT POINTS	^{60}Co	^{137}Cs
SURFACE	100	100
MEDIUM	250	510
BOTTOM	30	15

FIGURE 1

AMERICIUM-241 PEAK FOR THE FIRST MEASUREMENT POINT FOR PACKAGE N° 1901

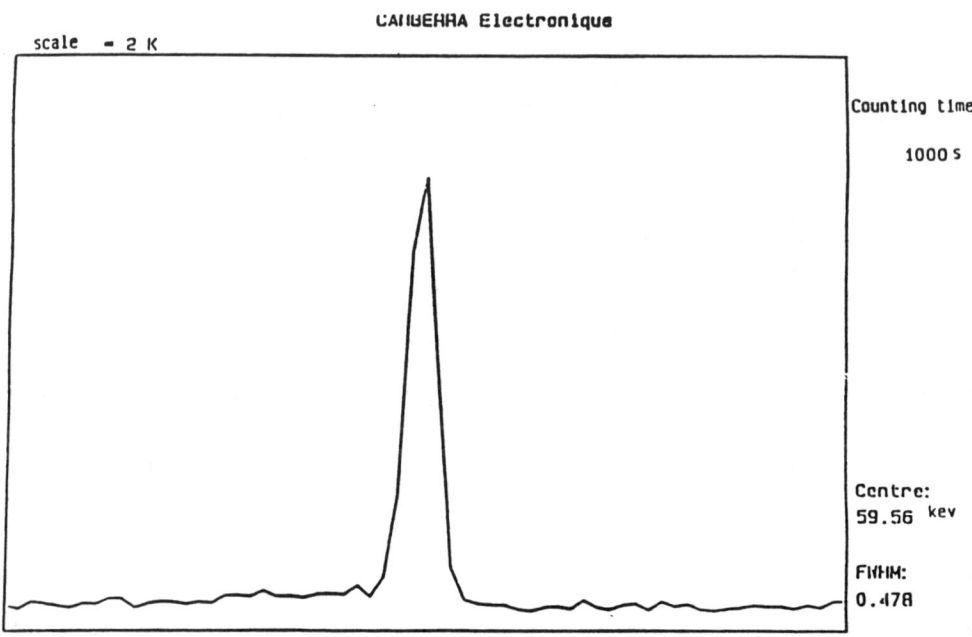

CANBERRA Electronique

scale = 2 K

Counting time

1000 s

Centre:
59.56 kev

FWHM:
0.478

TABLE XII

MEASUREMENTS OF DISMANTLING WASTE :

RAW WASTE MEASUREMENTS

REFERENCE OF DRUMS	ROUTINE CONTROL MEASUREMENTS	^{22}Na kBq	^{54}Mn kBq	^{60}Co kBq	^{134}Cs kBq	^{137}Cs kBq
4751	UNDETECTABLE	-	-	< 37	-	< 37
4752	1700	-	-	20190 ± 1700	-	2375 ± 236
4753	700	7.0 ± 1.4	337 ± 31	365 ± 31	98 ± 10	14393 ± 1400
4754	700	-	-	6016 ± 52	-	2915 ± 280
4756	3300	-	2527 ± 235	3197 ± 277	673 ± 68	101800 ± 9500

TABLE XIII

DISMANTLING WASTE

DENSITY - DOSE RATE - % OF ACTIVITY FOR DIFFERENT ISOTOPES

REFERENCE OF DRUMS CHECKING CART	4571	4752	4754	4753	4756
DENSITY	NOT MEASURED	0.22 g/cm^3	0.26 g/cm^3	0.24 g/cm^3	0.24 g/cm^3
ORIGIN	HOT CHANGING ROOM	DECONTAMINATION	DECONTAMINATION	REACTOR CONTAINMENT	REACTOR CONTAINMENT
MAXIMUM CONTACT DOSE RATE	< 0.1 mrad/h	40 mrad/h at mid-height	4 mrad/h at mid-height 10 mrad/h at bottom	4 mrad/h at mid-height 10 mrad/h at bottom	4 mrad/h at mid-height 10 mrad/h at bottom
RELATIVE ACTIVITY					
^{22}Na	-	-	-	0.05 %	
^{54}Mn	-	-	-	2.20 %	2.3 %
^{60}Co	50 %	90 %	68 %	2.50 %	3.1 %
^{134}Cs	-	-	-	0.60 %	0.6 %
^{137}Cs	50 %	10 %	32 %	94.65 %	94.0 %

TABLE XIV

DISMANTLING WASTE
CONVERSION TABLE : COUNT TO ACTIVITY IN kBq

TYPE OF WASTES	CONVERSION FACTORS
DECONTAMINATION	13.3 kg ± 20 % FOR 1 COUNT PER SECOND
REACTOR BARRIER	27.2 kBq ± 20 % FOR 1 COUNT PER SECOND

Although this kind of filter has a typical spectrum in normal operation, variations are observed when during plant starting or in case of serious cladding failure ; it is then necessary to measure filter activity before immobilizing them in concrete shells.

An example of routine measurement is given in table XV. This example shows the activity dispersion according to filter location, as well as the fluctuations of isotopic ratios :

e.g. $^{134}Cs/^{137}Cs$: 1.46 to 1.61 x 10^{-2}

$^{60}Co/^{137}Cs$: 1.74 x 10^{-2} to 1.7 x 10^{-3}

$^{60}Co/^{125}Sb$: 0.195 to 1.567.

5.4. Measurements on special waste

The number of nondestructive measurement needed for this type of waste increases with the need for storage and the requests to ANDRA for special packages.

Three examples illustrate this operating experience :

- measurements on a settling tank from a nuclear power plant ; this vessel was placed in a container for failed assemblies, surrounded by concrete shielding. Table XVI shows the results of γ and low-energy γ measurements,

- measurements on a control rod after replacement ; figure 2 shows the contents of this package. Table XVII shows both the measured spectrum including some minor radionuclides such as ^{124}Sb, and the activity of the two major isotopes : ^{58}Co and ^{60}Co (activity between 3800 and 4000 Ci),

- measurements on packages to be removed from storage because of radium α-activity. In addition to the overall activity per package, it was necessary to measure the distribution in the drum and in the absorbants placed at the top and bottom of the drum. Table XVIII shows an example of measurement for this type of package.

5.5. Industrial measurements performed when starting new facilities

Nondestructive measurements are performed on bitumized packages during the experimental operation periods.

The industrial targets are :

- validation of the activity evaluations made by the operator from measurements of raw waste,

- validation of the industrial measurement system based on dose rate measurement,

- verification that the results comply with the specifications and guaranteed process parameters.

TABLE XV

PROCESS WASTES (FILTERS)

(ACTIVITY MEASUREMENTS : MBq/FILTER)

SAMPLE NUMBER	^{60}Co	^{125}Sb	^{154}Eu	^{134}Cs	^{137}Cs
1	42.2 ± 7.0	106 ± 25	89 ± 21	99 ± 14	6150 ± 750
2	129 ± 17	82.3 ± 14.0	71 ± 10	113 ± 17	7400 ± 900
3	190 ± 32	950 ± 210	1260 ± 320	1273 ± 177	84700 ± 10700
4	218 ± 35	1117 ± 159	1140 ± 285	1560 ± 288	106300 ± 10700
5	130 ± 20	635 ± 77	571 ± 70	1143 ± 158	76700 ± 9700

TABLE XVI

ACTIVITY IN A SETTLING TANK

RADIONUCLIDE	ACTIVITY IN SETTLING TANK
^{241}Am	30.0 ± 7.5
^{125}Sb	28.8 ± 4.0
^{154}Eu	40.0 ± 6.0
^{134}Cs	42.0 ± 6.0
^{137}Cs	4220 ± 630
^{60}Co	13.6 ± 2.0

(ACTIVITY CALCULATED AT MEASURING TIME (05/09/89)

RADIONUCLIDE	ACTIVITY (TBq)
^{58}Co	69 ± 12
^{60}Co	73 ± 12

TABLE XVII

CONTROL ROD SPECTRUM

RADIONUCLIDE	ACTIVITY (TBq)
^{58}Co	69 ± 12
^{60}Co	73 ± 12

FIGURE 2

PACKAGE COMPOSITION

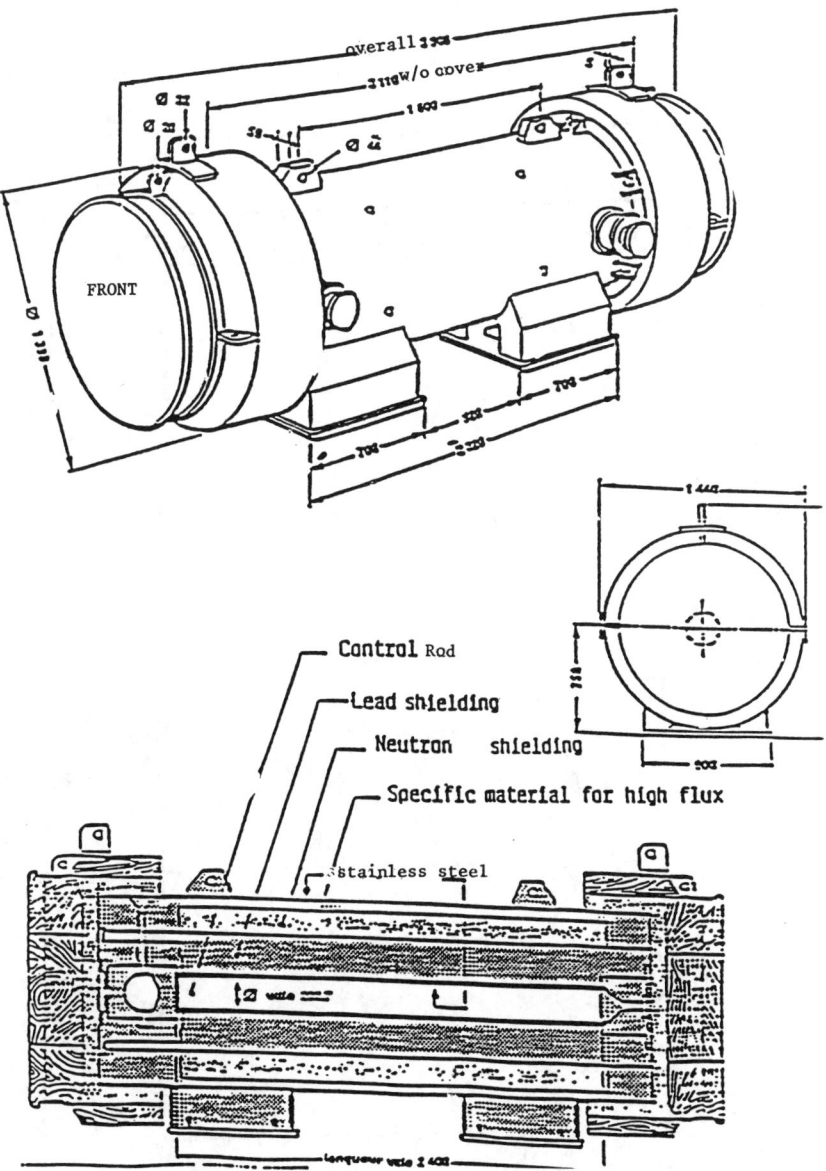

— Control Rod
— Lead shielding
— Neutron shielding
— Specific material for high flux
— stainless steel

TABLE XVIII

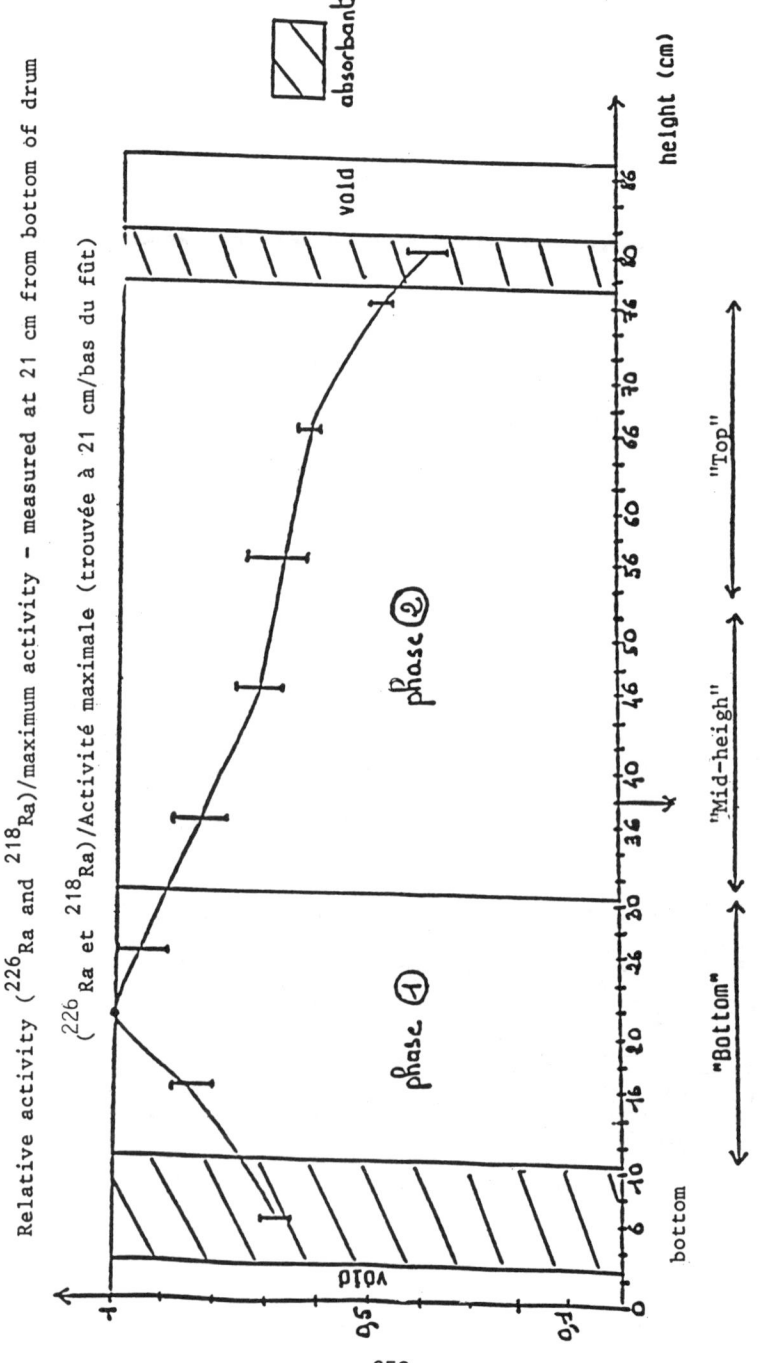

Relative activity (^{226}Ra and ^{218}Ra)/maximum activity – measured at 21 cm from bottom of drum

(^{226}Ra et ^{218}Ra)/Activité maximale (trouvée à 21 cm/bas du fût)

TABLE XVIII : Gamma-scanning of 200-liter drum for wastes containing radium

Table XIX shows an example of the measurements performed and of the evaluations given by the operator. The deviations between the evaluations, the full-scale nondestructive measurements and the dosimetries ranged from 15 to 20 % for packages containing 60 to 400 Ci of $\beta\gamma$ emitters.

6. CONCLUSIONS

These validation results and industrial measurement show that the measurements performed by the Laboratoire d'Expertise et de Caractérisation des Confinements comply with current ANDRA specifications for homogeneity and for mass activity of packages. Nevertheless, the only α-emitter that can be routinely measured at the present time is ^{241}Am. This determination will be used for relating with neutron measurements (active or passive) and for nondestructive validation of all the devices needed to assess mass activity for $\beta\gamma$ and α emitters.
Nondestructive $\beta\gamma$ measurements will remain necessary to determine pure β emitters and validate the destructive techniques employed for this type of isotope.
Acknowledgement : The validation of methods was carried out as part of a CEC contract (N° FI1W-0095).

TABLE XIX

NONDESTRUCTIVE MEASUREMENTS FOR STARTUP OF THE STE3 BITUMINIZATION FACILITY AT LA HAGUE

NONDESTRUCTIVE MEASURMENTS (CAD)
EVALUATION ACCORDING TO THE OPERATOR'S SPECIFICATIONS (LH)
VALIDATION OF ACCEPTANCE REQUIREMENTS (ACTIVITY CONTENT IN PACKAGES)

PACKAGE NUMBER	^{106}Ru + ^{106}Rh (Ci)		TOTAL $\beta\gamma$ (Ci)		DOSIMETRY (rad/h)	
	LH	CAD	LH	CAD	LH	CAD
1099	26.01	23.0	64.1	58.5	45.0	38.3 $\pm$ 1.8
1159	24.5	21.0	59.0	53.2	42	ND
992	147.1	122.1	340	314	102	101.3 $\pm$ 1.2
993	141.0	118.0	333	292	ND	96.7 $\pm$ 2.9

S E S S I O N   B 3   –   O P E R A T I O N A L   E X P E R I E N C E

Chairman : W.B. Bremner – AEA Technology, Dounreay
Secretary : K.P. Lambert – AEA Technology, Harwell

Opening statement

The "Operational Experience" session is primarily aimed at the application of NDA technology in operational facilities. It was recognised by the Organising Committee that many papers contained both development work and operational experience which resulted in some overlaps between previous sessions and this one. It is encouraging to find that many papers now indicate that the technology has firmly reached the stage where it is being routinely applied in a wide variety of measurement roles.

This session contains information on the application of NDA techniques to solid waste arisings from a wide cross-section of the nuclear industry. In some cases the experience has been accumulated for 8-10 years and clearly demonstrates the capability of routinely operating passive gamma, passive neutron and active neutron systems. In many cases a dual measurement approach is used which reduces some of the measurement uncertainties and gives increased confidence to the operator on the validity of the result. The implementation of improvements on the installed systems, which recognises progressive changes in the underlying technology, has also been a noticeable feature in some papers. It is often apparent that the ultimate performance of a selected measurement technique is not required for the solid waste stream concerned, either because the operational limits are not too stringent or the stream consistently contains quantities of fissile material which are relatively easy to detect. In one example a fully automated laboratory was detailed which is capable of 8000 measurements per year.

The use of variable geometry mobile system is a recent trend in the technology and several papers indicated that future systems (or more recent additions to an existing system) are likely to follow this philosophy in order to increase the flexibility of the device and to cater for a wider range of items. The use of computer linking with several NDA systems is also an emerging trend.

The provision of adequate Quality Assurance data is an important requirement for neutron measurements and has been the subject of further study in most facilities. It seems likely that this will become a more pressing requirement as both legislative and inspection procedures become more stringent.

8 YEARS UTILIZATION OF THE WASTE DRUMS COUNTING FACILITY AT THE UP1 REPROCESSING PLANT

R. BERNE, CEA (NUCLEAR STUDIES CENTER), CADARACHE, FRANCE
and
J.C. BATAILLES - C. BRESCHET, COGEMA - UP1, MARCOULE CENTER, FRANCE

Summary

Monitoring waste contaminated by α-emitting materials, especially plutonium, is essential for safety and waste management purposes. Since 1981, the COGEMA UP1 reprocessing plant at the Marcoule Center has included a facility for neutron and γ counting of α-contaminated 100 liter waste drums. During the past 8 years, more than 5,000 drums have been inspected with a satisfactory degree of accuracy. A new development is the neutron active method, which uses a ^{252}Cf source to activate the fissile material contained in the drums. This method uses detection and counting of both delayed γ and neutrons. It is hoped that detection limits will be improved using this technique.

1 - INTRODUCTION

The UP1 plant has introduced various measures aimed at counting α-contaminated waste.

In 1981, UP1 opened an installation designed to inspect 100 liter drums containing plutonium-contaminated waste. The measuring techniques initiated at the same time are of the "passive" type and are based on neutron counting and associated gamma spectrometry.

The installation has operated on an industrial basis since 1982 and has inspected more than 5,000 drums.

2 - MEASUREMENT PRINCIPLE

2.1 - "Neutron" Passive Method

The "neutron" passive method is based on counting the neutrons emitted by the contaminant contained in the drum. If the contaminant is PuO_2, it is known that two types of neutron are emitted:
- neutrons resulting from the reactions (α, n) of Pu α radiation on the associated oxygen nuclei;
- neutrons resulting from spontaneous fission of Pu

The following table gives the characteristic values of these two types of emission for the various actinides normally found in Plutonium Oxide contaminants:

actinide	$n_{(\alpha, n)} \cdot S^{-1} \cdot g^{-1}$	$n_{FS} \cdot s^{-1} \cdot g^{-1}$
^{238}Pu	1.2×10^4	2634
^{239}Pu	32	0.02
^{240}Pu	122	911
^{241}Pu	1.2	0
^{242}Pu	1.75	1712
^{241}Am	2330	1.3

For a PuO_2 contaminant of known isotopic composition, it is therefore possible, using the neutron emission count rate and the data given above, to establish the quantity of Pu by weight responsible for the emission.

2.2 - Passive "Gamma" Method

The passive γ method consists of simultaneously performing γ spectrometry and neutron counting on the drum. This spectrometry is only possible because the Pu contaminant is free from fission products normally found in irradiated fuels.

An analysis of the resulting spectrogram is used, by measuring the surface of a photoelectric peak, to correlate the number of representative pulses on this surface and the quantity of actinide present in the measured drum.

Measurement is performed during the neutron counting time (approximately 200 seconds) using an HP Ge detector which is moved up and down a generatrix of the drum. Since the drum rotates throughout the measurement, this technique provides γ spectrometry along a helix and also γ scanning to the height of the drum. This γ scanning is used to reveal the presence of an isolated source located in the volume of the drum. If such a source is present, γ scanning is used to correct the neutron count calibration coefficient, which is established for a Pu contaminant assumed to be uniform.

The HP Ge detector is collimated and designed so that the response from an isolated source present inside the drum is the same as that given by a source distributed evenly over the radial plane of the drum. This γ measurement should preferably be used for:

1) determining the isotopic composition of the Pu in order to correct, if necessary, the pre-established composition (neutron counting can only be interpreted to the extent that isotopic composition is known).

2) correcting the results of measurements obtained using neutron counting.

3 - BRIEF DESCRIPTION OF THE INSTALLATION

3.1 - Location

The UP1 drum counting station is located inside the building containing the reprocessing plant workshops.

In the immediate vicinity, there are:
- a permanent storage area for drums with a very high Pu content
- a temporary storage area for drums awaiting evacuation.

3.2 - Description of the Installation

The UP1 counting station consists of:

1) a turntable managed by a programmable controller on which the 100 liter drum to be counted is placed and rotated

2) 24 neutron counters, immersed in paraffin wax, positioned around the turntable, connected to both a dead time (DTNC) and a shift register (SRCS) electronic system

3) a γ radiation detector (hyper-pure Ge or similar) mounted on a movable support which is also managed by a programmable controller, which raises and lowers the detector while the drum rotates (a single up and down movement during the measurement)

4) a TRACOR TN 1710 computer used to analyze the data sent by the electronic counting systems and to provide results for isotopic composition and plutonium weight.

4 - CHARACTERISTICS OF WASTE COUNTED

The 100 liter waste drums come from the medium activity workshops of the plant. They have the following characteristics:

- physical

The waste is removed from glove boxes in packets and wrapped in a double layer of vinyl. In general it consists of fuel waste made up of PVC, natural or synthetic rubber, metallic parts, miscellaneous equipment and, rarely, solutions.

- physico-chemical

The plutonium may be in nitrate, oxalate, oxide, fluoride or metallic form. This obviously has a considerable influence on the self-absorption factor and on neutron emission.

- isotopic composition of plutonium

^{240}Pu content may be as high as 27% and, under exceptional circumstances, 45%

- γ activity

The γ dose rate of the drums is generally very low, of the order of several μSv.h^{-1}, measured using the babyline. Older drums containing ^{241}Pu (between 0.5 and 1% of total plutonium), or drums from other workshops, may however have much higher values (several hundred μSv.h^{-1}).

- quantity of Pu per drum

Approximately 80% of the drums contain less than 1.7 g total Pu and are therefore sent to the conditioning workshop. The others are stored in the medium activity workshop awaiting reprocessing. They contain an average weight of Pu of 20 g per drum. The total quantity in a drum may be as high as 250 g but in this situation, the constituent packets are measured individually.

5 - COUNTING CONDITIONS

It is clear that the counting station is subject to the influence of its immediate environment via the background radiation.

The background radiation of the permanent storage area is limited by the use of concrete storage cells and by the thickness and construction of the walls of the store.

The background radiation caused by the adjacent temporary storage area is limited, given the use to which the store is put (it contains no plutonium-rich drums).

In contrast however, the background radiation caused by Pu transfer operations in the immediate vicinity of the walls of the building is prohibitive and makes any counting operation impossible.

6 - EXPERIENCE FEEDBACK

6.1 - Baseline Criteria

For drums which are poor or very poor in plutonium, the sensitivity of the measurements obtained is the essential requirement.

In contrast, for plutonium-rich drums, measurement accuracy is crucial. This is necessary for drawing up evaluations and for respecting safety and criticality regulations.

6.2 - Development of the Facility

During the time it has been in operation, the counting facility has been modified in various ways to improve the sensitivity of the measurements produced.

6.2.1 - Gamma Spectrometry

Gamma spectrometry has, since 1983, been developed to give the best coverage of the radiation range of the drums. Various collimators have therefore been installed and calibrated with aperture dimensions of 2.5, 7.5, 15 and 30 mm.

Each collimator is allocated to counting a given corresponding band of radiation (limited values: 400 μSv/h, 30 μSv/h and 10 μSv/h).

Problems were experienced during calibration of the 2.5 mm collimator used for drums with the highest dose rate. The standard source was so strongly collimated that its activity was not high enough. As a result, this collimator is now only used for determining isotopic composition (see paragraph 2.2).

6.2.2 - Neutron Counting

For several years (1981 to 1987), the installation operated using a dead time electronics neutron counting system. This operating method only proved valid for ^{240}Pu weights of less than 30 grams. Above this threshold, there is an under-evaluation in the counting of spontaneous fission neutrons.

In 1987 a shift register electronic system was installed and calibrated. This has considerably widened the measurement range. The linearity of the response has been certified up to 300 g of ^{240}Pu in the form of plutonium oxide.

The spatial effect due to the exact location of the source in the drum is corrected for using the indications given by the gamma scanning.

Due to the limitations of the Tracor computer, the program uses one of the two neutron counting results available. The method selection threshold was set at 0.7 g of ^{240}Pu.

6.3 - Equipment and Environment Check
6.3.1 - Background Radiation Check

This check is performed at least once a week. The presence of "rich" drums in the temporary store next to the counting area has a slight influence on the background radiation. Therefore a check is carried out each time there is significant movement of drums in the store.

6.3.2 - Neutron Counting Output Check

This is carried out at least once a month. This check is used to detect any deficiency in an He^3 counter or in the electronics.

6.3.3 - γ Calibration Check

This check is performed after any operation is carried out on the electronics or on the processing program, using known PuO$_2$ sources.

6.3.4 - Other Checks

The counting parameters were set during testing and are not varied. The operator provides:
- the drum number
- the assumed ^{240}Pu content
- the calibration number, which corresponds to a given collimator. This automatically launches acquisition and, finally, expression of results.

6.4 - Installation Performance (see table 1)

Since the installation was commissioned, 3,700 100 liter drums, 1,400 packets and 140 containers of ash have been counted.

80% of the drums counted were sent to the solid waste processing station as containing less than 1.7 g of Pu.

6.4.1 - Selection of Counting Method

After counting, the installation supplies two groups of results:
- ^{240}Pu content, using γ spectrometry and the correlation tables included in the program
- weight of total Pu, expressed:
1) by γ spectrometry using the 413.7 keV peaks of ^{239}Pu and the 208 keV peaks of ^{241}Pu
2) by neutron counting using the two electronic systems and the isotopic composition:
- introduced by the operator
- established by the program using γ spectrometry and the correlation tables.

In neutron counting, the computer selects either the result given by DTNC or the SCRS. For a threshold set at 0.7 g of ^{240}Pu (see paragraph 6.2.2) it is seen that for drums containing more than 1.7 g total Pu, 80% of results selected are those given by the SCRS and 20% those from the DTNC.

The result given by the operator is the product of an interpretation of the results given above:
- ^{240}Pu content confirms or refines, in most cases, the content introduced by the operator, based on information concerning the origin of the waste
- the weight of Pu given:
1) by γ spectrometry is selected (in only a few per cent of cases) when the following conditions are met:
1/ measurement performed using the 7.5, 20 or 15 mm aperture
2/ weight of Pu given by γ spectrometry greater than that given by neutron counting (storage safety)
3/ dead time < 20% of counting time
4/ surface of photoelectric peak useable ($\geq$ 500 pulses for duration of counting).
2) by neutron counting using the isotopic composition:
- introduced by the operator is selected if:
1/ γ spectrometry has been carried out using the 2.5 mm aperture
2/ weight of total Pu obtained by neutron counting is greater than that obtained by spectrometry.

6.4.2 - Measurement Sensitivity

PuO$_2$ standards were counted in order to establish the sensitivity threshold of the equipment.

Results obtained with real, light vinyl matrix drums in which the contaminant is normally well-distributed, are normally more satisfactory. The detection limit is approximately:
- 4 mg of ^{240}Pu using neutron counting (spontaneous fission)
- 30 mg of total Pu using γ spectrometry (of the 208 keV ^{241}Pu peak) but always with overestimation of the value.

6.4.3 - Measurement Accuracy

Measurement accuracy is made up of several factors:
- statistical accuracy of the measurement
- accuracy linked to the spatial effect in the matrix
- accuracy linked to the nature and density of the matrix
- accuracy linked to the counting method selected

The higher the quantity of Pu present, the more important accuracy becomes.

1) Error linked to statistical accuracy is more or less repeated from one counting method to the other and has little influence on the result, whichever method is selected. Tests and calibrations carried out on drums justify the quality of the counting. However, in Pu-rich drums containing old waste containing large amounts of americium, measurement repeatability remains hard to guarantee, even today, given the large number of accidental coincidences in relation to real coincidences taken into account by SCRS.

2) Error linked to the spatial effect has most influence on measurement accuracy. This effect is corrected for using γ scanning but, as was explained above, correction is imperfect.

3) Error linked to the matrix effect due to absorption by the matrix can be significant to the extent that the matrix density of the drum counted is different from that used for calibration. At UP1, a single matrix density was taken into account: $d_H = 0.15$. In addition, this matrix is always of the light vinyl type (gloves, vinyl, cottons) which limits error due to the matrix effect perfectly.

4) In general, experience has shown that error due to the counting method selected is less when quantitative results are given using neutron counting and when γ spectrometry is only used to analyze isotopic composition and for γ scanning.

Overall measurement error is approximately 20% for a 100 liter vinyl matrix drum, density $d_H = 0.15$ for a counting time of approximately 30 minutes.

6.5 - Future Prospects

Given the performances achieved using the shift register electronic system, this will certainly remain, in the short term, the sole neutron counting method used, without deterioration in the detection limits.

Improvements must be made in the repeatability of measurements carried out on old, Pu-rich drums.

The Tracor computer must be replaced by a microcomputer with a greater capacity and using a simpler computer language than that used by the Tracor.

Lastly, it is planned to introduce methods using neutron interrogation of fissile material in the medium term.

TABLE 1

Pu O_2 enriched with ^{240}Pu (%)	Weight of ^{240}Pu (g)	Weight obtained by:					
		γ spectrometry		dead time neutron counting		Shift register neutron counting	
		^{240}Pu (g)	Deviation (%)	^{240}Pu (g)	Deviation (%)	^{240}Pu (g)	Deviation (%)
8	0.63	0.63	0	0.6			
8	1.43	1.31	-8	1.25	-13		
8	2.23	2.05	-8	1.8	-19		
8	3.03	2.8	-8	2.4	-20		
8	3.83	3.6	-6	2.8	-27		
8	4.6	4.4	-4	3.5	-24		
8	5.4	5.3	-2	4.1	-24		
8	6.2	5.7	-8	4.7	-24		
8	7.0	6.2	-11	5.2	-26		
8	7.8	7.1	-9	2	-74	7	-10
25	12.5	7.7	-38	9.4	-25	10	-20
25	25	30	+20	17	-31	23	-8
25	37.5	41	+9	23	-39	32,5	-13
25	50	63	+26	26	-48	44	-12
25	62.5	-	-	28	-55	50.5	-19

OPERATIONAL EXPERIENCE OF NDA TECHNIQUES AT DOUNREAY IN SOLID WASTE MEASUREMENT FROM REPROCESSING 13t FBR FUEL

BY

W B BREMNER AND D W ADAWAY
AEA TECHNOLOGY, DOUNREAY

Summary

An integrated solid waste measurement system has been in operation at Dounreay since 1978. In most cases the systems were purpose-designed for the particular waste stream. One of the priorities of the integrated system is to route waste to the appropriate storage facility and achieve the most economic use of engineered storage capacity. The integrated system as a whole is discussed.

The design and operation of a Californium Shuffler is described which is used to assay head-end waste. A new design of Shuffler is outlined which has been installed in a PIE facility.

An integrated subsystem consisting of a segmented gamma scanner and passive neutron counters is described along with a computerised data storage approach. These systems are mainly used to assay intermediate level waste and low level waste arising from laboratories.

The use of mobile slab counters is finding increasing application in non-standard waste streams, and streams where a dedicated counter cannot be justified economically.

The application of mathematical modelling techniques to improving the understanding of counter response in the Californium Shuffler, passive neutron counters, and the segmented gamma scanner is discussed.

1.   INTRODUCTION

The accurate measurement of nuclear materials is an essential requirement for the efficient operation of nuclear fuel reprocessing and the associated waste treatment plant. Conventional destructive analysis techniques are difficult to apply, particularly to solid wastes, because of the difficulty in obtaining representative samples from material which is not of a uniform nature. Non-destructive assay (NDA) techniques allow the determination of uranium and plutonium in nuclear materials, including the plutonium isotopic composition, without altering the chemical and physical form of the material. There are considerable cost incentives in using NDA techniques.

At AEA Technology, Dounreay the reprocessing of some 13t of fuel from the Prototype Fast Reactor (PFR) has made extensive use of an integrated solid waste system comprising a range of NDA devices which use passive gamma, passive neutron and active neutron techniques to provide adequate measurement capability for the various waste arisings (fig 1). The highest loss of fissile material is from head-end arisings such as leached hulls (0.3% of plant throughput) which has put great emphasis on the active neutron technique using a sealed-tube generator or a Californium Shuffler, in the delayed neutron counting mode.

It is becoming increasingly important to obtain accurate and verifiable data for these systems, and to ensure that the range of matrices in the stream can be adequately covered. A mathematical modelling programme has provided an essential ingredient to this requirement, and in some cases has enabled the fundamental calibration data to be applied to several different matrices.

The fundamental approach to the design and selection of the integrated measurement system has been found to be sound but some changes have been introduced such as the use of variable geometry portable devices.

2. INTEGRATED SYSTEM DESCRIPTION

The principle systems employed are described below with their locations.

SYSTEM	LOCATION	DESCRIPTION
NDA 2	High alpha low beta-gamma waste store	Segmented gamma scanner for waste in 270mm La Calhène containers
NDA 3	High alpha low beta-gamma waste store	Passive neutron coincidence counter for waste in 270mm La Calhène containers
NDA 4	High alpha low beta-gamma waste store	Passive neutron coincidence counter for bulked waste in 220l drums
NDA 5	Disassembly cave PFR fuel reprocessing plant	Californium shuffler
NDA 6	High alpha high beta-gamma waste store	Passive neutron counter
NDA 7	Waste posting cell in PIE facility	Active neutron interrogation using neutron generator, about to be replaced by Californium shuffler
NDA 10	Laboratories	Passive neutron counter for sentencing waste in 220l and 100l drums to burial
NDA 17	portable counter	Passive neutron, variable geometry coincidence counter for general use (flasks, filters, drums etc.)

3. WASTE ARISINGS

The waste arisings dealt with by the major systems since PFR reprocessing began are detailed below.

SYSTEM NUMBER	NUMBER OF CONTAINERS MEASURED	MATRIX	ASSOCIATED ACTIVITY	DENSITY g/cc
NDA 2	4033	paper/PVC/ polythene	0-5µSv/hr	0.1-0.3
NDA 3	8010	paper/PVC/ metal	0-0.75mSv/hr	0.1-1.0
NDA 4	17007	paper/PVC/ metal	0-0.75mSv/hr	0.1-0.3
NDA 5	2302	stainless steel	37GBq	1.5
NDA 6	287	stainless steel	75GBq	1.3
NDA 7	11500	polythene/ paper/steel	37GBq	0.1-1.3
NDA 10	3195	paper/metal	0-5mSv/hr	0.1-0.3

4.   SYSTEM AVAILABILITY

It is important in a fuel reprocessing facility that the supporting NDA systems have good availability throughout the campaigns. Most of the systems have been available for greater than 80% of the time with some systems achieving 95%. The systems which have had lower availability (70%) are NDA 6 and NDA 10. NDA 6 which assays head-end waste consigned to the high alpha, high beta-gamma waste store has given problems with noise primarily due to dampness in the working environment. The system has recently been extensively modified and is currently undergoing further modifications to update the computer control. NDA 10 has been unavailable due to extensive refurbishment of the counting assembly but it was possible throughout this period to use alternative counting systems, notably the variable geometry slab counter NDA 17.

5.   HEAD-END MEASUREMENTS

5.1   Selection of Technique

The choice of an active interrogation technique was dictated by the high neutron emission of ^{242}Cm and ^{244}Cm present in irradiated PFR fuel. A typical isotopic composition of PFR plutonium at reprocessing is given in table I. The neutron emission contribution for the various isotopes for irradiated PFR plutonium is given in table II. Typically the passive neutrons from spontaneous fission of ^{242}Cm and ^{244}Cm exceed the passive neutrons from spontaneous fission of ^{240}Pu by a factor of ten. This does not preclude the use of passive neutron counting techniques but makes the accuracy of the result highly dependant on precise knowledge of the fuel and its history. In practice a passive count is made routinely and the result is corrected for neutron emission from other sources (a,n and sf) based on fuel history and FISPIN calculations. This result acts as a useful check on the active result and, if wide and repeatable discrepancies exist, can act as an indication of system fault conditions or of the presence of fuel not consistent with the expected input.

5.2   Description of Technique

The system is shown in fig 2. A typical measurement sequence is described, as follows. The dissolver basket is positioned at the top of the counting well and then automatically lowered for interrogation of the first segment. A background count is made for 120 sec before the ^{252}Cf source (2.89mg, 6.69E9 n/sec in 1984) is used to irradiate the fissile material for 8 seconds. The source is withdrawn to the storage position (transfer time about 0.8 sec) and the delayed neutrons from fission of the ^{239}Pu are counted for 6 sec. The irradiation - count sequence is repeated 10 times. The dissolver basket is automatically raised 90mm and the above procedure repeated. A total of 10 segments are counted. The number of delayed neutrons counted is corrected for source decay and a result calculated by comparison with the calibration produced from synthetic standards. The dissolver basket is then positioned centrally with respect to the counting tubes and a passive count is made for 10 min. This result is also compared with a calibration produced from synthetic standards and is corrected for neutrons from (α,n) production and from spontaneous fission of ^{242}Cm and ^{244}Cm.

5.3   Description of Installed Equipment

5.3.1 Counting Well and Counting System

The counting well, which is constructed to take a dissolver basket 1000mm high by 200mm diameter, is positioned in the floor of the fuel disassembly cell on a vertical axis. Movement of the dissolver basket is made by an in-cell hoist which is controlled by the main control computer. The well is surrounded by 150mm lead and 300mm wax. In the original counter the nine ^{3}He tubes were equispaced round the well parallel to the well axis in re-entrant tubes within the wax. This arrangement posed several problems. Access to the tubes was via the cell interior and led to extensive surface contamination of the tubes. Cable connection to the tubes was also via the cell interior and led to radiation embrittlement of the cable insulation. The counting geometry was redesigned in 1987. Holes were bored horizontally from the exterior face of the cell floor to positions tangential to the well axis at the same distance from the well centre as the original tubes. A total of eight holes were bored, four on each side of the well. The holes were bored to allow a lead sleeve 25mm thick to surround each tube reducing the gamma dose. Additional lead shielding was also installed in the cell as an annulus round the well above the tubes. This arrangement has proved entirely satisfactory. The tubes cannot become contaminated and they can be withdrawn from the front face of the cell. The cabling is also no longer subject to radiation damage.

5.3.2 Shuffler

The Shuffler mechanism was obtained from Los Alamos National Laboratory (LANL) under the US/UK LMFBR exchange agreement. The source was encapsulated in a special container designed by Savannah River Laboratories. A polythene plug was designed to house the shuffler mechanism and guide tube. The source is moved on a Teleflex cable by a gear wheel controlled by a stepping motor. Proximity sensors are used to indicate correct movement of the source. Source movement is controlled by a Stepping Motor Controller which is in turn controlled by a Commodore PET computer via a Harwell MOUSE unit. The MOUSE unit, in addition to providing I/O lines for control, contains a timer and scalers which are used to collect the counts. The counts are transferred to the computer using the IEEE-488 bus.

5.3.3 Computer Control
The main control programme contains three routines:

a) Active Count
b) Passive Count
c) Change Parameters

The Active Count sequence is the main routine for fissile material assay and controls movement of the in-cell hoist, movement of the ^{252}Cf source, and collection and evaluation of data.

The Passive Count sequence is similar though simpler and includes an empty well background count.

The Change Parameters routine enables the operator via password control to change number of segments, segment size, delay time between segments, background counting time per segment, number of assay cycles per segment, sample irradiation time, and delayed neutron counting time. It also permits the calibration coefficients to be entered for a polynomial equation.

5.4   Experience with Detector System
Initial problems with gamma sensitivity were addressed by setting each detector tube operating voltage on the knee of the response plateau. As the gamma dose rate increases the slope of the plateau increases and its length decreases. If the operating point is set midway on the original plateau, obtained during calibration in low radiation conditions, the detector response is greatly changed in high radiation conditions. By setting the operating point on the knee of the plateau the normal benefits of being midway on the plateau are lost (ie. reduced sensitivity to changing tube voltage) but much reduced sensitivity to increased gamma dose rate ensues.

Problems were also encountered with the tube voltage. The high voltage supply to the detectors is not removed during the irradiation period. After the irradiation sequence, particularly at high gamma dose rates, it was found that the voltage dropped by about 20 volts and took in excess of two seconds to recover. This effect was removed by substituting a different model of power supply and modifications to the pre-amplifiers.

With these modifications the counting system operates well under all circumstances.

5.5   Overall Performance of System
The Shuffler has made 2 300 determinations on leached hulls, centrifuge bowls, and standards since 1984. It is estimated that the source has been shuffled a total of 250 000 times. On the actual Shuffler the Stepping Motor Controller has been the only item to fail and has been replaced once. The Teleflex cable has once bypassed the stop sensor which led to modifications on the Dounreay system and other LANL systems. Some problems have occurred with the in-cell hoist.

6.   PCM MEASUREMENTS

The PCM measurements are made using either passive neutron coincidence counters or a passive gamma system. In the passive neutron counters conventional detectors employing BF$_3$ or ^{3}He tubes are used. At the start of PFR reprocessing the coincidence electronics were Variable Dead Time but all have now been changed to Shift Register Units.

The most important waste stream is associated with materials being sent to the high alpha, low beta-gamma waste store. This stream uses three NDA devices to assist in decategorisation and economic storage of waste. The systems used are NDA 2, 3 and 4. NDA 3 is a passive neutron counter and is used to measure La Calhène containers where the density of the material is too high for passive gamma or where the plutonium level is too low (< 0.5g). NDA 2 which assays laboratory waste in 270mm La Calhène containers is a segmented gamma scanner. The container is placed on a turntable and a HPGe detector is used to scan the container in 22 segments. The 414 kev photopeak is used to determine the ^{239}Pu content, and a correction for matrix density of each segment is made using a ^{75}Se transmission source. The plutonium spectrum is also used to determine the isotopic composition and the final result is corrected to total plutonium. NDA 4 is a conventional passive neutron counter which has the capability of measuring 200l drums in the range 0.01 to 50g plutonium and thus fulfils both an accountancy and decategorisation role.

The use of different NDA techniques to measure both the individual La Calhène containers and the same material after bulking into 200l drums gives an opportunity to make a comparison of the respective assays (fig 3).

7. CALIBRATION

7.1 Hulls

The calibration for leached hulls was made by preparing simulated hull standards. These consist of (U,Pu)O2 granules, of similar composition to that used in PFR, dispersed on clean hulls. Twelve standards were made covering the range 0.5-15g Pu. By combining standards calibrations up to 63g Pu were produced in both active and passive modes. A basket of leached hulls which gave an active interrogation result of 134g was split into four batches and each batch interrogated separately. This produced results of 28, 32, 34 and 37g confirming the calibration up to about 130g. Certain standards are combined to produce a 30g standard which is regularly used to provide a system control. From approximately 100 determinations the mean was 31g with a standard deviation of 5.8g.

7.2 Centrifuge Bowls

Similar standards were made up to simulate centrifuge bowls using the same (U,Pu)O2 granules. The calibration has been demonstrated to be linear up to 300g in the passive mode. In active mode a decision has to made regarding the geometry of the material in the bowl as different calibrations are used for normal on-wall material and slumped material (see section 8).

7.3 PCM

A wide range of standards to simulate PCM waste were prepared using (U,Pu)O2 granules distributed on latex. The latex sheets were rolled up and inserted in aluminium cans, which in turn were placed in drums containing a simulated waste matrix which is characteristic of the waste stream. In some cases absolute calibration were carried out using small discrete masses of plutonium.

8. MATHEMATICAL MODELLING

A collaborative programme on mathematical modelling has operated between Dounreay and JRC Ispra since 1977. Work on the Shuffler has concentrated on:

a) calculation of errors which may arise if the plutonium in leached hulls is not evenly distributed

b) calculation of errors which may arise if the plutonium in centrifuge bowls changes from the normal on-wall distribution to a slumped distribution in the bottom of the bowl. Experiments have been conducted using the cadmium ratio technique to provide experimental data for this work but results have yet to be evaluated.

Modelling work has also been successfully carried out on response characteristics for different density matrices in NDA 4, and on evaluation of results from the segmented gamma scanner NDA 2. In both cases the algorithms deriving from the modelling work have been incorporated in the software packages controlling the systems.

9.   FUTURE WORK AT DOUNREAY

The sealed tube neutron generator in NDA 7 has recently been replaced by another Shuffler system in the Post-Irradiation Facility at Dounreay. This system, based on the LANL principle, has a source transfer mechanism designed by UKAEA staff at Dounreay, Harwell and Risley. An STE-bus based computer will be used for controlling the source movement instead of a Stepping Motor Controller. An IBM-PC will be used both to control the STE-bus computer and to process the data. This system is expected to be commissioned later in 1989 and if successful it will form the basis for updating the system in the PFR fuel reprocessing plant.

10.   CONCLUSIONS

The integrated system has met all of the criteria for both fissile material measurement and decategorisation of solid waste at Dounreay. It has sufficient flexibility to allow systems to be modified as improved detectors or electronics become available. The mathematical modelling work has given further confidence in the ability of the system to meet the required measurement performance for various matrices and in some cases has provided on-line correction procedures.

Future systems are likely to be based on variable geometry portable systems rather than fixed geometry dedicated systems. Three year's experience with the variable geometry high efficiency slab counter has demonstrated its versatility in counting waste in a variety of containers up to 500l drums and lead flasks, and a second counter of the same type is being constructed. The modular design of the counter lends itself to empirical mathematical design of counters for particular applications which can be readily constructed from existing modules.

11.   REFERENCES

1. W B BREMNER, M L CLARK and B W SPENCE. Operational experience relating to measurement of plutonium in solid waste streams from Fast Breeder Reactor reprocessing. IAEA-SM-260/68 International Meeting on Nuclear Safeguards, Vienna, 1982.

2. O PUGH. Fuel Cycle Operations at Dounreay. The Nuclear Engineer, Vol 26 No 4 July/August 1985.

3. W B BREMNER. Methodology of solid waste measurement and the role of an integral experiment for calibration and interpretation of NDA systems at DNPDE. Seminar on measurement of plutonium in solid waste and application of NDA techniques to FBR fuel reprocessing. Universite Claude Bernard, Lyon, France, May 1983.

4. B J MCDONALD, G FOX and W B BREMNER. Non-destructive measurement of plutonium and uranium in process wastes and residues. IAEA-SM-201/61. International Meeting on Nuclear Safeguards, Vienna, 1976.

5. G W ECCLESTON and others. Fast facility spent waste assay instrument. ANS/INMM Conference on Safeguards Technology: The Process Safeguards Interface, South Carolina USA, Dec 1983.

6. G W ECCLESTON and others. Dounreay Shuffler diagnostic software operations manual. Los Alamos report LA-10470-M July 1985.

7. W B BREMNER and others. An integral experiment relating to calibration and interpretation of solid waste measurements at DNPDE. Commission of the European Communities Report EUR 8020 En 1984.

8. G BIRKHOFF and W B BREMNER. Calibration and interpretation of plutonium waste measurements. International Symposium on Management of Alpha-contaminated Wastes: Vienna 1980.

TABLE I

TYPICAL ISOTOPIC COMPOSITION
OF Pu IN PFR FUEL AT REPROCESSING

	Isotopic Composition (wt%)
^{238}Pu	0.3
^{239}Pu	70.0
^{240}Pu	25.7
^{241}Pu	3.0
^{242}Pu	1.0

TABLE II

NEUTRONS FROM IRRADIATED PFR FUEL
FUEL HISTORY   BURN-UP   8%
COOLING TIME   1 YEAR

NUCL	MODE	n/s/kg	%ages excluding delayed neutrons	%ages including delayed neutrons
Pu	sf	3E5	11	6
Pu	α,n	1E5	4	2
Cm	sf	2E6	74	43
Cm	α,n	3E5	11	6
Pu	del	2E6		43

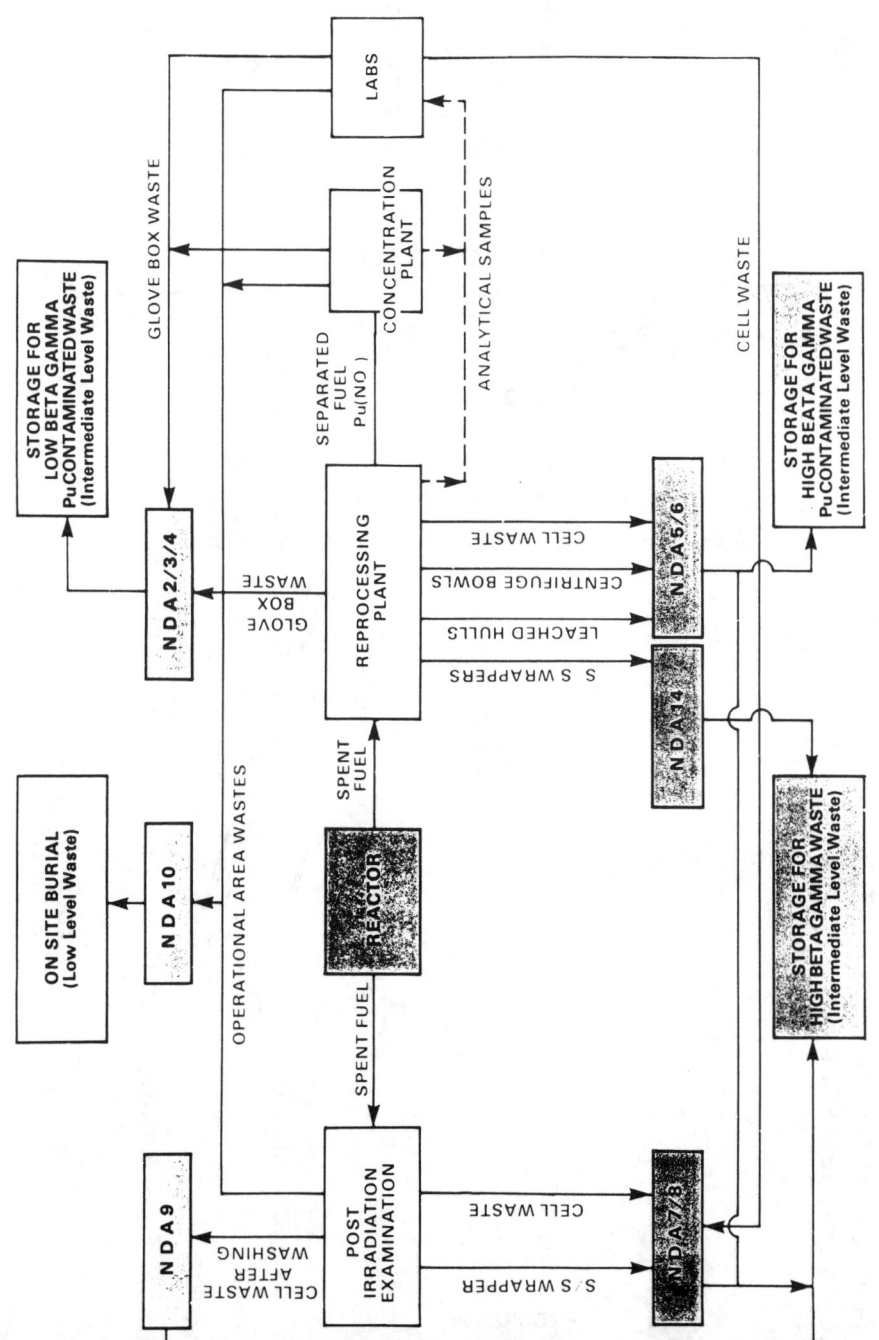

FIGURE 1. NDA FLOW DIAGRAM.

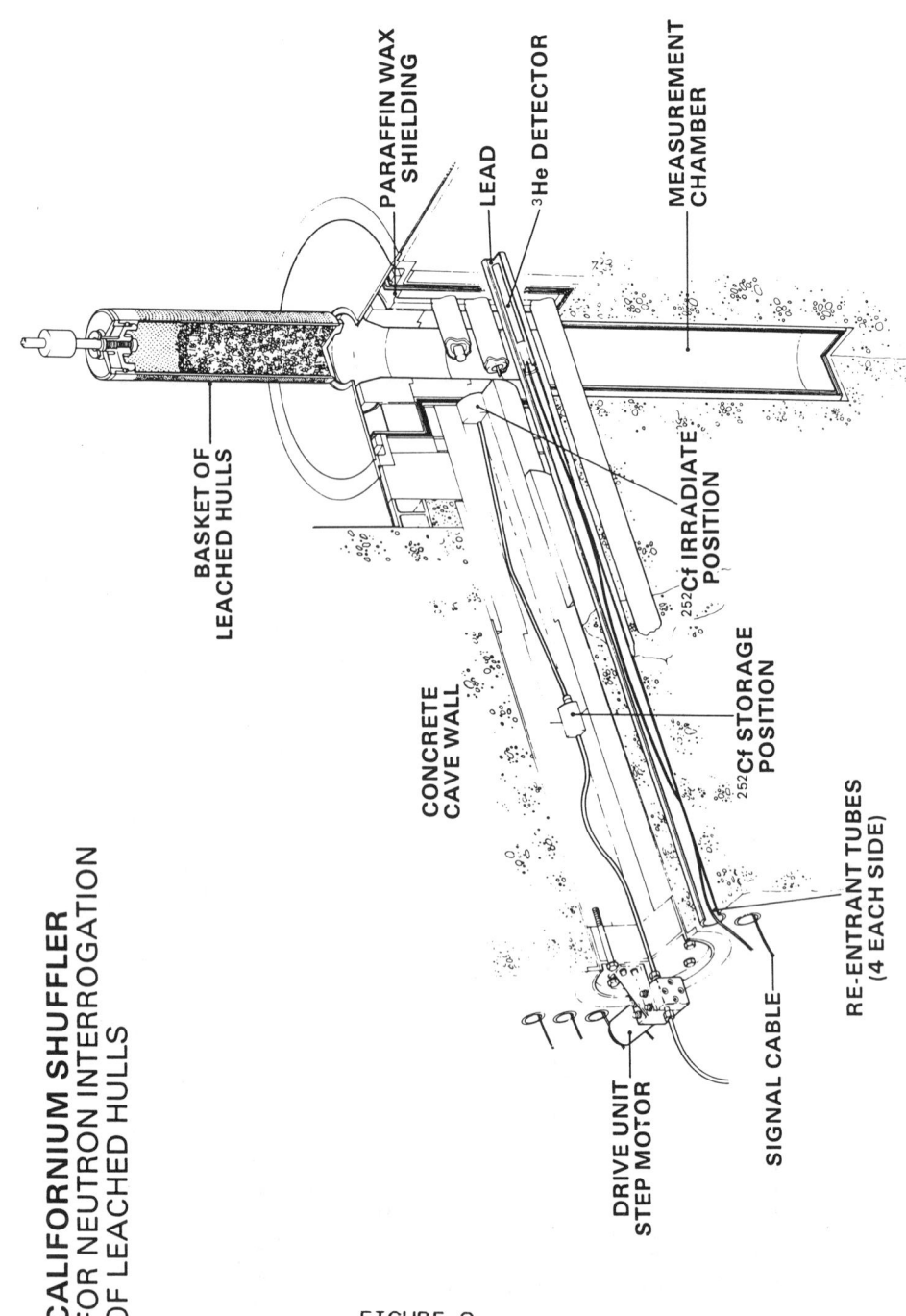

CALIFORNIUM SHUFFLER
FOR NEUTRON INTERROGATION
OF LEACHED HULLS

BASKET OF
LEACHED HULLS

PARAFFIN WAX
SHIELDING

LEAD

^{3}He DETECTOR

MEASUREMENT
CHAMBER

^{252}Cf IRRADIATE
POSITION

CONCRETE
CAVE WALL

^{252}Cf STORAGE
POSITION

RE-ENTRANT TUBES
(4 EACH SIDE)

DRIVE UNIT
STEP MOTOR

SIGNAL CABLE

FIGURE 2

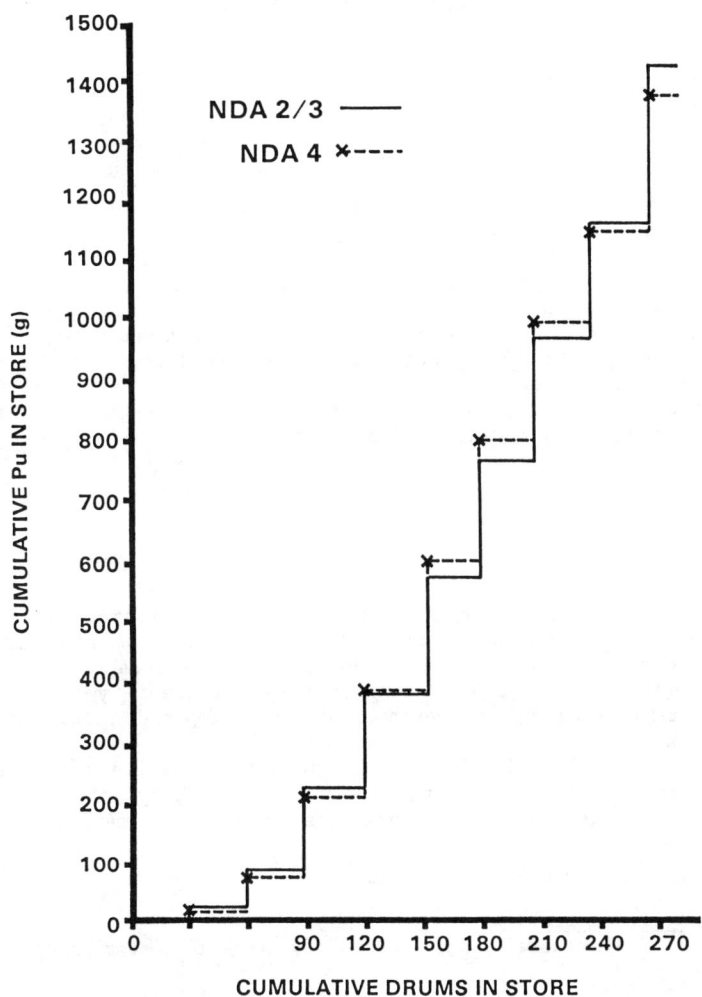

COMPARISON OF MEASUREMENTS
ON NDA2/3 AND NDA4

NDA 2/3 ———
NDA 4 ✖----

CUMULATIVE Pu IN STORE (g)

CUMULATIVE DRUMS IN STORE

FIGURE 3

FULLY AUTOMATED LABORATORY FOR THE ASSAY OF PLUTONIUM IN WASTES AND RECOVERABLE SCRAPS

Ph. GUIBERTEAU, F. MICHAUT and C. BERGEY
C.E.A. Centre d'Etudes de Valduc - 21120 IS-sur-TILLE, France

Th. DEBRUYNE
C.E.A. Centre d'Etudes de Bruyères-le-Châtel
91680 Bruyères-le-Châtel, France

Summary

To determine the plutonium content of wastes and recoverable scraps in intermediate size containers (ten liters) an automated laboratory has been carry out. Two passive methods of measurement are used. Gamma ray spectrometry allows plutonium isotopic analysis, americium determination and plutonium assay in wastes and "poor" scraps. Calorimetry is used for accurate (± 3 %) plutonium determination in "rich" scraps. A full automation was realized with a barcode management and a supply robot to feed the eight assay set-ups. The laboratory works on a 24 hours per day and 365 days per year basis and has a capacity of 8,000 assays per year.

1. INTRODUCTION

To perform the plutonium entrance balance of a reprocessing unit, a laboratory has been specially studied and equipped for non destructive determination of plutonium in wastes and recoverable scraps.

The requirement to have to measure very different types of products and the techniques used (passive methods) have imposed a full automation of the laboratory.

We will describe, in first, the problem to solve and the chosen solution. Then we will give details on the laboratory with emphasize on the main characteristics and some examples of the performances obtained.

2. PROBLEM TO SOLVE AND CHOSEN SOLUTION

2.1 The data of the problem

The non destructive assay laboratory has to answer very different requirements :

- measurements, in the same room of "rich" (MPu > 100 g) and "poor" (MPu < 1 g) products

- wastes and scraps have very different shapes, weights and matrices (metal, oxide, graphite moulds, gloves ...)

- presence of several plutonium isotopic compositions (2 % < ^{240}Pu < 15 at.%) and americium contents (10^2 to 10^4 ppm) but no fission products

- necessity to manage the laboratory with only two persons in normal day hours (8 hours a day). And this, for an annual flow of 2,000 "rich" and 6,000 "poor" parcels (samples).

2.2 The existing solutions

Active neutron interrogation methods have not been considered because the installation cost and the state of development in France when this study was launched.

Three passive methods of determination were tested (1) : gamma-ray spectrometry, calorimetry and passive neutron assay.

In our situation every technique has advantages and limitations.

2.2.1 Gamma-ray spectrometry

Allows a simultaneous determination of plutonium masses and isotopic compositions. Gamma-ray spectrometry is useable in a wide range of concentrations but precision (typically ± 5 %) and accuracy are limited by sample self-attenuation of the gamma-rays. The method can be used neither on large metallic samples nor on strongly absorbant matrix.

2.2.1 Calorimetry

In the case of low burn-up plutonium, calorimetry is the more precise non destructive method for plutonium determination. All types of samples and matrices are accepted, but calorimetry needs to know the isotopic composition and americium concentration. The determination times are long (t ≥ 6 h) and the equipment expensive.

2.2.3 Passive neutron assay

Used for determination of medium quantities of plutonium (0,5 g ≤ Mpu ≤ 150 g), the passive neutrons counting presents in our conditions (low burn-up, important quantities of low Z matrix), two major draw backs : the necessity to know with precision the ^{240}Pu percentage and the sensitivity of the technique to the α-n reactions.

2.3 - The chosen solution

It is clear that among the three passive methods, only the use of combined gamma-ray spectrometry and calorimetry will give a correct answer to our two main requirements :

- knowledge of isotopic composition
- precise determination (± 3 %) of plutonium in "rich" scraps.

But, these methods need very long counting times (2 to 12 h) and we have to realize a complete automation of the laboratory to be able to test 8,000 samples per year.

3. EXPERIMENTAL

3.1 The laboratory (fig. 1)
The laboratory consists of two different parts :

- an operator room where the multichannel analyzer system and the micro-computer which allows automatic data acquisition analysis and management of all the hardware are located

- a measurement room including :
    - 7 gamma-ray spectrometry set-ups
    - 1 calorimeter (a second one will be purchased in 1990)
    - 2 sites for the parcels transportations racks
    - 1 supplier robot which is an interface between the transportation racks and the assay set-ups.

3.2 The assay set-ups

3.2.1 Calorimeter (fig.2)
The calorimeter (SETARAM FRANCE) is a Tian-Calvet type with two cells (one reference and one measurement cell).
The dimensions of the cells are : L = 360 mm ; l : 215 mm ; h = 268mm
The e.m.f. is measured with a millivoltmeter interfaced with a micro-computer. The software allows an automatic drawing of the v = f/t) curve (fig. 3) and the automatic determination of the equilibrium. The calorimeter can be loaded by the robot.

3.2.2 The gamma-ray spectrometry set-ups
Each is composed of :

- a counting chamber with five centimeters lead shields
- a sample rotating table (one rotation table allows vertical displacements for segmented gamma-ray scanning)
- a detector NaI(1) or GeHP(6)) with pulse processing electronics
- ^{152}Eu gamma-ray sources used to determine the self attenuation correction
The seven set-ups are connected with a IN1200 (Intertechnique-France) multichannel analyzer.

3.3 Automation

3.3.1 The supply robot (fig. 4)
Scraps and wastes are put into P.V.C. containers (Ø : 315, h : 320) which are disposed on the transportation racks. The racks are used to collect the wastes from the plant storage and then in the assay laboratory, to play the part of an interface between the men and the robot.
The supply robot (CYBERNETIX), a three axis arm with an articulated hand is hanged up to a rail. Fited with several safety devices, the robot allowed a safe transfer of the containers from the racks to the assay set-ups, following a procedure described in 3.4, and under computer control.

3.3.2 Computer NETWORK (fig. 5)

A central microcomputer (PC COMPAQ 386) controls all the different functions of the laboratory :

- data acquisition (references of the parcels, results of the measurements)
- management of the robot and analysers
- edition of final results on stickers

3.3.3 Barcode management (fig. 6)

To minimize the human errors of transcription a barcode management is used.

Every parcel is characterized by a barcode (isotopic composition, nature, origin). This "parcel barcode" is then, associated to a "container barcode" and, finally, to a "rack site barcode".

All these data are transfered into the central microcomputer, before the measurement, by an optical pen.

3.4 Measurement procedure

A general procedure illustrated by the fig.7 allows the assay of very different samples with an optimal utilization of the time. Every "poor" parcel is measured first by the NaI detector spectrometer. After ten minutes of counting the first results allow the automatic guidance of the parcel to one of the six GeHP spectrometer set-ups. All this procedure is fully automated under the control of the central microcomputer.

4. PERFORMANCES AND CHARACTERISTICS

4.1 Calorimetry

4.1.1 Standardization

To reach better quality control two different methods of standardization are used : standardization by Joule effect and utilization of plutonium standards.

Ten plutonium reference samples allow to cover all the different isotopic composition used. The two methods of standardization are in good agreement and the value of the conversion factor has been determined to $k = 77,1 \pm 0,6$ mv/w

4.1.2 Measurement time

The measurement time is a function of sample composition and packing. A preheating of the sample at the equilibrium temperature (38° C) allows a reduction of this measurement time which is typically about 6 hours for metal plutonium and 14 hours for calciothermic slags (19 h without preheating).

4.1.3 Precision - Accuracy

In calorimetry, precision is composed of three main components :

- precision on the knowledge of the isotopic composition of plutonium and ^{241}Am concentration
- precision on the conversion factor K
- precision on the calorimetric measurement (mV)

In our case, the first is the major component, so, the precision is typically ± 3 % in the case of gamma-ray isotopic measurement and ± 1 % if thermoionisation mass spectrometry is used.

Figure 8 gives some examples of the accuracy of our calorimetry set-ups.

4.2 Gamma-ray spectrometry

4.2.1 Gamma-ray attenuation correction

The gamma-ray attenuation correction is the main problem in gamma-ray assay.

For wastes or scraps the origin of this attenuation could be the matrix and/or the plutonium itself.

The method of correction used (2) is a combination of the two classical methods : the "infinite energy method" (3) and the transmission measurement of an external ^{152}Eu source placed behind the sample. Figure 9 summarizes the principle of the calculations.

This combined method gives good results for every types of matrices (low or high Z).

4.2.2 Isotopic analysis

The isotopic composition of plutonium is calculated with the gamma-ray lines in the region from 120 to 414 kev. The sophisticated peak fitting algorithms have been established earlier by T. DEBRUYNE (4).

4.2.3 Precision - Accuracy

Figure 10 gives results obtained by gamma-ray spectrometry on very different samples. The comparison with calorimetry or destructive analysis shows only little deviations (< 10 %) for masses of plutonium higher than 1 g.

For the wastes the results of an internal round-robin (figure 11) show the usefulness of the gamma-ray attenuation correction even in the case of a light Z matrix (cellulose, vinyl ...).

5. CONCLUSION

The combined use of optimized classical passive non destructive methods and a manual feeding have allowed in the last six months, the plutonium assay of more than 500 intermediate size parcels of recoverable scraps. Typical precision was 3 to 10 % according to the matrix and the plutonium concentration.

Now, the assembly of the supply robot comes to an end. The fully automated working of the laboratory, expected for the beginning of the next year, will give some advances in several fields :

- Safety :
The barcode management will allow a better identification of the parcels and will suppress the transcription mistakes.

- Flexibility and productivity :
Two chemists in legal day hours (8 h per day) will manage the laboratory working on a 24 h day and 365 days per year basis. In this condition, 2,000 parcels of recoverable scraps and 6,000 parcels of wastes will be measured in one year.

- Precision and accuracy :
The work on a 24 h a day basis will allow a better precision by an increase of the counting time (gamma ray spectrometry) and a better accuracy by a multiplication of the measurements of the standard reference materials.

REFERENCES

1. J.M. JEHAN, F. MICHAUT et C. BERGEY
    Comparaison de trois méthodes de détermination des masses de plutonium dans les déchets
    Proceedings - Réunion technique sur l'analyse des déchets
    CADARACHE (1987)

2. F. MICHAUD, J.J. RADECKI
    Etude de la correction d'atténuation dans une matrice homogène en spectrométrie gamma
    Proceedings - Réunion technique sur l'analyse des déchets
    CADARACHE (1987)

3. J. MOREL, M. VALLEE
    Note Technique 86 CEA/LMRI/175 (1986)

4. Th. de BRUYNE, M. SILLY, Al. ADAM, J. LAUREC
    CEA R 5290 (1985)

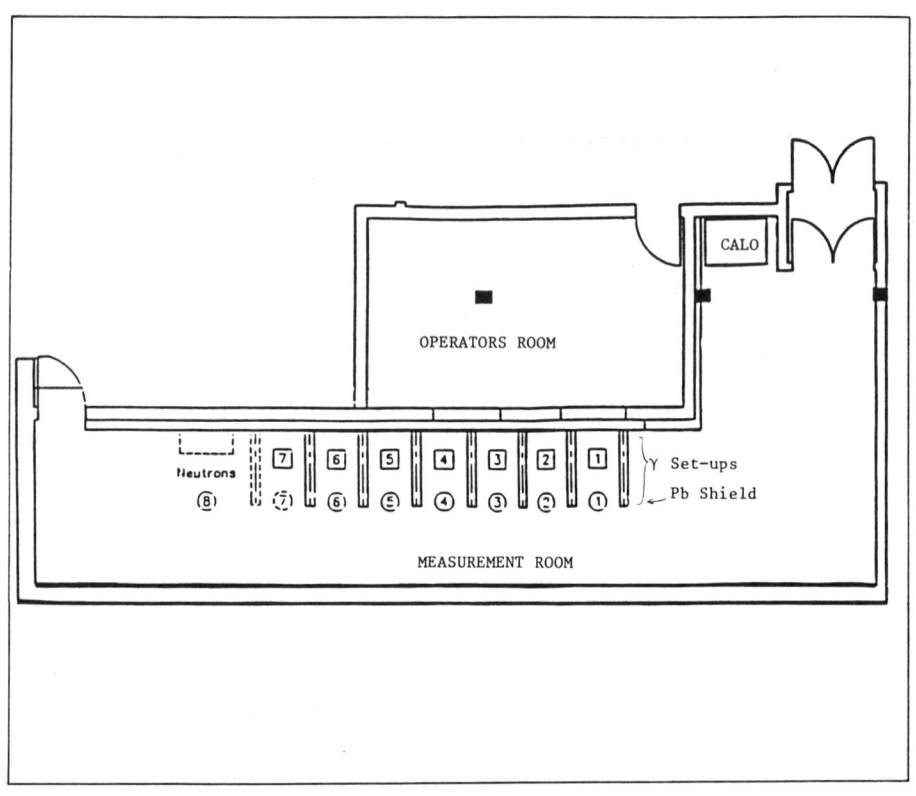

FIG. 1 - GENERAL VIEW OF THE LABORATORY

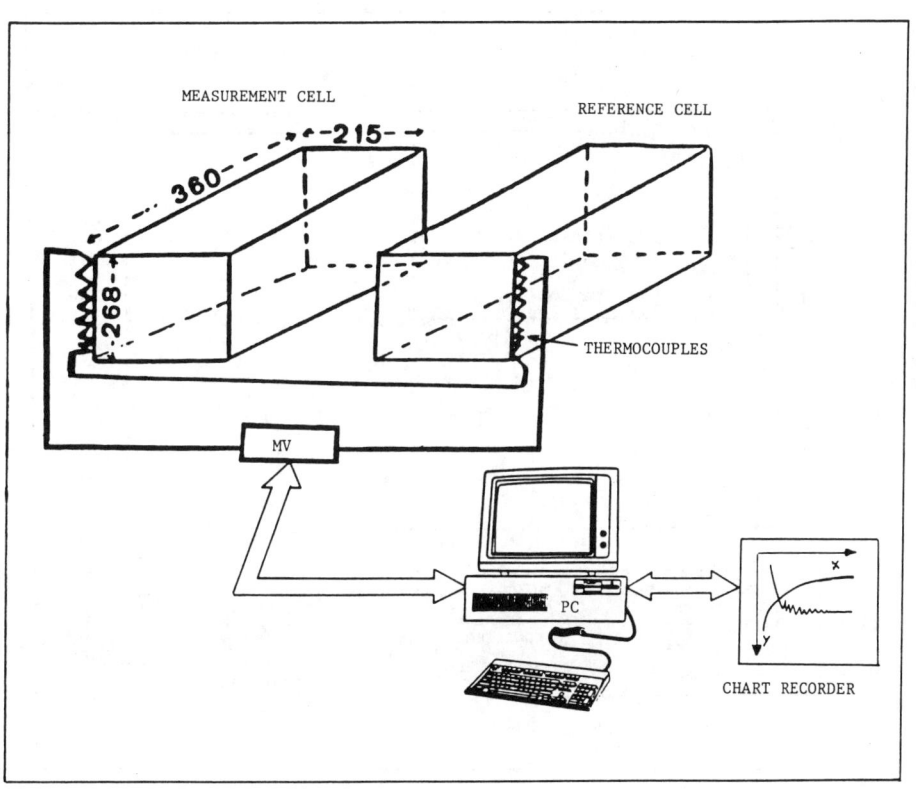

MEASUREMENT CELL

REFERENCE CELL

215

360

268

THERMOCOUPLES

MV

PC

CHART RECORDER

FIG. 2 - SHEME OF THE CALORIMETER (DOUBLE CELL FLUXMETRIC TYPE)

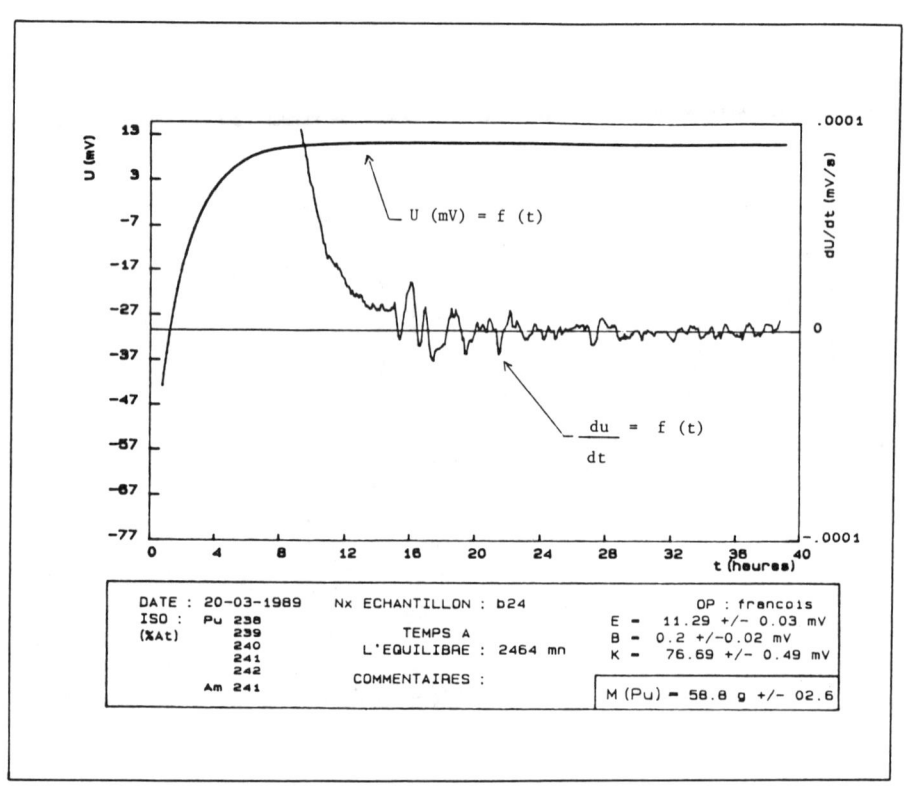

FIG. 3 - CALORIMETRIC DATA

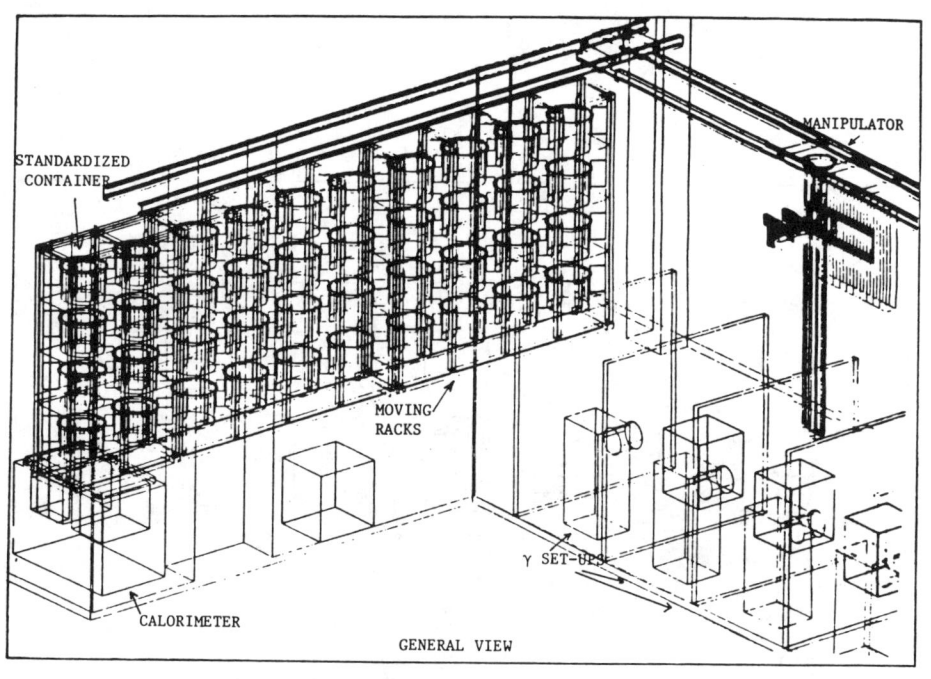

STANDARDIZED
CONTAINER

MANIPULATOR

MOVING
RACKS

γ SET-UPS

CALORIMETER

GENERAL VIEW

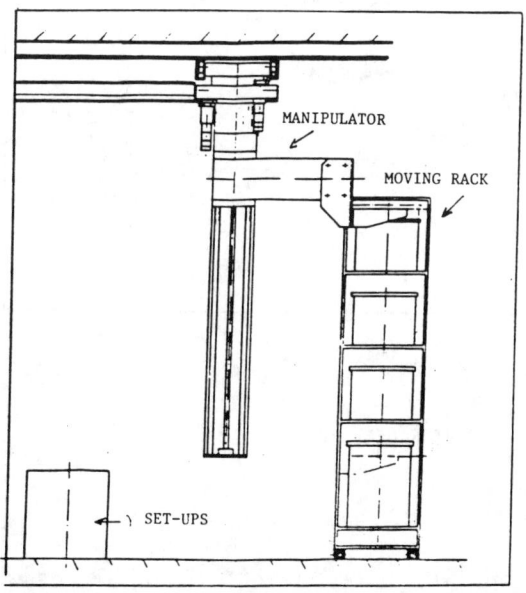

MANIPULATOR

MOVING RACK

γ SET-UPS

FIG. 4 - THE SUPPLY ROBOT

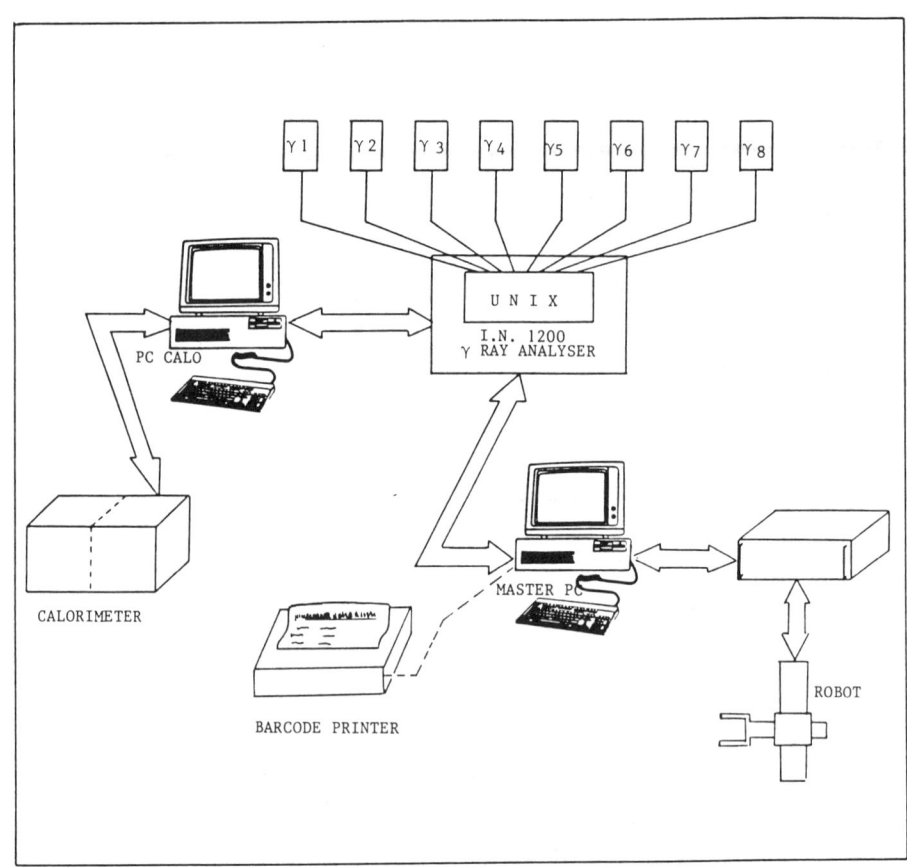

FIG. 5 - COMPUTER NETWORK

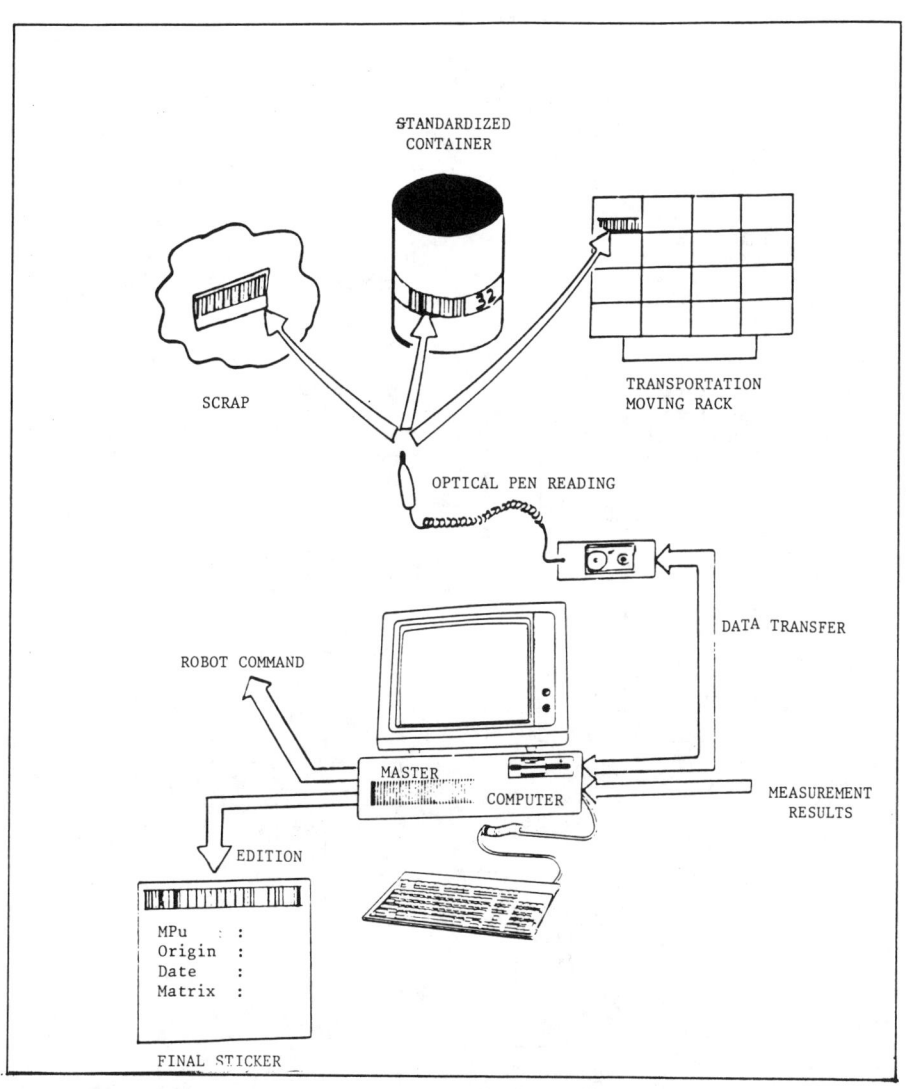

FIG. 6 - BAR CODE MANAGEMENT

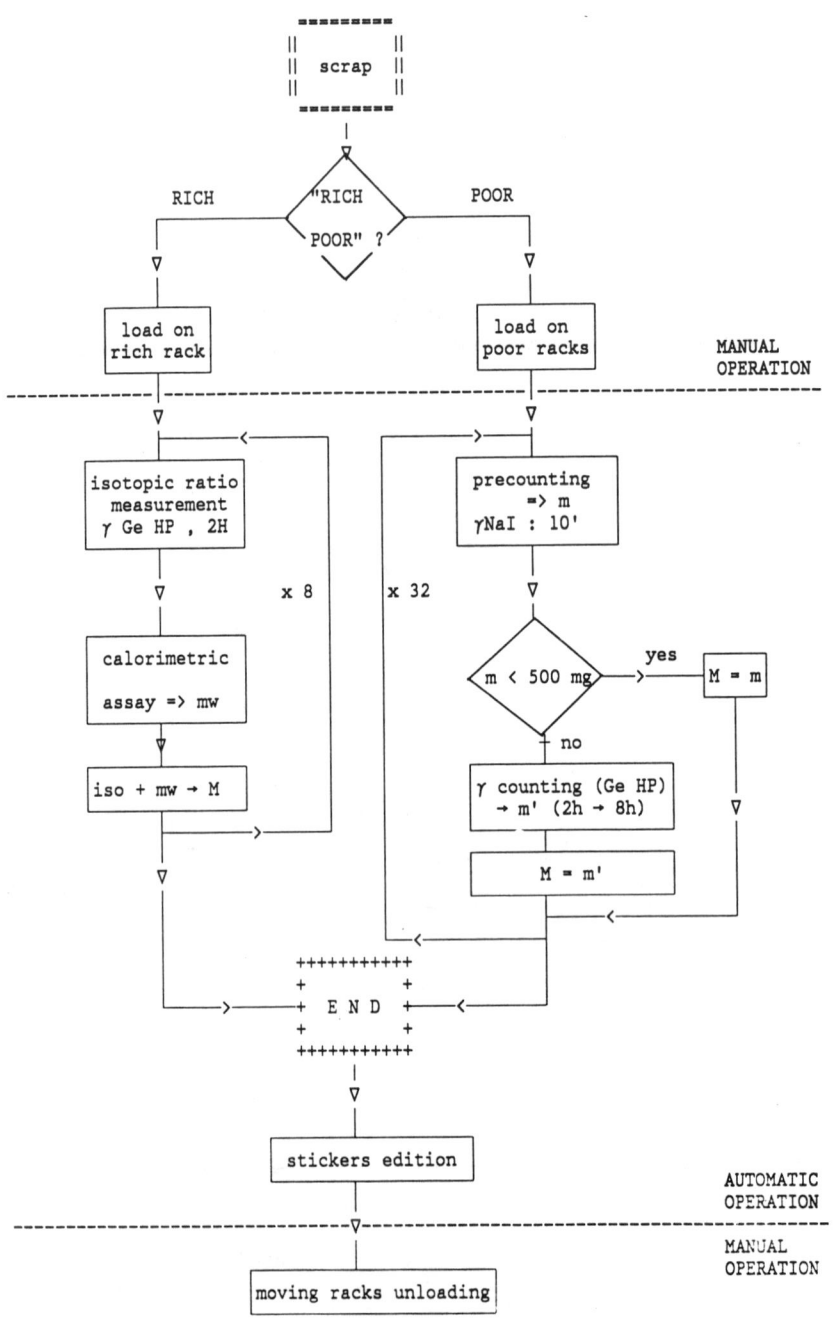

FIG. 7 MEASUREMENT PROCEDURE

| STANDARD | ISO | | MPu (g) | P (mw) | MPu (CALORIMETRIC MEASUREMENT) | | | |
| | % Pu^{240} | ppMAm^{241} | (certified) | (measured) | ISO BY TIMS (*) | | ISO BY GAMMA SPEC. | |
					MPu (g)	Δ %	MPu (g)	Δ %
1	6.5	3000	39.5	108	39.4	+ 0.25 %	39.6	+ 0.3 %
2	6.5	3000	134.4	369	134.1	+ 0.2 %	134.6	+ 0.2 %
3	8	3900	58.9	173	58.6	+ 0.5 %	60,1	+ 2 %
4	8	3900	117.9	347	117.3	- 0.5 %	117.8	- 0.1 %
5	10.2	5500	40.31	137	40.28	- 0.1 %	40.3	0 %
6	10.2	5500	135.2	447	134.8	- 0.3 %	134.9	- 0.2 %

(*) Thermal ionisation mass spectrometry.

FIG. 8 - ACCURACY OF CALORIMETRY - SOME EXAMPLES

P R O C E D U R E   (SEVEN STEPS)

1. ^{152}Eu $\longrightarrow$ [DETECTOR]

2. ^{152}Eu   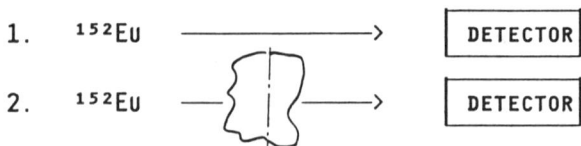 $\longrightarrow$ [DETECTOR]

3.   ── PLOTTING OF $\lambda\Delta = F(\text{LOG} E)$

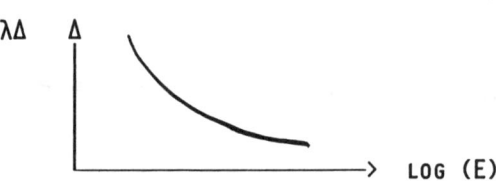

4. ^{239}Pu ACTIVITY MEASUREMENT $\rightarrow$ A'

129.3	keV	E1
203.4	keV	E2
345	keV	E3
393	keV	E4
413.7	keV	E5
451.5	keV	E6

5.   CORRECTION BY $A'' = \dfrac{A'.B\,\lambda\Delta}{1 - E^{-0.8\lambda\Delta}}$

(CYLINDER GEOMETRICAL APPROXIMATION)

6.   INFINITE ENERGY EXTRAPOLATION $\rightarrow A_0''$

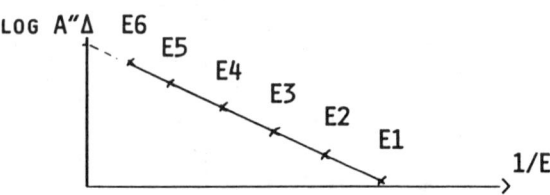

7.   ISOTOPIC MEASUREMENT $\longmapsto$ MPu
        $+ A_0''$

FIG. 9 – GAMMA RAY ATTENUATION CORRECTION

MATRIX	RESULTS (g)		$\dfrac{a - b}{a}$ (%)
	GAMMA-RAY SPECTROMETRY	CALORIMETRY OR DESTRUCTIVE ANALYSIS	
	(a)	(b)	
BULK Pu	20,7	20,68	+ 0,1
Ta	53,5	54,7	− 2,2
Pu OXYDE	142	139	+ 2,2
	12	11,5	+ 4,3
CALCIOTHERMIC SLAGS	14	14,5	− 3,4
	25	24,5	+ 2
	14	15	− 6,7
GRAPHITE	3,2	3,0	+ 6,7
	7	7,1	− 1,4
CELLULOSE	18,6	18,5	+ 0,5
	23	22,0	+ 4,6

FIG. 10 - TYPICAL RESULTS OBTAINED BY GAMMA-RAY SPECTROMETRY

REFERENCE VALUE (mg)	EXPERIMENTAL RESULTS (mg)	
	WITH ATTENUATION CORRECTION	WITHOUT ATTENUATION CORRECTION
15	13	10,4
39	31	20
57	59	52
240	250	210
510	500	480
872	900	830

FIG. 11 - GAMMA SPECTROMETRY

RESULTS OF AN INTERNAL ROUND ROBIN ON WASTES

(LIGHT MATRIX)

NONDESTRUCTIVE DETERMINATION OF RESIDUAL FUEL ON LEACHED HULLS AND DISSOLVER SLUDGES FROM LWR FUEL REPROCESSING

H. Würz, K. Wagner, H.J. Becker *)
Nuclear Research Centre Karlsruhe
*)WAK Betriebsgesellschaft

Summary

In reprocessing plants leached hulls and dissolver sludges represent rather important intermediate level α-waste streams. A control of the Pu content of these waste streams is desirable.
The nondestructive assay method to be preferred would be passive neutron counting. However, before any decision on passive neutron monitoring becomes possible, a characterization of hulls and sludges in terms of Pu content and neutron emission is necessary. For the direct determination of Plutonium on hulls and in sludges, as coming from reprocessing, an active neutron measurement is required. A simple, and sufficiently sensitive active neutron method which can easily be installed uses a stationary ^{252}Cf neutron source. This method was used for the characterization of hulls and sludges in terms of Plutonium content and total neutron emission in the Karlsruhe reprocessing plant WAK.

1. INTRODUCTION

In reprocessing plants leached hulls and dissolver sludges represent rather important intermediate level α-waste streams. Reprocessing of 1 ton heavy metal yields 300 l uncompacted hulls and up to 6kg dissolver sludges.Control of the dissolution process requires monitoring of leached hulls. This monitoring has to be performed directly at the dissolver basket.
For plant safety, monitoring of the dissolver sludges in terms of Plutonium could become necessary, depending on the size of the sludge collecting tanks. Nondestructive measurements (NDA = Non Destructive Assay) can contribute to control achievement of termination criteria.
The preferable NDA method, would be passive neutron counting. This method is rather sensitive and allows the assay of large sample sizes. The measurement is simple and reliable performance is achieved also under real plant conditions. However, before any decision on passive neutron monitoring of these wastes becomes possible, a characterization of hulls and sludges in terms of Pu content and neutron emission is necessary.
Hulls and sludges are Cm bearing wastes. The neutron emission is dominated by Cm. Therefore Plutonium can be determined indirectly only by a passive neutron measurement. If the Cm/Pu ratio for hulls and sludges is known, the amount of Pu can be determined from a passive neutron measurement of Cm.
Activities in characterization of hulls and sludges from LWR reprocessing are reported from different laboratories | 1-4 |. NDA measurements have been performed and are maintained at the reprocessing plants of La Hague, Windscale and Tokai Mura. At Karlsruhe, chemical analysis was used for the determination of the amount of residual fuel on hulls and wave length dispersive X-ray fluorescence (XRF) for the analysis of sludge samples | 5 |. In Table I a comparison

of published results of the different laboratories in terms of residual fuel is presented. The values are mean values over several batches. Results of NDA measurements are available for hulls only.

Passive NDA-measurements on hulls showed, that the amount of residual fuel on the hulls is rather small. For the direct determination of Plutonium on hulls and in sludges an active neutron measurement is required.

Active neutron measurements hitherto have been carried out by measuring the delayed fission neutrons using a pulsed neutron generator or an oszillating ^{252} Cf neutron source and by measuring delayed fission neutrons | 1 |. Due to the small fraction of delayed neutrons, these methods are cumbersome and insensitive or in presence of neutron emitters such as Curium are requiring rather high neutron source strenghts. Moreover, these active methods until now have been applied to the assay of hulls from fast reactor fuel reprocessing only.

For the active assay of hulls and dissolver sludges from LWR fuel reprocessing an attractive approach with respect to sensitivity, accuracy and volume of analysis would be the differential die away time method | 6 |. However, the remote installation of such a complex measurement system in an existing plant is difficult. A simple, and sufficiently sensitive active neutron method which can easily be installed uses a stationary ^{252}Cf neutron source. This method was used here for a first characterization of hulls and dissolver sludges in terms of Plutonium content and total neutron emission.

2. THE METHOD

The assay is done using an active and passive neutron measurement. For the active measurement a stationary ^{252}Cf-neutron source is used. The passive measurement is done on the same batch with removed neutron source. The principle is shown in Fig. 1. The n-source and the n-detectors are arranged externally to the assay volume. The fast source neutrons are slowed down and finally become thermalized in the water of the item to be assayed, thereby causing induced fission at the fissile material. This additional neutron production is detected as an increase in the detector countrate. The total countrate CR_{tot} will be measured. It is given by three different contributions.

$$CR_{tot} = CR_{dir} + CR_{ne} + CR_{ind} \qquad (1)$$

with CR_{dir} : countrate due to direct source neutrons; reference countrate of the fissile free batch
CR_{ne} : countrate due to neutron emission from the batch, it is determined in the passive measurement with the external neutron source removed
CR_{ind} : countrate due to induced fission

The dominating contribution to the total countrate CR_{tot} comes from direct source neutrons. In order to reduce CR_{dir} the batches are immersed in water. But in that case the reference countrate CR_{dir} becomes batch dependent and is given by

$$CR_{dir}(batch) = K(batch)\, CR_{dir}(H_2O) \qquad (2)$$

with $CR_{dir}(H_2O)$: countrate without batch

The factor K(batch) describes the influence of the fissile free batch. This factor is determined in the laboratory. $CR_{dir}(H_2O)$ is determined before and after each measurement thus, changes in moderator temperature and in detector sensitivity are eliminated. CR_{ind} the quantity to be determined by the active measurement is a function of the amount of the fissile material.

For the same reason as for eq(2) we define

$$K(batch + fuel) = \frac{CR_{tot} - CR_{ne}}{CR_{dir}(H_2O)} \qquad (3)$$

All countrates in this term are measured.
From eq. (3) together with eq. (1) and (2) the CR_{ind} is given according to

$$CR_{ind} = CR_{dir}(H_2O) \ [K(batch + fuel) - K(batch)] \qquad (4)$$

Moreover CR_{ind} depends on the fissile material concentration and is given according to

$$CR_{ind} = A \cdot C_{fiss} \qquad (5)$$

with
C_{fiss} = concentration of fissile material
The fissile material is assumed to be homogeneously distributed and is given as effective ^{239}Pu concentration $^{239}Pu_{eff}$ with:

$$^{239}Pu_{eff} = {}^{239}Pu + 1,38 \ {}^{241}Pu + 0,66 \ {}^{235}U \qquad (6)$$

where the isotope symbols stand for the concentration.
The weight factors are determined by the ratios

$$\frac{(v\sigma_f)^{isotope}}{(v\sigma_f)^{239}}$$

with
v   number of fission neutrons
σ_f fission cross section
From CR_{ne} the neutron emission ne is determined according to

$$CR_{ne} = B \cdot ne \qquad (7)$$

The neutron emission ne is mainly due to spontaneous fission neutrons from Curium. Minor contributions are due to spontaneous fission neutrons from the even numbered Pu isotopes and from (a,n) neutrons due to the Pu a-activity.
The passive measurement is not only used for correcting the result of the active measurement, but can be used independently to determine the fractional fuel content by correlating the measured neutron emission rate with that of the spent fuel prior to dissolution, if the ratio Cm/Pu is known.

3. MEASUREMENT SYSTEM

The measurement system is schematically shown in Fig. 2 for the measurement of hulls. The assay volume is defined by the dissolver basket with a diameter of 22,2 cm. The basket is put into the water tank. The tank is surrounded by a PE reflector. There the stationary neutron source and on the opposite side 3 neutron detectors are embedded. Boron-10 lined detectors are used. The detector shielding against n-background is not optimized. Especially top and bottom of the system are shielded only weakly.

In Fig. 3 the arrangement for the measurement of dissolver **sludges is shown.** The filter bag containing sludge is put in a cylindrical vessel with **a diameter of 15** cm and a height of 35 cm made from stainless steel and flooded with water. For the measurement the vessel is located in the center of the water tank at half height.

The monitor is installed in the fuel element disassembly cell of the Karlsruhe pilot reprocessing plant WAK. Due to the water flooding of the batches, the active method is only applicable to basket diameters of up to 50 cm. For larger baskets the inner region will increasingly be selfshielded and is therefore not assayable by the active measurement.

4. THE CALIBRATION

The calibration in the laboratory had the aim to determine the factors A and B (see eq (5) and (7)) and to quantify the batch dependent factor K(batch) (eq(2)). The geometry and the materials used for calibration were identical or similar to those for the real measurements. As hulls original PWR zircaloy tubes cut into pieces of 5 cm length were used. The sludge was simulated by SiO_2 powder in an original filter bag.

Calibration of the active measurement was done using Plutonium samples and U and MOX fuel pins having active lengths comparable to the height of the monitor for the hulls and of the height of the container for the sludges. The selfshielding of the samples was corrected. In Fig. 4 the assayable volume for the active measurement is shown in axial direction. It is restricted to a segment with 18 cm above and below the source position. Assaying the whole basket, thus requires scanning. This is possible only by lifting the basket. In our case two active measurements per basket were done in the lower part at heights of 15 cm and 45 cm.

Relative detection sensitivities for fissile material in the water tank filled with hulls and immersed in water are shown in Fig. 5. By using three detectors azimutally displaced by 45° the detection sensitivity becomes rather flat over the cross section of the basket and is constant over the cross section of the sludge containing vessel. The circles are indicating the position of the fuel pin used for the calibrations.

The detection sensitivity for neutron emission is rather flat in axial direction due to the active length of the detectors. However, the three detectors only are able to monitor the adjacent part of the basket. For the passive measurement therefore the basket is turned by 180° and a second measurement is performed. For the sludge only one active and passive measurement without movement of the item is necessary.

The calibration factors A and B were determined to be

$$A_{hulls} = (833 \pm 90) \ \frac{c/5 \ min}{mg \ ^{239}PU_{eff}/1}$$

$$A_{FCS} = (222 \pm 25) \ \frac{c/5 \ min}{mg \ ^{239}PU_{eff}/1}$$

$B_{hulls} = (4 \pm 0,3) \ cps/n/cm^3 s$
$B_{FCS} = (0,7 \pm 0,05) \ cps/n/cm^3 s$

The error in the calibration data is due to uncertainties in the self shielding of the samples and due to errors in the positioning of the samples in the assay volume.

5. SENSITIVITY AND ACCURACY

Reference measurements without fuel were performed before and after each item measurement, demonstrating that reproducibility in laboratory was within $\pm$ 5σ. Due to the rather high countrate contribution of direct source neutrons (CR_{dir}) the 5σ values for the active measurement correspond to a signal to background ratio s/b of 0,5%, thus permanent quality assurance of the monitor system is essential.

In the reprocessing plant due to handling of spent fuel assemblies in the neighbourhood of the monitor, the neutron background is rather high and may change from measurement to measurement, thus requiring permanent background monitoring. Before and after each item measurement, reference measurements are done with and without external neutron source.

Moreover the detector electronics showed some unexpected dependence on the γ-activity at the detector position. This effect has been determined in laboratory and was taken into account in the evaluations of the data. Both effects and the influence of the batch handling caused a deterioration of the reproducibility of the monitor reference countrate to $\pm$ 1% for the hot measurements.

Signal to background ratios as function of the fissile material concentration are shown in Fig. 6. The minimum detectable amount of residual fuel is 30 mg $239Pu_{eff}/l$ for hulls and 90 mg $239Pu_{eff}/l$ for sludges.

The accuracy of the active measurement mainly is influenced by the batch dependence of the reference countrate $CR_{dir}(batch)$. For flooded hulls this is caused by the volume ratio hulls/water, which differs from batch to batch and has to be measured, and from the possible entrainment of an air film on the hulls surfaces or air bubbles in the hulls. Air entrainment is increasing the reference countrate and, if not corrected, leads to an overestimation of the result of the fissile values.

The dependence of K(batch) on the volume ratio hulls/water is shown in Fig. 7 for laboratory batches and for inactive batches handled under real conditions in the reprocessing plant. In laboratory a tensid was added to the water to prevent air entrainment. As is seen from Fig. 7, the air entrainment increases K(batch) by about 2% and doubles the error $\Delta K/K$. From the countrate increase the entrained air volume is estimated to be about 1% of the total batch volume.

For evaluation of the assay measurements it is assumed, that no air entrainment on the active hulls is occuring due to the dissolution process in boiling nitric acid. For future hot batch measurements a tensid will be added to the water to definitly exclude air entrainment.

For sludges the batch dependence of the reference countrate is caused by the mass of the dissolver sludge and by air entrainment in the filter bag. To exclude air entrainment, detailed batch handling procedures were taken into account by the operators.

6. RESULTS AND CONCLUSIONS

6.1. Hulls
6.1.1. Results

Up to now 28 batches of PWR hulls have been measured. The results for PWR hulls are shown in Fig. 8. Signal to background ratios up to 5% are obtained. For 50% of the batches the ratio hulls/water was not measured. These cases are indicated by larger error bars. In Fig. 9 the results of the active and the passive measurements are shown. If there exists a constant ratio Cm/Pu the points belonging to the same burnup should show a linear relation between CR_{ind} and

CR_{ne}. This actually is the case, as can be seen from a comparison of measured and calculated values.

However for obtaining agreement between calculation and measurement, the neutron emission and hence the amount of Cm has to be reduced by a factor of 2.7. Also indicated in Fig. 9 is the maximum amount of residual Pu left on the hulls. This amounts up to 0,5% what results in 37 g Pu_{tot} left on the hulls per ton heavy metal reprocessed. The mean value is about 0,3% or about 24g Pu_{tot}/tHM.

6.1.2. Conclusions

On the assumption, that no air entrainment occured with the hulls, the amount of residual fuel is clearly above the sensitivity limit of the method. The ratio Cm/Pu then is a factor of 2.7 lower than in the undissolved fuel. This factor is independent from burn up. Therefore the amount of residual fuel could be determined by a passive measurement.

If air entrainment is occuring, the batch factor K(batch) has to be increased by about 2% (see Fig. 7) and therefore the signal to background ratio s/b decreases (see chap. 5). In this case the ratio Cm/Pu more closely corresponds to that for the undissolved fuel, and the distance of the results to the sensitivity limit of the method becomes smaller. The then obtained results are mean amount of residual Pu about 0,1% what corresponds to about 8 g Pu per 1 ton heavy metal reprocessed. This value is comparable to what has been found for two batches during the Coquenstock exercise /4/.

6.2. Sludge
6.2.1. Results

Up to now 22 batches of dissolver sludges have been assayed. The results are shown in Fig. 10 together with the batch factor K. The signal to background ratio is considerably lower as for the hulls and taking into account the error bars, in most cases is in the range of the fissile free K-value. This is valid also for 2 MOX batches. Some outliers are observed presumably due to air entrainment in the filter bag.

The results hitherto obtained do not show any relation between CR_{ind} and CR_{ne}. Therefore no information on Cm/Pu is gainable from the hitherto obtained results. The maximum amounts of Pu_{eff} are 2 g/batch or about 14 g Pu_{tot} per 1 ton heavy metal reprocessed.

7. SUMMARY

The results as discussed are valid for WAK batches from LWR fuel reprocessing. Due to the specific filtering technique, the WAK sludge batches contain particle sizes above 15 μm only. The amount of Pu contained in the sludge fraction having smaller particle sizes presently is unknown.

For the hulls, due to individual pin cutting instead of assembly cutting, homogeneous mixing of hulls, coming from different axial pin positions and thus from different axial core regions, is guaranteed. Therefore mean values for the amount of residual fuel are obtained without requiring time consuming axial scanning of the basket.

The neutron method with stationary n-source is sensitive enough for characterization of hulls and dissolver sludges. The amount of Plutonium on hulls as determined from the active measurement is higher than for the passive measurement, that means, the weight ratio Cm/Pu for hulls is smaller than for the fuel prior to dissolution. The factor has been determined to be 2,7.

If this is taken into account, passive n-monitoring of hulls is adequate also on an item to item basis.

8. REFERENCES

/1/   N. Hagano et al., Evaluation of a leached hull monitoring system at the Tokai Reprocessing Plant, Proc. 6th Annual Symp. on Safeguards and Nucl. Mat. Management, Venice 14.-18. May 1984

/2/   B.J. Mc Donald et al., Nondestructive measurement of Pu and U in process wastes and residues. IAEA Conf. on Safeguarding Nuclear Materials, Vienna, 20.-24. Oct. 1975, IAEA-SM 201/61

/3/   G. Frejaville et al., Nondestructive measurements: Hull monitoring and burnup determinations. Conf. Safeguards Technology, November 28 - December 2, 1983, Hilton Head Island. Trans ANS 45 (1983)

/4/   J.P. Gue et al Operation Coquenstock, Determination de la composition et de la radiocativité des coques provenant du traitement industriel des combustibles des réacteurs à l'eau légère. Rep. EUR 10923 FR 1986

/5/   H. Wertenbach, Determination of nuclear fuel residues in dissolver sludges and on hulls after reprocessing. Int. Conf. on Nuclear and Radiochemistry, Lindau, Oct. 8.-12., 1984

/6/   H. Würz, The pulsed neutron method - a new method for determination of small quantities of Plutonium in waste, Proc. Jahrestagung Kerntechnik, München , 21.-23.05.1985, P. 317/320

ACKNOWLEDGEMENTS

The technical support of the operation staff of the pilot reprocessing plant WAK (batch preparation and remote handling) has been essential for the measurements. We thank therefore Mr. Kunstmann and Mr. Ratzke for their continuous and kind help. Further acknowledged is the support of Mr. Hespeler, who kindly assisted in performing the measurements.

	hulls		sludges
	γ (1)	n (2)	
La Hague (3)	0.14 ±0.1 0.25±0.1	0.06±0.1 0.15 ±0.1	
Tokai mura (3)	0.17 ±0.04		
Karlsruhe (4)	DA 0.08 ±0.065 (5)		XRF: 0.16 ±0.13 (6)

(1) : passive γ-measurements
(2) : passive n-measurements
(3) : no determination of Cm/Pu
(4) : Cm/Pu ratios available
(5) : 10 hulls/batch as sample
(6) : 2 probes/batch as sample
DA : Destructive Analysis
XRF : X-Ray Fluorescence

Comparison of results given as residual fuel in percent of total fuel.

Table 1

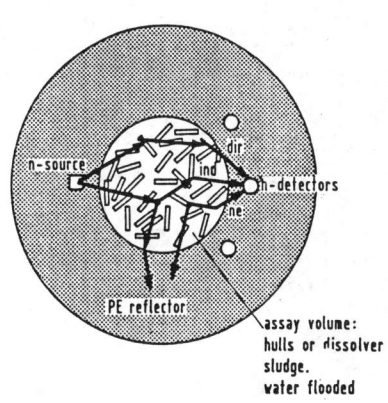

Fig.1 <u>active measurement</u>

$$CR_{tot} = CR_{dir} + CR_{ne} + CR_{ind}$$

$$CR_{ind} = A \times C_{fiss}$$

Masses:
$$^{239}Pu_{eff} = {}^{239}Pu + 1.38 \cdot {}^{241}Pu + 0.66 \cdot {}^{235}U$$

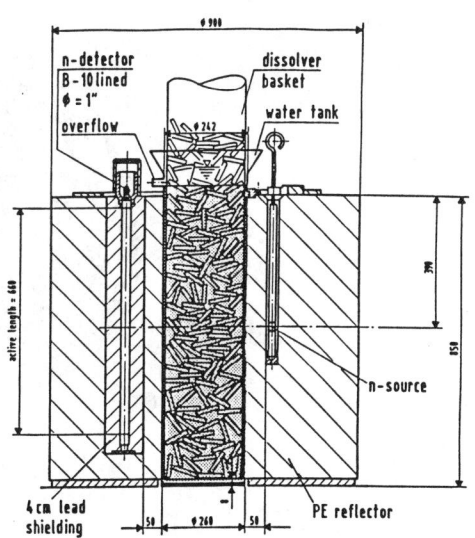

Fig.2 **Arrangement for measurement of hulls in the dissolver basket**

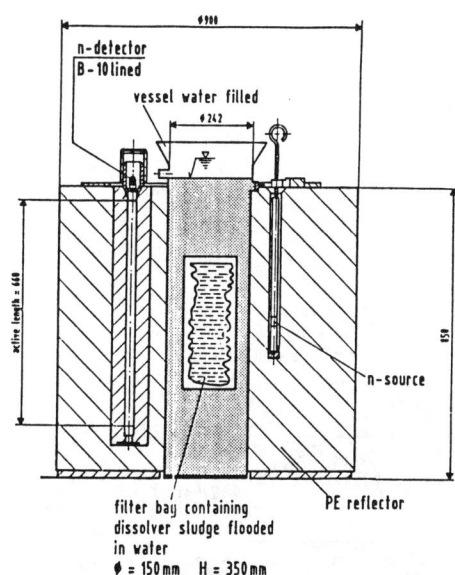

Fig.3 **Arrangement for measurement of dissolver sludge in filter bag**

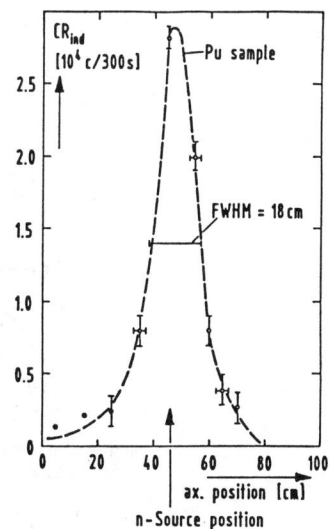

Fig.4 **Detection sensitivity in axial direction for small Pu sample**

— 307 —

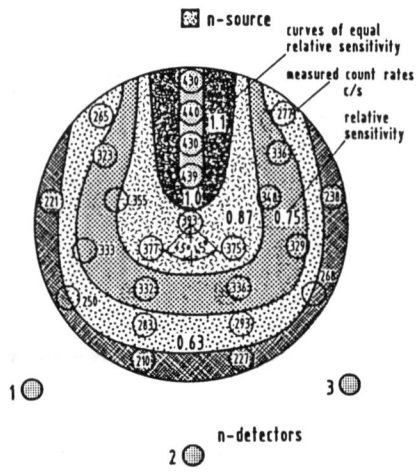

Fig.5 **Detection sensitivities for 3 detectors**
Zircaloy hulls water flooded, diameter
of assay volume 24 cm.

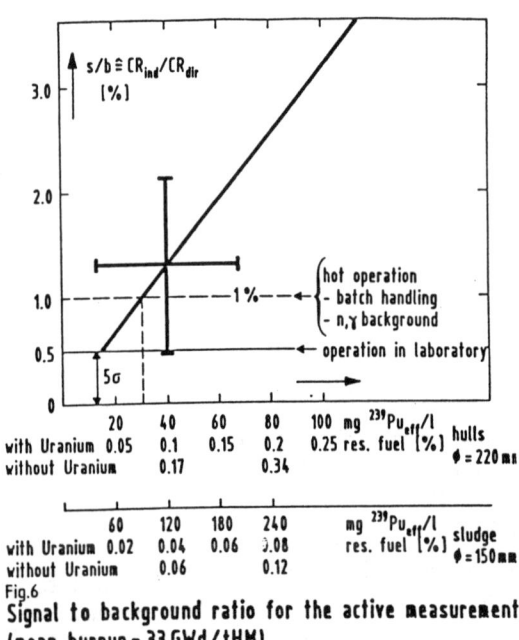

Fig.6
Signal to background ratio for the active measurement
(mean burnup = 33 GWd/tHM)

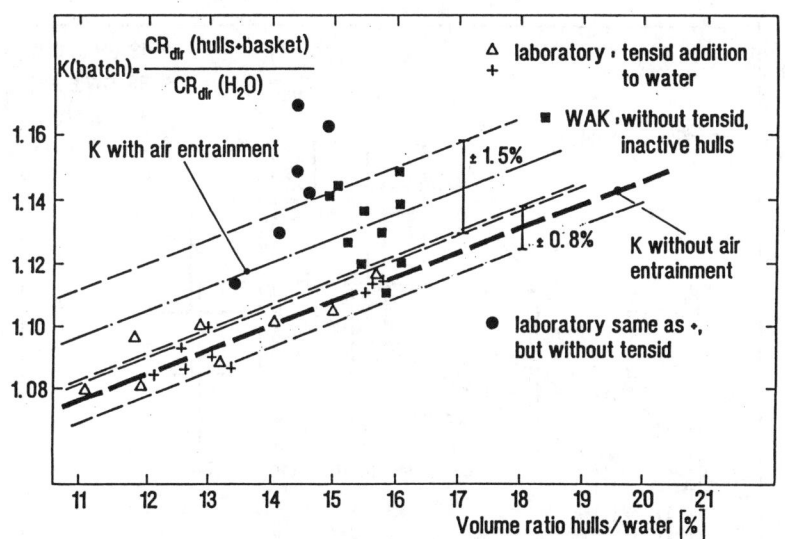

Fig. 7   Dependence of the batch factor K from volume
ratio hulls/water and air entrainment

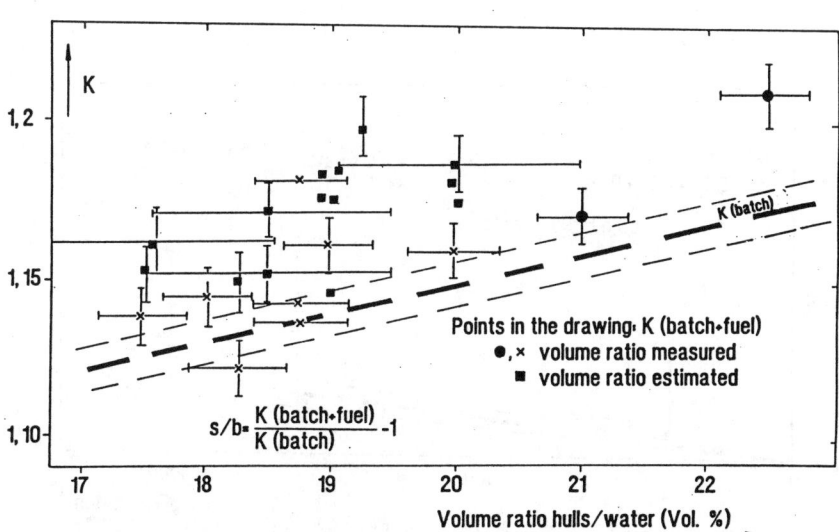

Fig. 8   Comparison of results for hulls K (batch+fuel) with
the batch factor K (batch)

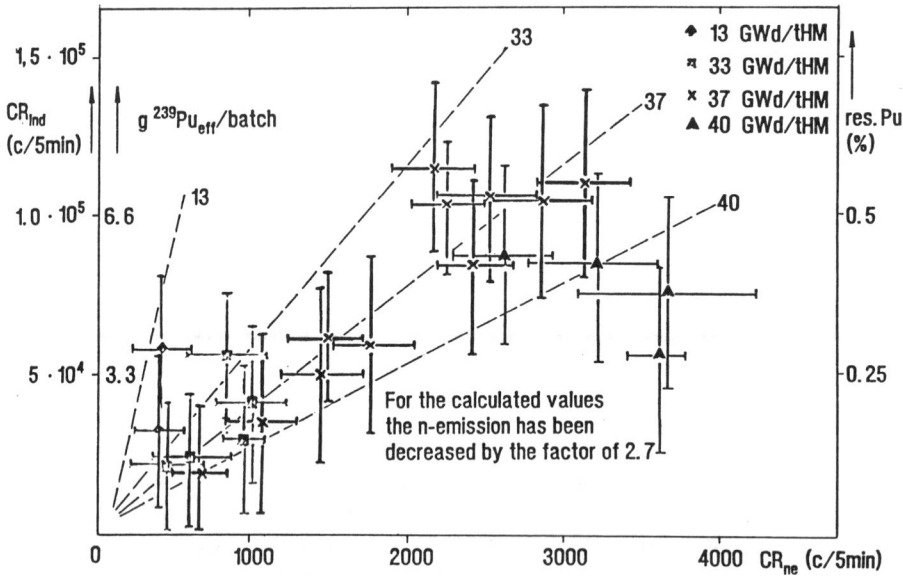

Fig. 9 Results of the active and passive measurements and comparison with calculated values.

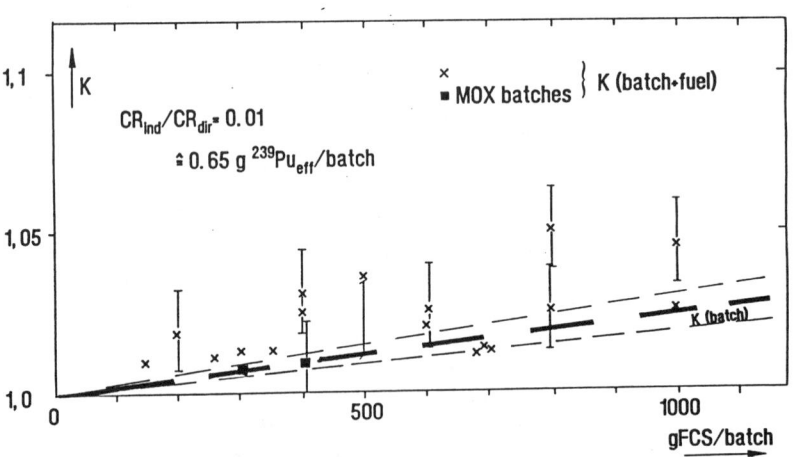

Fig. 10 Results for sludges

FINAL DISCUSSION

FINAL DISCUSSION

Chairman : J. LEFEVRE
           CEA Fontenay-aux-Roses

During the final discussion, led by Messrs LEVEFRE, BREMNER, SCHULTZ, SIMON, SWENNEN and WURZ, the following main points were addressed :

. Accuracy and representativity of measurement methods (introduced by Mr WURZ),

. Quality Assurance and guaranteed appraisal of the radioactive content of waste and associated standards (indroduced by Mr SWENNEN),

. Application of source or final package measurements (introduced by Messrs BREMNER and SCHULTZ),

. Problem of very low detection limits for decommissioning (introduced by Mr LEFEVRE),

. Balance of fissile materials (introduced by Mr SIMON).

The following principles resulted from discussions :

- Participants expressed the need to compare and improve harmonization of the points of view of the various people involved in the field: producers, measurers, systems and methods developers, organizations responsible for storage, official authorities, etc. The following points were raised with respect to this subject :

  . More information is needed on available methods (routine, qualification, development), their scope and recommendations for their application.

  . It is important for specialists to speak the same language (especially as regards accuracy and representativity).

  . Common standards and approaches are necessary (in spite of the fact that regulations differ from one country to another).

  . Meetings should be more specialized.

- Accuracy and representativity:

The following points were emphasized:

. Measurements are interpreted on the basis of assumptions concerning the matrix or the contaminant of waste (density, homogeneity, granulometry, presence of absorbents, etc.).

. The notion of representativity is extremely important (this is related to how sensitive the interpretation of the measurement is to the failure to consider matrix-and contaminant-related data).

. Combined methods (active-passive) offer a number of advantages.

. With respect to alpha activity criteria for storage, it is important to have measurements combining accuracy (and low detection limit) and representativity.

The following particular points were raised with respect to the application of source measurements to the final package:

. There was a consensus of opinion that the sorting and measuring of waste at source (better knowledge and sorting of waste, small volumes, etc.) considerably enhance accuracy and representativity (the further measurements are taken downstream, the more accuracy and representativity deteriorate).

. "Measurers" considered that the measurement of packages which are inadequately known, or where information is lacking, should be refused.

. If, on the same site, there are one or more systems upstream or downstream, the comparison of the data provided by these systems must be well controlled.

. It was considered important, however, to possess the means required to measure existing processed packages.

- Quality Assurance, Guaranteed Values and Standards

. Actions depend on the objective (classification, determination of alpha activity for storage, materials accounting).

. It is important to identify tools (measuring methods and systems) and error sources (including information related to the origin [isotopies, chemical form] and waste composition).

. As regards standards, the discussion concerned the advantages of a restricted number of specific standards for each method and the complementary use of Monte Carlo codes.

- Decommissioning

. Regulations have not yet been fixed but could result in very low values (a few Bq/g). It is important to develop and perfect suitable measuring methods.

. With regard to dismantling, which involves measurements of large volumes and the use of sampling methods, the difficulties relating to the representativity of samples were stressed.

- In conclusion, attention was drawn to the fact that the main objective was to reach the measurement performance levels required by the safety authorities, corresponding to very low measurement limit values. It was therefore considered essential to continue R & D work towards this goal.

LIST OF PARTICIPANTS

LIST OF PARTICIPANTS

BELGIUM

DE BAERE P.

SCK/CEN
BOERETANG 200
B - 2400 MOL

EID Ch.

COMMISSION DES COMMUNAUTES
EUROPEENNES
200, RUE DE LA LOI
DG XII/D-2 (ARTS 2/13)
B - 1049 BRUXELLES

HENDRICKX J.P.

BELGOPROCESS
GRAVENSTRAAT 73
B - 2480 DESSEL

REYNDERS R.

BELGOPROCESS
GRAVENSTRAAT 73
B - 2480 DESSEL

SIMON R.

COMMISSION DES COMMUNAUTES
EUROPEENNES
200, RUE DE LA LOI
DG XII/D-2 (ARTS 2/14)
B - 1049 BRUXELLES

SUZUKI Y.

SCK/CEN
BOERETANG 200
B - 2400 MOL

SWENNEN R.

NIRAS/ONDRAF
1, PLACE MADOU
BTES 24/25
B - 1030 BRUXELLES

FRANCE

ANTHONI S.

CEA/CEN CADARACHE
BP 1
F - 13108 ST-PAUL-LEZ-DURANCE CEDEX

AUDE G.

ANDRA
CEA - SIEGE
31-33 RUE DE LA FEDERATION
F - 75752 PARIS CEDEX 15

BERNARD P.

CEA/CEN CADARACHE
BP 1
F - 13108 ST-PAUL-LEZ-DURANCE CEDEX

BERGEY C.	CEA/CEN VALDUC DAM BP 14 F - 21120 IS-SUR-TILLE
BERNE R.	CEA/CEN CADARACHE BP 1 F - 13108 ST-PAUL-LEZ-DURANCE CEDEX
BEROUD Y.	S.G.N. 1, RUE DES HERONS MONTIGNY-LE-BRETONNEUX F - 78182 ST-QUENTIN-EN-YVELINES CEDEX
BESLU P.	CEA/CEN CADARACHE BP 1 F - 13108 ST-PAUL-LEZ-DURANCE CEDEX
BIGNAN G.	CEA/CEN CADARACHE BP 1 F - 13108 ST-PAUL-LEZ-DURANCE CEDEX
BOSSER R.	CEA/CEN FONTENAY-AUX-ROSES BP 6 F - 92265 FONTENAY-AUX-ROSES CEDEX
BOUCHARD J.	ANDRA CEA - SIEGE 31-33 RUE DE LA FEDERATION F - 75752 PARIS CEDEX 15
BUTEZ M.	CEA/CEN CADARACHE BP 1 F - 13108 ST-PAUL-LEZ-DURANCE CEDEX
CANCE M.	CEA/CEN BRUYERES-LE-CHATEL DAM F - 91680 BRUYERES-LE-CHATEL
CHABERT J.	S.G.N. F - 50440 BEAUMONT-HAGUE
CHASSEUR Ph.	CEA/CEN VALRHO F - 30205 BAGNOLS-SUR-CEZE CEDEX
CHOPLIN	CEA/CEN BRUYERES-LE-CHATEL DAM F - 91680 BRUYERES-LE-CHATEL
CLOUE J.	CEA/CEN CADARACHE BP 1 F - 13108 ST-PAUL-LEZ-DURANCE CEDEX

COUROUBLE J.M.	CEA/CEN FONTENAY-AUX-ROSES BP 6 F - 92265 FONTENAY-AUX-ROSES CEDEX
DE BRUYNE T.	CEA/CEN BRUYERES-LE-CHATEL DAM F - 91680 BUYERES-LE-CHATEL
DELARUE G.	CEA/CEN FONTENAY-AUX-ROSES BP 6 F - 92265 FONTENAY-AUX-ROSES CEDEX
DE SAINTE MARIE N.	COGEMA - MARCOULE BP 170 F - 30205 BAGNOLS-SUR-CEZE CEDEX
DESPRETS A.	CEA/CEN CADARACHE BP 1 F - 13108 ST-PAUL-LEZ-DURANCE CEDEX
DEYSON M.	ANDRA CEA/CEN CADARACHE F - 13108 ST-PAUL-LEZ-DURANCE
DOLLO R.	E.D.F. 6, RUE AMPERE F - 92203 SAINT DENIS CEDEX 1
DUCOS O.	COGEMA - MARCOULE BP 170 F - 30205 BAGNOLS-SUR-CEZE CEDEX
DOUTRELUINGNE P.	COGEMA - MARCOULE BP 170 F - 30205 BAGNOLS-SUR-CEZE CEDEX
GUIBERTEAU Ph.	CEA/CEN VALDUC DAM BP 14 F - 21120 IS-SUR-TILLE
JEANDIDIER C.	ANDRA CEA - SIEGE 31-33 RUE DE LA FEDERATION F - 75752 PARIS CEDEX 15
JOSSO F.	CEA/CEN FONTENAY-AUX-ROSES BP 6 F - 92265 FONTENAY-AUX-ROSES CEDEX
JOURDE P.	CEA/CEN FONTENAY-AUX-ROSES BP 6 F - 92265 FONTENAY-AUX-ROSES CEDEX

LE GAC A.	EDF/SPT/DSRE 6, RUE AMPERE F - 92203 SAINT DENIS CEDEX 1
LAMARQUE G.	CEA/CEN VALRHO F - 30205 BAGNOLS-SUR-CEZE CEDEX
LAMBERMONT G.	SODERN 1, AVENUE DESCARTES BP 23 F - 94451 LIMEIL-BREVANNES CEDEX
LEFEVRE J.	CEA/CEN FONTENAY-AUX-ROSES BP 6 F - 92265 FONTENAY-AUX-ROSES CEDEX
LE GUILLOU G.	CEA/CEN CADARACHE BP 1 F - 13108 ST-PAUL-LEZ-DURANCE CEDEX
LYON F.	CEA/CEN CADARACHE BP 1 F - 13108 ST-PAUL-LEZ-DURANCE CEDEX
LEROUGE B.	CEA/CEN FONTENAY-AUX-ROSES BP 6 F - 92265 FONTENAY-AUX-ROSES CEDEX
MALFONDET A.M.	CEA/CEN VALDUC DAM BP 14 F - 21120 IS-SUR-TILLE
MARMONIER P.	CEA/CEN FONTENAY-AUX-ROSES BP 6 F - 92265 FONTENAY-AUX-ROSES CEDEX
MARTIN-DEIDIER L.	CEA/CEN CADARACHE BP 1 F - 13108 ST-PAUL-LEZ-DURANCE CEDEX
MEGY J.	CEA/CEN CADARACHE BP 1 F - 13108 ST-PAUL-LEZ-DURANCE CEDEX
MICHAUD F.	CEA/CEN VALDUC DAM BP 14 F - 21120 IS-SUR-TILLE
MOREL J.	CEA/CEN SACLAY F - 91191 GIF-SUR-YVETTE CEDEX

Mlle NEUILLY M.   CEA/CEN CADARACHE
                                   BP 1
                                   F - 13108 ST-PAUL-LEZ-DURANCE CEDEX

PEROLAT J.P.   CEA/CEN FONTENAY-AUX-ROSES
                                   BP 6
                                   F - 92265 FONTENAY-AUX-ROSES CEDEX

REGIMBEAU P.   ANDRA
                                   CEA - SIEGE
                                   31-33 RUE DE LA FEDERATION
                                   F - 75752 PARIS CEDEX 15

Mme ROCCA-SERRA N.   ANDRA
                                   CEA/CEN CADARACHE
                                   F - 13108 ST-PAUL-LEZ-DURANCE CEDEX

ROMEYER-DHERBEY J.   CEA/CEN CADARACHE
                                   BP 1
                                   F - 13108 ST-PAUL-LEZ-DURANCE CEDEX

SAAS A.   CEA/CEN CADARACHE
                                   BP 1
                                   F - 13108 ST-PAUL-LEZ-DURANCE CEDEX

SALA T.   TECHNICATOME
                                   BP 34
                                   F - 13762 LES MILLES CEDEX

Mme SIMONET G.   CEA/CEN SACLAY
                                   F - 91191 GIF-SUR-YVETTE CEDEX

TACHON M.   CEA/CEN VALRHO
                                   F - 30205 BAGNOLS-SUR-CEZE CEDEX

THAUREL B.   CEA/CEN FONTENAY-AUX-ROSES
                                   BP 6
                                   F - 92265 FONTENAY-AUX-ROSES CEDEX

Mme THOREZ M.   CEA/CEN CADARACHE
                                   BP 1
                                   F - 13108 ST-PAUL-LEZ-DURANCE

VIDALIE M.   CEA/CEN BRUYERES-LE-CHATEL
                                   DAM
                                   F - 91680 BRUYERES-LE-CHATEL

VIGREUX B.   S.G.N.
                                   1, RUE DES HERONS
                                   MONTIGNY-LE-BRETONNEUX
                                   F - 78182 ST-QUENTIN-EN-YVELINES
                                   CEDEX

F.R.G.

KOPP D.A.
NIEDERSÄCHSISCHES UMWELTMINISTERIUM
ARCHIVSTRASSE 2
D - 3000 HANNOVER 1

MARTENS B.R.
BUNDESAMT FÜR STRAHLENSCHUTZ
ALBERT-SCHWEITZER-STRASSE 18
D - 3320 SALZGITTER 1

ODOJ R.
FORSCHUNGSZENTRUM JÜLICH
ICT/PKS
POSTFACH 1913
D - 5170 JÜLICH

SOHNIUS B.
NUKEM
POSTFACH 11 00 80
D - 6450 HANAU 11

WÜRZ H.
NUCLEAR RESEARCH CENTER
POSTFACH 3640
D - 7500 KARLSRUHE 1

VOLG K.
INSTITUT FÜR RADIOCHEMIE DER
TECHNISCHEN UNIVERSITÄT MÜNCHEN
WACHTER-MEISSNER-STRASSE 3
D - 8046 GARCHING

ITALY

CRESTI P.R.
E.N.E.A.
CRE - CASACCIA
SP. ANGUILLARESE KM 1+300
I - 00060 ROMA

CHIODI P.L.
NUCLECO
SP. ANGUILARESE KM 1+300
I - 00060 ROMA

DIERCKX R.
J.R.C. ISPRA
I - 21027 ISPRA (VARESE)

FAZIO A.
E.N.E.A.
CRE - CASACCIA
SP. ANGUILLARESE KM 1+300
I - 00060 ROMA

FRAZZOLI F.V.
DIPARTIMENTO DI ENERGETICA
UNIVERSITA DEGLI STUDI DI ROMA
"LA SAPIENZA"
VIA A SCARPA 14
I - 00161 ROMA

HAGE W.	J.R.C. ISPRA I - 21027 ISPRA (VARESE)
PEDERSEN B.H.	J.R.C. ISPRA ESSOR EO 83 I - 21027 ISPRA

SPAIN

PINA LUCAS G.	CIEMAT BUILDING 17 AVENIDA COMPLUTENSE 22 E - 28040 MADRID
SUAREZ GONZALEZ DEL REY J.A.	CIEMAT BUILDING 18 AVENIDA COMPLUTENSE 22 E - 28040 MADRID

SWEDEN

TOVEDAL H.	STUDSVIK AB S - 61182 NYKOPING

SWITZERLAND

ALDER J.C.	NAGRA PARKSTRASSE 23 CH - 5401 BADEN
KURTZ K.	NAGRA PARKSTRASSE 23 CH - 5401 BADEN

U.K.

ADAWAY D.	AEA TECHNOLOGY DOUNREAY NUCLEAR POWER DEVELOPMENT ESTABLISHMENT THURSO UK - CAITHNESS KW14 7TZ
ALLDRED K.	BRITISCH NUCLEAR FUELS plc SELLAFIELD WORKS SEASCALE UK - CUMBRIA CA20 1PG
ARMITAGE B.	AEA TECHNOLOGY BUILDING 721 HARWELL LABORATORY DIDCOT UK - OXFORDSHIRE OX11 ORA

BREMNER W.   AEA TECHNOLOGY
                                    DOUNREAY   NUCLEAR POWER DEVELOPMENT
                                    ESTABLISHMENT
                                    THURSO
                                    UK - CAITHNESS KW14 7TZ

FARREN J.   AEA TECHNOLOGY
                                    HARWELL
                                    UK - OXFORDSHIRE OX11 OLW

FINDLAY D.J.S.   AEA TECHNOLOGY
                                    BUILDING 418
                                    HARWELL LABORATORY
                                    DIDCOT
                                    UK - OXFORDSHIRE OX11 ORA

HARVEY C.   NATIONAL POWER NUCLEAR
                                    TECHNOLOGY DIVISION
                                    CANAL ROAD
                                    GRAVESEND
                                    UK - KENT DA1Z ZRS

LAMBERT K.   AEA TECHNOLOGY
                                    HARWELL
                                    UK - OXFORDSHIRE OX11 OLW

Mc DONALD Ch.R.   HMIP/DOE
                                    ROMNEY HOUSE
                                    ROOM A502
                                    43 MARSHAM STREET
                                    UK - LONDON SW1P 3PY

MOLESWORTH T.V.   TAYLOR   WOODROW   MANAGEMENT   AND
                                    ENGINEERING LTD
                                    ALPHA HOUSE
                                    WESTMOUNT CENTRE
                                    DELAMERE ROAD
                                    HAYES
                                    UK - MIDDLESEX UB4 OHD

ORR Ch.   BRITISH NUCLEAR FUELS plc
                                    SELLAFIELD WORKS
                                    SEASCALE
                                    UK - CUMBRIA CA20 1PG

PACKER T.   AEA TECHNOLOGY
                                    BUILDING 721
                                    ROOM G10
                                    HARWELL LABORATORY
                                    DIDCOT
                                    UK - OXFORDSHIRE OX11 ORA

U.S.

CALDWELL J.
PAJARITO SCIENTIFIC CORPORATION
322 KIMBERLY
USA - LOS ALAMOS NM 87544

ESTEP R.
PAJARITO SCIENTIFIC CORPORATION
322 KIMBERLEY
USA - LOS ALAMOS NM 87544

SCHULTZ F.J.
OAK RIDGE NATIONAL LABORATORY
P.O. Box X
USA - OAK RIDGE TN 37831

RINARD P.
SAFEGARDS ASSAY
MAIL STOP E540
USA - LOS ALAMOS NM 87544

SPRINKLE J.
LOS ALAMOS NATIONAL LABORATORY
MAIL STOP E540
USA - LOS ALAMOS NM 87544

European Communities — Commission

EUR 12890 — Non-destructive assay of radioactive waste.
Topical meeting

Edited by: *C. Eid, P. Bernard*

Luxembourg: Office for Official Publications of the European
Communities

1990 – XII, 327 pp., tab., fig. – 16.2 × 22.9 cm

Nuclear science and technology series

EN

ISBN 92-826-1636-3

Catalogue number: CD-NA-12890-EN-C

Price (excluding VAT) in Luxembourg: ECU 18.75

The purpose of the meeting reported in this volume was to review the experience gained in the laboratories of different countries on the measurement methods that accompany radioactive waste management, from the source of the waste to its disposal. Non-destructive assay finds applications in the initial sorting of waste, declassification after contamination, the dosing of waste during processing and conditioning, and the certification of compliance with disposal criteria. The meeting, jointly organized by the Commission of the European Communities and the French Commissariat à l'Energie Atomique, was divided into sessions on active methods, passive methods, integrated and/or combined systems, performance, correlation, and operating experience.